Control of Linear Systems
Observability and Controllability
of General Linear Systems

Control of Linear Systems

Observability and Controllability of General Linear Systems

Lyubomir T. Gruyitch

CRC Press
Taylor & Francis Group
Boca Raton London New York

CRC Press is an imprint of the
Taylor & Francis Group, an **Informa** business

CRC Press
Taylor & Francis Group
6000 Broken Sound Parkway NW, Suite 300
Boca Raton, FL 33487-2742

Printed on acid-free paper
Version Date: 20180927

International Standard Book Number-13: 978-1-138-35315-2 (Hardback)

Visit the Taylor & Francis Web site at
http://www.taylorandfrancis.com

and the CRC Press Web site at
http://www.crcpress.com

Contents

Preface

On the state of the art

This book concerns observability and controllability as two of the fundamental topics of the ***time*-invariant continuous-*time* linear control systems**, in the sequel called for short ***control systems*** or just ***systems***. It shows that in the framework of the linear systems there are still problems left untouchable but solvable.

All dynamical systems have their (internal and/or output) dynamics that determines their (internal and/or output) dynamical situation called *state* and *state variables*, i.e., its (internal and/or output) *state vector* all that *regardless of the existence or the nonexistence of the input derivatives*. The basis of the existing concept of state is that the input vector derivatives do not influence the system (internal and/or output) dynamics, or at least that they do not appear in the system mathematical model. It has become very useful to develop the effective mathematical machinery for the related studies. The theory and the practice have been very well developed only for one class of the physical control systems. In order to treat other classes of the physical control systems, their mathematical models should be transformed formally mathematically with the full loss of the physical sense of the new (mathematical) variables.

The concept of the state, the state variables, the state vector and the state space is well defined and widely effectively directly used only in the framework of the dynamical systems in general, and of the control systems in particular, described by the first order vector linear differential state equation and by the algebraic vector linear output equation, which are called **Input - State - Output** (abbreviated: *ISO*) (or: *state-space*) (**control**) **systems.** Their mathematical models do not contain any derivative of the input vector function $\mathbf{I}(.)$.

There is a fundamental lacuna in the control theory due to the nonexistence of the clear, well defined, concept of the state also for the systems subjected to the influence of the input vector derivatives so that the physical meaning and sense of the system variables is preserved.

On the book

The author, in addition to the analysis of the scientific papers listed in the bibliography D.15, consulted in particular the books by the following authors: B. D. O. Anderson and J. B. Moore [1], P. J. Antsaklis and A. N. Michel [2], [3], S. Barnett [4], A. Benzaoiua, F. Mesquine and M. Benhayoun [5]; further by L. D. Berkovitz [6], L. D. Berkovitz and N. G. Medhin [7], S. P. Bhattacharyya, A. Datta and L. H. Keel [9], D. Biswa [10], P. Borne, G. Dauphin - Tanguy, J.- P. Richard, F. Rotella and I. Zambettakis [11], W. L. Brogan [13], F. M. Callier and C. A. Desoer [15], [16], C.-T. Chen [18], H. Chestnut and R. W. Mayer [19], M. J. Corless and A. E. Frazho [20], J. J. D'Azzo and C. H. Houpis [21], J. J. D'Azzo, C. H. Houpis and S. N. Sheldon [22], C. A. Desoer [24], C. A. Desoer and M. Vidyasagar [25], F. W. Fairman [27], F. R. Gantmacher [30], [31], G. C. Goodwin [34], Ly. T. Gruyitch [40], [41], [42], M. Haidekker [46], J. P. Hespanh [50], C. H. Houpis and S. N. Sheldon [52], D. G. Hull [53], M. K.-J. Johansson [55], T. Kaczorek [56], T. Kailath [57], D. E. Kirk [64], B. Kisačanin and G. C. Agarwal [65], B. C. Kuo [66], [67], H. Kwakernaak and R. Sivan [68], P. Lancaster and M. Tismenetsky [69], J. M. Maciejowski [71], J. L. Melsa and D. G. Schultz [75], R. K. Miller and A. N. Michel [77], K. Ogata [81], [82], D. H. Owens [83], H. M. Power and R. J. Simpson [86], H. H. Rosenbrock [88], A. Sinha [89], R. E. Skelton [90], D. M. Wiberg [95], R. L. Williams II and D. A. Lawrence [96], W. A. Wolovich [97], W. M. Wonham [98] and B.-T. Yazdan [99]. This book is complementary to them and/or extends, broadens, and generalizes inherently their parts that are related to the dynamical system state concept, observability, and controllability.

The book treats observability and controllability for the following five classes of the **systems**.

- *The Input-Output (IO) systems* described by the ν-*th* order linear vector differential equation expressed in terms of the output vector $\mathbf{Y} \in \Re^N$. This class of the control systems has been only partially studied; it has been studied by formally mathematically transforming its mathematical model into the form of the *ISO* systems. The variables of such transformations lose the physical sense of the system variables if the system is subjected to actions of the derivative(s) of the input vector. The book resolves this lacuna of the control theory.

- *The (first order) Input-State-Output (ISO) systems* determined by the first order linear vector differential equation expressed in terms of the state vector $\mathbf{X}$, which is *the state equation*, $\mathbf{X} \in \mathfrak{R}^n$, and by the algebraic vector equation expressed in terms of the output $\mathbf{Y}$, which is *the output equation*. They are well known as *the state-space systems*. They contain only one derivative that is the first derivative of the state vector. They do not contain any derivative of the input vector.

- *The (first order) Extended-Input-State-Output (EISO) systems* determined by the first order linear vector differential equation expressed in terms of the state vector $\mathbf{X}$, which is *the state equation*, $\mathbf{X} \in \mathfrak{R}^n$, and by the algebraic vector equation expressed in terms of the output $\mathbf{Y}$, which is *the output equation*. They contain the first order derivative $\mathbf{X}^{(1)}$ of the state vector $\mathbf{X}$ and μ derivatives of the input vector $\mathbf{I}$ only in the state vector equation, $\mu > 0$ (because for $\mu = 0$ the *EISO* system becomes the *ISO* system). The *EISO* systems have not been studied so far.

- *The Higher order Input-State-Output (HISO) systems* characterized by the α-*th* order linear vector differential equation expressed in terms of the vector $\mathbf{R}$, which is *the state equation*, $\mathbf{R} \in \mathfrak{R}^p$, and by the linear vector algebraic equation of the output vector $\mathbf{Y}$, which is *the output equation*. This class of the control systems has not been studied so far.

- *The Input-Internal and Output state (IIO) systems* characterized by the α-*th* order linear vector differential equation expressed in terms of the internal dynamics vector $\mathbf{R}$, which is *the internal dynamics, i.e., the internal state, equation*, and by the ν-*th* order linear vector (differential if $\nu > 0$, algebraic if $\nu = 0$) equation expressed in terms of the output vector $\mathbf{Y}$, which is *the output (state if $\nu > 0$) equation*. The books [36], [40] introduced and initiated the study of this class of the dynamical systems. However, the *IIO control systems* have not been studied so far.

The existence of the actions of the input vector derivatives on the system is the reason, justification, and need **to extend and to generalize the state concept of dynamical, hence of control, systems**.

Various subsidiary statements, results, exercises, and rigorous detailed proofs form the Appendix.

The goal of the book is to contribute to the advancement of the linear control systems theory and the corresponding university courses, to open new directions for research and for applications in the framework of *time*-invariant continuous-*time* linear control systems. It represents a further development of the existing linear control systems theory that is not repeated herein. *The contributions of the book largely and crucially are beyond the existing control theory.*

In gratitude

The author expresses his gratitude to:

Ms. Nora Konopka, Global Editorial Director – Engineering, for her formidable, exceptionally careful leadership of the publication process during which she proposed to divide the original manuscript of 633 pages into two books, this one and the accompanying book [37], under their titles as they are published.

Ms. Michele Dimont, Project Editor, for her devoted leadership of the book's editing.

Ms. Vanessa Garrett, Editorial Assistant – Engineering, for her careful and effective administrative work.

All of CRC Press/Taylor & Francis.

The author is grateful to Mr. George Pearson with MacKichan Company for his very kind and effective assistance to improve my usage of the excellent Scientific Work Place software for scientific works.

Belgrade, September 28, 2017, March 13, April 23, May 18 - 20, 2018.

Lyubomir T. Gruyitch

Part I

SYSTEM CLASSES

Chapter 1

Introduction

1.1 *Time*

All processes, motions and movements, all behaviors of the systems and their responses, as well as all external actions on the systems, occur and propagate in *time*. It is natural from the physical point of view to study the systems directly in the temporal domain. This requires to be clear how we understand what *time* is and what its properties are, which we explain in brief as follows (for the more complete analysis see: [39], [40], [41], [42], [43], [44]).

Definition 1 *Time*
   *Time (i.e., **the temporal variable**) denoted by t or by τ is an independent scalar physical variable such that:*
   *- Its value called **instant** or **moment** determines uniquely **when** somebody or something started/interrupted to exist,*
   *- Its values determine uniquely **since when and until when** somebody or something existed/exists or will exist,*
   *- Its values determine uniquely **how long** somebody or something existed/exists or will exist,*
   *- Its values determine uniquely whether an event E_1 occurs **then when** another event E_2 has not yet happened, or the event E_1 takes place just **then when** the event E_2 happens, or the event E_1 occurs **then when** the event E_2 has already happened,*
   *- Its value **occupies (covers, encloses, imbues, impregnates, is over and in, penetrates) equally** everybody and everything (i.e., beings, objects, energy, matter, and space) **everywhere and always**, and*

3

- Its value has been, is, and will be **permanently changing smoothly, strictly monotonously continuously, equally** *in all spatial directions and their senses, in and around everybody and everything,* **independently** *of everybody and everything (i.e., independently of beings, objects, energy, matter, and space),* **independently** *of all other variables, independently of all happenings, movements, and processes.*

Time is a basic and elementary constituent of the existence of everybody and of everything [42], [43], [44].

All human trials during millenniums have failed to explain, to express, the nature, the phenomenon, of *time* in terms of other well-defined notions, in terms of other physical variables and phenomena [42], [43, Axiom 25, p. 52], [44, Axiom 25, p. 53]. The nature of *time*, the physical content of it, cannot be explained in terms of other basic constituents of the existence (in terms of energy, matter, space) or in terms of other physical phenomena or variables. *Time* has its own, original, nature that we can only call it *the nature of time*, i.e., *the temporal nature* or *the time nature* [40], [41], [42], [43], [44].

An arbitrary value of *time* t (τ), i.e., **an arbitrary instant** or **moment**, is denoted also by t (or by τ), respectively. It is an **instantaneous (momentous)** and **elementary** *time* value. It can happen exactly once and then it is the same everywhere for, and in, everybody and everything (i.e., for, and in, beings, energy, matter, objects, and space), for all other variables, for all happenings, for all movements, for all processes, for all biological, economical, financial, physical, and social systems. It is not repeatable. Nobody and nothing can influence the flow of instants [42], [43], [44].

The physical dimension of *time* is denoted by $[T]$, where T stands for *time*, t $[T]$. It cannot be expressed in terms of the physical dimension of another variable. Its physical dimension is one of the basic physical dimensions. It is used to express the physical dimensions of most of the physical variables. A selected unity 1_t of *time* can be arbitrarily chosen and then fixed. If it is *second s* then $1_t = s$, which we denote by $t\langle 1_t\rangle = t\langle s\rangle$.

There can be assigned *exactly one* (which is denoted by $\exists!$) real number to every moment (instant), and vice versa. The numerical value **num** t of the moment t is a real number and dimensionless, $num\ t \in \Re$ and $num\ t$ $[-]$, where $\Re$ is the set of all real numbers.

Theorem 2 *Universal time speed law, [42], [43], [44]*

Time is the unique physical variable such that the speed v_t (v_τ) of the evolution (of the flow) of its values and of its numerical values:

a) *is invariant with respect to a choice of a relative zero moment* t_{zero}, *of an initial moment* t_0, *of a time scale and of a time unit* 1_t, *i.e., invariant relative to a choice of a time axis, invariant relative to a selection of spatial coordinates, invariant relative to everybody and everything, and*

b) *its value (its numerical value) is invariant and equals one arbitrary time unit per the same time unit (equals one), respectively,*

$$v_t = 1[TT^{-1}]\left\langle 1_t 1_t^{-1} \right\rangle = 1[TT^{-1}]\left\langle 1_\tau 1_\tau^{-1} \right\rangle = v_\tau, \ numv_t = numv_\tau = 1, \tag{1.1}$$

relative to arbitrary time axes T *and* T_τ, *i.e., its numerical value equals 1 (one) with respect to all time axes (with respect to any accepted relative zero instant* t_{zero}, *any chosen initial instant* t_0, *any time scale and any selected time unit* 1_t*), with respect to all spatial coordinate systems, with respect to all beings and all objects.*

The uniqueness of *time*, the constancy and the invariance of the *time* speed determine that *time* itself is not relative and cannot be relative [38], [39], [42], [43], [44], [80].

Time set $\mathcal{T}$ is the set of all moments. It is open, unbounded, and connected set. It is in the biunivoque (one-to-one) correspondence with the set $\mathfrak{R}$ of all real numbers,

$$\mathcal{T} = \{t : num\ t \in \mathfrak{R},\ dt > 0,\ t^{(1)} \equiv 1\},$$
$$\forall t \in \mathcal{T},\ \exists! x \in \mathfrak{R} \Longrightarrow x = num\ t$$
$$and\ \forall x \in \mathfrak{R},\ \exists! t \in \mathcal{T} \Longrightarrow num\ t = x,$$
$$num\ inf\mathcal{T} = num\ t_{\inf} = -\infty \notin \mathcal{T}\ and\ num\ sup\mathcal{T} = num\ t_{\sup} = \infty \notin \mathcal{T}. \tag{1.2}$$

The rule of the correspondence determines an **accepted relative zero numerical *time* value** t_{zero}, a ***time* scale** and a ***time* unit** denoted by 1_t (or by 1_τ). The *time* unit can be ... , millisecond, second, minute, hour, day, ... , *which Newton explained by clarifying the sense of **relative time*** [80, I of Scholium, p. 8]. Unfortunately, this fact has been ignored in the modern physics and science.

Note 3 *Choice of the relative zero moment* t_{zero} *and the initial moment* t_0

We accept herein the relative zero moment t_{zero} *to have the zero numerical value,* $num\ t_{zero} = 0$ *because we deal with the time-invariant systems. Besides, we adopt it for* t_{zero} *to be also the initial moment* t_0, $t_0 = t_{zero}$, *num* $t_0 = 0$, *in view of the time-invariance of the systems to be studied. This determines the subset* $\mathfrak{T}_0$ *of the time set* $\mathfrak{T}$,

$$\mathfrak{T}_0 = \{t : t \in \mathfrak{T}, \ num\,t \in [0, \infty[\}.$$

Sometimes, we will denote the initial moment explicitly by t_0 *but it will mean that num* $t_0 = 0$.

Note 4 *We usually use the letters* t *and* τ *to designate time itself and an arbitrary moment, as well as the numerical value of the arbitrary moment with respect to the chosen zero instant, e.g.,* $t = 0$ *is used in the sense num* $t = 0$. *From the physical point of view this is incorrect. The numerical value num* t *of the instant* t *is a real number without a physical dimension, while the instant* t *is a temporal value that has the physical dimension, which is the temporal dimension* T *of time. We overcome that by using the normalized, dimensionless, mathematical temporal variable, denoted by* $\bar{t}$ *and defined by*

$$\bar{t} = \frac{t}{1_t}[-],$$

so that the time set $\mathcal{T}$ *is to be replaced by*

$$\overline{\mathcal{T}} = \{\bar{t}[-] : \bar{t} = num\,\bar{t} = num\ t \in \mathfrak{R}, \ d\bar{t} > 0, \ \bar{t}^{(1)} \equiv 1\}.$$

With this in mind we will use in the sequel the letter t *also for* $\bar{t}$, *and* $\mathcal{T}$ *also for* $\overline{\mathcal{T}}$. *Hence,*

$$t[-] = num\,t[-].$$

Between any two different instants $t_1 \in \mathcal{T}$ and $t_2 \in \mathcal{T}$ there is a third instant $t_3 \in \mathcal{T}$, either $t_1 < t_3 < t_2$ or $t_2 < t_3 < t_1$. The *time* set $\mathcal{T}$ is **continuum**. It is called also **the continuous-*time* set**. This book is on *continuous-time* systems and their control.

1.2 Notational preliminaries

Lower case ordinary letters denote scalars, bold (lower case and capital, Greek and Roman) letters signify vectors, capital italic letters stand for matrices, and we use capital 𝔉𝔯𝔞𝔨𝔱𝔲𝔯 letters for sets and spaces. For example, the identity matrix of the dimension i is denoted by I_i,

$$I_i = diag\,\{1 \quad 1 \quad ... \quad 1\} \in \mathfrak{R}^{i \times i}, \ I_n = I \in \mathfrak{R}^{n \times n}. \tag{1.3}$$

The variables in the mathematical models are dimensionless because their values are normalized relative to their characteristic values. Throughout the book we accept the following condition to hold:

Condition 5 *Normalized variables*

The value of every variable Z appearing in a system mathematical model is dimensionless normalized physical variable Z_{Ph} relative to some its characteristic value Z_{PhCh} (e.g., nominal value Z_{PhN} or the unit value 1_Z):

$$Z \ [-] = \frac{Z_{Ph} \ [Z_{Ph}]}{Z_{PhCh} \ [Z_{Ph}]}. \tag{1.4}$$

Note 6 *Useful simple vector notation [36], [40]*

Instead of using, for example,

$$\mathbf{Y}^{\mp}(s) = F(s)\bullet$$

$$\bullet \begin{bmatrix} I^{\mp^T}(s) & I^T(0^{\mp}) & .. & I^{(\mu-1)^T}(0^{\mp}) & Y^T(0^{\mp}) & .. & Y^{(\nu-1)^T}(0^{\mp}) \end{bmatrix}^T,$$

the following simple vector notation enabled us to define and use effectively the system full transfer function matrix $F(s)$:

$$\mathbf{Y}^{\mp}(s) = F(s)\mathbf{V}(s),$$

$$\mathbf{V}(s) = \begin{bmatrix} \mathbf{I}^{\mp}(s) \\ \mathbf{C}_0^{\mp} \end{bmatrix}, \ \mathbf{C}_0^{\mp} = \begin{bmatrix} \mathbf{I}^{\mu-1}(0^{\mp}) \\ \mathbf{Y}^{\nu-1}(0^{\mp}) \end{bmatrix},$$

$$\mathbf{I}^{\mu-1}(0^{\mp}) = \begin{bmatrix} I(0^{\mp}) \\ I^{(1)}(0^{\mp}) \\ ... \\ I^{(\mu-1)}(0^{\mp}) \end{bmatrix}, \ \mathbf{Y}^{\nu-1}(0^{\mp}) = \begin{bmatrix} Y(0^{\mp}) \\ Y^{(1)}(0^{\mp}) \\ ... \\ Y^{(\nu-1)}(0^{\mp}) \end{bmatrix},$$

by introducing the general compact vector notation

$$\mathbf{Y}^k = \begin{bmatrix} \mathbf{Y} \\ \mathbf{Y}^{(1)} \\ ... \\ \mathbf{Y}^{(k)} \end{bmatrix} = \begin{bmatrix} \mathbf{Y}^{(0)} \\ \mathbf{Y}^{(1)} \\ ... \\ \mathbf{Y}^{(k)} \end{bmatrix} \in \mathfrak{R}^{(k+1)N}, \ k \in \{0, 1, ...\}, \ \mathbf{Y}^0 = \mathbf{Y}. \tag{1.5}$$

It is different from the k-th derivative $\mathbf{Y}^{(k)}$ of $\mathbf{Y}$:

$$\mathbf{Y}^{(k)} = \frac{d^k \mathbf{Y}}{dt^k} \in \mathfrak{R}^N, \ k \in \{1, ...\}, \ \mathbf{Y}^k \neq \mathbf{Y}^{(k)}.$$

This permits to express $\sum_{i=0}^{i=\nu} A_i \mathbf{Y}^{(i)}(t)$ as follows,

$$A_i \in \mathfrak{R}^{N \times N}, \ \sum_{i=0}^{i=\nu} A_i \mathbf{Y}^{(i)}(t) = \begin{bmatrix} A_0 & \vdots & A_1 & \vdots & ... & \vdots & A_\nu \end{bmatrix} \begin{bmatrix} \mathbf{Y}^{(0)}(t) \\ \mathbf{Y}^{(1)}(t) \\ ... \\ \mathbf{Y}^{(k)}(t) \end{bmatrix},$$

i.e., in the compact form by introducing the extended system matrix $A^{(\nu)}$
composed of the system matrices $A_i \in \mathfrak{R}^{N \times N}$, $i \in \{0, 1, ..., \nu\}$,

$$A^{(\nu)} = \begin{bmatrix} A_0 & \vdots & A_1 & \vdots & ... & \vdots & A_\nu \end{bmatrix} \in \mathfrak{R}^{N \times (\nu+1)N}, \tag{1.6}$$

$$A^{(\nu)} \neq A^\nu = \underbrace{AA...A}_{\nu-times} \in \mathfrak{R}^{N \times N} \tag{1.7}$$

so that

$$\sum_{i=0}^{i=\nu} A_i \mathbf{Y}^{(i)}(t) = A^{(\nu)} \mathbf{Y}^\nu(t). \tag{1.8}$$

Other notation is defined at its first appearance in the text and in Appendix A.

1.3   Compact, simple, and elegant calculus

The introduction and definition of:

- The extended vector $\mathbf{Y}^k \in \mathfrak{R}^{(k+1)N}$ (1.5), which is composed of the vector $\mathbf{Y}$ and its derivatives up to the order k,

- The extended matrix $A^{(\nu)} \in \mathfrak{R}^{N \times (\nu+1)N}$ (1.6), the entries of which are submatrices A_i, $i = 0, 1, 2, .., \nu$,

enable us to develop a compact, simple, and elegant calculus.
The matrix differential equation:

$$\sum_{i=0}^{i=\nu} A_i \mathbf{Y}^{(i)}(t) = \sum_{i=0}^{i=\mu \leq \nu} B_i \mathbf{I}^{(i)}(t),$$

$$A_i \in \mathfrak{R}^{N \times N}, \ \mathbf{Y} \in \mathfrak{R}^N, \ B_i \in \mathfrak{R}^{N \times M}, \ \mathbf{I} \in \mathfrak{R}^M, \tag{1.9}$$

has the equivalent compact form in the time domain [36], [40]:

$$A^{(\nu)} \mathbf{Y}^\nu(t) = B^{(\mu)} \mathbf{I}^\mu(t),$$

$$A^{(\nu)} \in \mathfrak{R}^{N \times (\nu+1)N}, \ \mathbf{Y}^\nu \in \mathfrak{R}^{(\nu+1)N}, \ B^{(\mu)} \in \mathfrak{R}^{N \times (\mu+1)M}, \ \mathbf{I}^\mu \in \mathfrak{R}^{(\mu+1)M}. \tag{1.10}$$

Comment 7 *Compact form of the linear differential equation*
Equation (1.10) is a differential, not algebraic, equation that is the compact form of the original differential Equation (1.9).

If $N \leq M$ and rank $A^{(\nu)} = N$ then the matrix $\left(A^{(\nu)} \left(A^{(\nu)} \right)^T \right)$ is nonsingular and the right inverse $\left(A^{(\nu)} \right)^T \left(A^{(\nu)} \left(A^{(\nu)} \right)^T \right)^{-1}$ of $A^{(\nu)}$ is well defined. If Equation (1.10) had been algebraic and treated as algebraic then we would have been formally able to solve it for $\mathbf{Y}^{\nu}(t)$:

$$\mathbf{Y}^{\nu}(t) = \left(A^{(\nu)} \right)^T \left(A^{(\nu)} \left(A^{(\nu)} \right)^T \right)^{-1} B^{(\mu)} \mathbf{I}^{\mu}(t),$$

but this would not have been a solution to Equation (1.10) because it is a differential, not algebraic, equation.

The compact, simple, and elegant calculus is the basis for all calculations in the book. It is effectively applicable not only to linear continuous-time systems [36], [40], but also to linear discrete-time systems [14] and to nonlinear dynamical systems [41].

1.4   System behavior

Time is a basic constituent of the environment of every dynamical physical system. *The time field is the temporal environment, i.e., the time environment, of the system* [42], [43], [44].

A *time*-dependent variable will be denoted for short by the corresponding letter, e.g., scalar variables by $D, I, R, S, U, Y, ...$ and vector variables by $\mathbf{D}, \mathbf{I}, \mathbf{R}, \mathbf{S}, \mathbf{U}, \mathbf{Y}, ...$. From the mathematical point of view they are functions, e.g., $D = D(.) : \mathfrak{T} \longrightarrow \mathfrak{R}^1$, $\mathbf{D} = \mathbf{D}(.) : \mathfrak{T} \longrightarrow \mathfrak{R}^d$.

A variation of the value of every *time*-dependent variable is in *time*.

As usual, $\mathfrak{R}_+$ is the set of all nonnegative real numbers, $\mathfrak{R}^+$ is the set of all positive real numbers, $\mathfrak{R}^k$ is the k-dimensional real vector space, the elements of which are k-dimensional real valued vectors, where k is any natural number. Notice that $\mathfrak{R}^1 \neq \mathfrak{R}$.

There are three substantial characteristic groups of the variables that are associated with the dynamical system in general. Their definitions follow by referring to [18, Definition 3-6, p. 83], [40], [41], [66, p. 105], [78, 2. Definition, p. 380], [79, 2. Definition, p. 380], [81, p. 4], [82, p. 664].

Note 8 *The capital letters D, I, R, S, U, Y (and $\mathbf{D}, \mathbf{I}, \mathbf{R}, \mathbf{S}, \mathbf{U}, \mathbf{Y}$) denote the total scalar (vector) values of the variables $D(.), I(.), R(.), S(.), U(.),$*

Y (.) (of the vector variables $\mathbf{D}$ (.), $\mathbf{I}$ (.), $\mathbf{R}$ (.), $\mathbf{S}$ (.), $\mathbf{U}$ (.), $\mathbf{Y}$ (.)) relative to their total zero scalar (vector) value, if it exists, or relative to their accepted zero scalar (vector) value, respectively.

A characteristic of the dynamical systems is their *dynamical behavior.* The dynamical system can possess the explicit internal dynamics and the implicit output dynamics or explicit both the internal and output dynamics.

A special family of the dynamical systems is that of *plants (objects).*

Definition 9 *Plant (object)*

*A **plant** $\mathcal{P}$ (i.e., an **object** $\mathcal{O}$) is a system that should under specific conditions called **nominal (nonperturbed)** realize its demanded dynamical behavior and under other (nonnominal, perturbed, real) conditions should realize its dynamical behavior sufficiently close to its demanded dynamical behavior over some (bounded or unbounded) time interval.*

The physical nature of a plant can be anyone.

Definition 10 *Input variables, input vector, and input space*

*A variable that acts on the system and its influence is essential for the system behavior is the **system input variable** denoted by $I \in \mathfrak{R}$. The system can be under the action of several mutually independent input variables I_1, I_2, ..., I_M. They compose the **system input vector** (for short, **input**)*

$$\mathbf{I} = [I_1 \ \ I_2 \ \ ... \ \ I_M]^T \in \mathfrak{R}^M, \tag{1.11}$$

*which is an element of the **input space** $\mathfrak{R}^M$.*

*The **instantaneous values** of the variables I_i and $\mathbf{I}$ at an instant $t \in \mathfrak{T}$ are $I_i(t)$ and $\mathbf{I}(t)$, respectively.*

The capital letters I and $\mathbf{I}$ denote the total (scalar, vector) values of the variable I and the vector $\mathbf{I}$ relative to their total zero (scalar, vector) value, if it exists, or relative to their accepted zero (scalar, vector) value, respectively.

Definition 11 *Disturbance variable and disturbance vector*

*An input variable D of a system that acts on the system without using any information about the system demanded dynamical behavior or by using it in order to perturb the system behavior is the **disturbance variable** (for short: **disturbance**) for the system.*

*If there are several, e.g., d, disturbance variables D_1, D_2, ... , D_d, then they are entries of the **disturbance vector** (for short: **disturbance**) $\mathbf{D}$,*

$$\mathbf{D} = \left[D_1 \vdots D_2 \vdots ... \vdots D_d \right]^T \in \mathfrak{R}^d. \tag{1.12}$$

The instantaneous values of the variables D_i and $\mathbf{D}$ at an instant $t \in \mathfrak{T}$ are $D_i(t)$ and $\mathbf{D}(t)$, respectively.

A disturbance action on a system most often is not rejectable. The disturbance acts on the system at best independently of the system behavior, because if the disturbance exploits the information about the system demanded behavior in order to perturb the system behavior then it is **an enemy disturbance**. In order to interrupt its action on the system its source should be destroyed, which is rarely possible. The physical nature of disturbances can be anyone.

The system output behavior is determined by the temporal evolution of its *output variables and their derivatives,* in the sense of the following definitions:

Definition 12 *Output variables, output vector, output space, and response*

*A variable $Y \in \mathfrak{R}$ is an **output variable** of the system if and only if its values result from the system behavior, they are (directly or indirectly) measurable, and we are interested in them.*

*The number N is the maximal number of linearly independent output variables Y_1, Y_2, ..., Y_N on $\mathfrak{T}$ of the system. They form the **output vector** $\mathbf{Y}$ of the system, which is an element of the **output space** $\mathfrak{R}^N$:*

$$\mathbf{Y} = [Y_1 \ Y_2 \ ...Y_N]^T \in \mathfrak{R}^N. \tag{1.13}$$

The time evolution $\mathbf{Y}(t)$ of the output vector $\mathbf{Y}$ takes place, i.e., the output vector $\mathbf{Y}$ propagates, in the integral output space $\mathcal{I}$,

$$\mathcal{I} = \mathfrak{T} \times \mathfrak{R}^N. \tag{1.14}$$

The instantaneous values of the variables Y_i and $\mathbf{Y}$ at an instant $t \in \mathfrak{T}$ are $Y_i(t)$ and $\mathbf{Y}(t)$, respectively.

*The time variation $\mathbf{Y}(t)$ of the system output vector $\mathbf{Y}$ is the **system (output) response**.*

***The plant desired output behavior** is denoted by $\mathbf{Y}_d(t)$.*

Note 13 *There are systems, the output variable of which is fed back to the system input. Such output variable is also the system input variable, and such system has its own (local) feedback.*

A (physical and a mathematical) dynamical system can be subjected to the action of the input vector derivatives $\mathbf{I}^{(l)}(t)$, $l \in \{1, 2, ...\}$. The system internal and output dynamical behavior depend then not only on the input vector $\mathbf{I}(t)$ but also on all its derivatives acting on the system. This is reality that inspires us, justifies and demands us to generalize the concept of the dynamical system *state* as follows.

Definition 14 *State of a dynamical system*
The (internal, output) state of a physical dynamical system at **a moment** $\tau \in \mathfrak{T}$ *is, respectively, the system (internal, output) dynamical physical situation at the moment τ, which, together with the input vector and its derivatives acting on the system at any moment $(t \geq \tau) \in \mathfrak{T}$, determines uniquely the system behavior, [i.e., the system (internal, output) state and the system output response], for all $(t > \tau) \in \mathfrak{T}$, respectively.*
The (internal, output) state of a mathematical dynamical system at a moment $\tau \in \mathfrak{T}$ *is, respectively, the minimal amount of information about the system at the moment τ, which, together with information about the action on the system (about the system input vector and its derivatives acting on the system) at any moment $(t \geq \tau) \in \mathfrak{T}$, determines uniquely the system behavior (i.e., the system (internal, output) state and its output response) for all $(t > \tau) \in \mathfrak{T}$, respectively.*
The minimal number $n_{(.)}$ of linearly independent variables $S_{(.)i}$ on $\mathfrak{T}$, $i = 1, 2, ... , n_{(.)}$, the values $S_{(.)i}(\tau)$ of which are at every moment $\tau \in \mathfrak{T}$ in the biunivoque correspondence with the system (internal: $(\cdot) = I$, output: $(\cdot) = O$) state at the same moment τ, is **the state dimension** *and the variables $S_{(.)i}$, $i = 1, 2, . . ., n_{(.)}$, are, respectively,* **the (internal: $(\cdot) = I$, output: $(\cdot) = O$) state variables of the system.** *They compose, respectively,* **the (internal: $(\cdot) = I$, output: $(\cdot) = O$) state vector** $S_{(.)}$ **of the system,**

$$\mathbf{S}_{(.)} = \begin{bmatrix} S_{(.)1} & S_{(.)2} & ...S_{(.)n_{(.)}} \end{bmatrix}^T \in \mathfrak{R}^n, \ (.) = \ , I, \ O. \qquad (1.15)$$

The space $\mathfrak{R}^{n_{(.)}}$ is, respectively, **the (internal: $(.) = I$, output: $(\cdot) = O$) state space of the system.**
The state vector function $\mathbf{S}(.) : \mathfrak{T} \longrightarrow \mathfrak{R}^n$ *is* **the motion of the system.**
The instantaneous value of the (internal, output) state vector function $\mathbf{S}_{(.)}(.)$ *at an instant $t \in \mathfrak{T}$ is the instantaneous (internal, output) state vector $\mathbf{S}_{(.)}(t)$ at the instant t, respectively*
The plant desired state behavior *is denoted by* $\mathbf{S}_d(t)$.

This definition broadens and generalizes the well known and commonly accepted definition of the state of the dynamical in general, control in particular, systems.

In what follows the term *mathematical system* denotes the accepted mathematical model (description) of the corresponding physical system.

The system explicit internal dynamics variable is its *internal (dynamics) state variable* S_I. This is typical for the *ISO*, *EISO*, and *HISO* systems.

The *IO* and *IIO* systems possess the explicit output dynamics, too. The internal dynamics of the *IO* systems has not been well-studied directly. The system output dynamics variable is its *output (dynamics) state variable* S_O.

The *IO* system internal dynamics is simultaneously its output dynamics so that $S_I = S_O = S$, where S is the system *full state variable* S_F, $S_F = S$.

The internal dynamics of the *ISO*, *EISO*, and *HISO* systems determines completely their output dynamics in the free regime so that for them $S_I = S_O = S_F = S$, too.

The *IIO* system internal dynamics and output dynamics are explicit and different so that $S_I \neq S_O$ and the full state variable is the vector variable

$$\mathbf{S}_F = \mathbf{S} = \left[\mathbf{S}_I^T \vdots \mathbf{S}_O^T\right]^T .$$

The properties of the system determine the form and the character of the system state vector $\mathbf{S}$:

- **The Input-Output (*IO*) systems** are described by the ν-th order linear vector differential *input-output, i.e., the output state,* equation of the output vector $\mathbf{Y} \in \mathfrak{R}^N$,

$$\mathbf{Y} = \left[Y_1 \vdots Y_2 \vdots ... \vdots Y_N\right]^T \in \mathfrak{R}^N, \ Y_i \in \mathfrak{R}, \ i = 1, 2, ..., N. \quad (1.16)$$

Their extended output vector $\mathbf{Y}^{\nu-1}$,

$$\mathbf{Y}^{\nu-1} = \left[\mathbf{Y}^T \vdots \mathbf{Y}^{(1)^T} \vdots ... \vdots \mathbf{Y}^{(\nu-1)^T}\right]^T \in \mathfrak{R}^{\nu N}, \ n = \nu N, \quad (1.17)$$

is their state vector $\mathbf{S}_{IO}$, which is also their internal state vector $\mathbf{S}_{IOI}$, their output state vector $\mathbf{S}_{IOO}$, and their full state vector $\mathbf{S}_F$,

$$\mathbf{S}_{IOI} = \mathbf{S}_{IOO} = \mathbf{S}_{IOF} = \mathbf{S}_{IO} = \mathbf{Y}^{\nu-1} \in \mathfrak{R}^n, \ n = \nu N. \quad (1.18)$$

- **The Input-State-Output** (*ISO*) **systems** are determined by the first order linear vector differential equation in the vector **X** (1.20), which is *the (internal) state equation,* by *the algebraic output vector equation* of the output vector **Y**, and the only derivative in them is the first derivative of the state vector. Their state vector $\mathbf{S}_{ISO}$ is the vector **X**, which is also their internal state vector $\mathbf{S}_{ISOI}$ and their full state vector $\mathbf{S}_{ISOF}$:

$$\mathbf{S}_{ISOI} = \mathbf{S}_{ISOF} = \mathbf{S}_{ISO} = \mathbf{X} \in \mathfrak{R}^n. \tag{1.19}$$

They do not possess the output state vector $\mathbf{S}_O$ because they do not have an independent output dynamics. Their output equation does not contain any derivative of the output vector.

- **The Extended Input-State-Output** (*EISO*) **systems** are determined by the first order linear vector differential equation in the vector **X** (1.20),

$$\mathbf{X} = \left[X_1 \vdots X_2 \vdots \ldots \vdots X_n \right]^T \in \mathfrak{R}^n,\ X_i \in \mathfrak{R},\ \forall i = 1, 2, \ldots, n, \tag{1.20}$$

which is *the (internal) state equation,* by *the algebraic output vector equation* of the output vector **Y**, and, in addition to the first derivative of the state vector, there are derivatives of the input vector only in the state equation. Their state vector $\mathbf{S}_{EISO}$ is the vector **X** (1.20) that is also the internal state vector $\mathbf{S}_{EISOI}$, and the full state vector $\mathbf{S}_{EISOF}$:

$$\mathbf{S}_{EISOI} = \mathbf{S}_{EISOF} = \mathbf{S}_{EISO} = \mathbf{X} \in \mathfrak{R}^n, \tag{1.21}$$

They do not possess the output state vector $\mathbf{S}_O$ for the same reason for which the *ISO* systems do not have the output state vector.

Note 15 *On the highest derivative of the input vector*

In order to avoid the problem of the appearance of impulse discontinuities in the system behavior the systems theory and the control theory restrict the order of the highest derivative of the input vector to be at most equal to the system order. However, the problem of the appearance of impulse disconti- nuities in the system behavior does not exist if the input vector function is defined and continuously differentiable μ−times, where μ is the order of the highest input vector derivative acting on the system. For its physical origin see in the sequel Note 50 (Subsection 2.1.1).

- **The Higher Order-Input-State-Output** $(HISO)$ **systems** are characterized by the α-*th* order linear vector differential equation, i.e., *the α-th order (internal) state equation*, in *the substate vector* $\mathbf{R}$,

$$\mathbf{R} = \left[R_1 \vdots R_2 \vdots ... \vdots R_\rho \right]^T \in \mathfrak{R}^\rho, \ R_i \in \mathfrak{R}, \ i = 1, 2, ..., \rho, \quad (1.22)$$

and are additionally determined by *the algebraic output vector equation* of the output vector $\mathbf{Y}$. Their internal state vector $\mathbf{S}_{HISOI}$ is the extended vector $\mathbf{R}^{\alpha-1}$,

$$\mathbf{R}^{\alpha-1} = \left[\mathbf{R}^T \vdots \mathbf{R}^{(1)^T} \vdots ... \vdots \mathbf{R}^{(\alpha-1)^T} \right]^T \in \mathfrak{R}^{\alpha\rho}, \ n = \alpha\rho, \quad (1.23)$$

which is also their full state vector $\mathbf{S}_f$,

$$\mathbf{S}_{HISOI} = \mathbf{S}_{HISOf} = \mathbf{S}_{HISO} = \mathbf{R}^{\alpha-1} \in \mathfrak{R}^n, \ n = \alpha\rho. \quad (1.24)$$

They do not possess the output state vector $\mathbf{S}_O$. The derivatives of the input vector can exist only in the state equation.

- **The Input-Internal and Output state** (IIO) **systems** are characterized by the α-*th* order linear vector differential equation, i.e., by *the α-th order internal state equation*, in *the substate vector* $\mathbf{R}$, and by the linear output vector ν-*th* order differential equation, i.e., by *the output state equation* of the output vector $\mathbf{Y}$. Their extended vector $\mathbf{R}^{\alpha-1}$ (1.24) is their internal state vector $\mathbf{S}_{IIOI}$ (1.25),

$$\mathbf{S}_{IIOI} = \mathbf{R}^{\alpha-1} \in \mathfrak{R}^{n_I}, \ n_I = \alpha\rho, \quad (1.25)$$

and their output state vector $\mathbf{S}_{IIOO}$ is the extended output vector $\mathbf{Y}^{\nu-1}$,

$$\mathbf{S}_{IIOO} = \mathbf{Y}^{\nu-1} = \left[\mathbf{Y}^T \vdots \mathbf{Y}^{(1)^T} \vdots ... \vdots \mathbf{Y}^{(\nu-1)^T} \right]^T \in \mathfrak{R}^{n_O}, \ n_O = \nu N.$$
$$(1.26)$$

Their full state vector $\mathbf{S}_{IIOf}$, which is their state vector $\mathbf{S}_{IIO}$, is composed of their internal state vector $\mathbf{S}_{IIOI} = \mathbf{R}^{\alpha-1}$ and of their output state vector $\mathbf{S}_{IIO} = \mathbf{Y}^{\nu-1}$,

$$\mathbf{S}_{IIOf} = \left[\begin{array}{c} \mathbf{S}_{IIOI} \\ \mathbf{S}_{IIOO} \end{array} \right] = \left[\begin{array}{c} \mathbf{R}^{\alpha-1} \\ \mathbf{Y}^{\nu-1} \end{array} \right] = \mathbf{S}_{IIO} \in \mathfrak{R}^n, \ n = \alpha\rho + \nu N. \quad (1.27)$$

Comment 16 *The state variables and the state vectors defined by (1.16)-*
(1.27) have the full physical sense (for more details, see Note 28 in Section
2.1 and Note 42 in Section 3.1).

Definition 17 *System state, motion, and response*
   The system state vector $S(t)$ at a moment $t \in \mathfrak{T}$ is the vector value of
the system motion $S(.;t_0;S_0;\mathbf{I})$ at the same moment t:

$$\mathbf{S}\left(t\right) \equiv \mathcal{S}(t;t_0;\mathbf{S}_0;\mathbf{I}) \Longrightarrow \mathbf{S}\left(t_0\right) \equiv \mathcal{S}(t_0;t_0;\mathbf{S}_0;\mathbf{I}) \equiv \mathbf{S}_0.$$

1.5   Control

Definition 18 *Control variable and control vector*
   An input variable U of a system (e.g., of a plant) that acts, together
with its μ derivatives, on the system by using information about the sys-
tem demanded behavior in order to force the system to realize its demanded
behavior under the system nominal conditions and to force the system real
behavior to be sufficiently close to the system demanded behavior under per-
*turbed conditions is **the control variable** for the system.*
   If and only if there are several, e.g., r, control variables U_1, U_2, ... , U_r,
*then they form **the control vector** (for short: **control**) $\mathbf{U}$,*

$$\mathbf{U} = \left[U_1 \vdots U_2 \vdots ... \vdots U_r \right]^{T} \in \mathfrak{R}^r, \tag{1.28}$$

and together with their μ derivatives that act on the system form the extended
control vector $\mathbf{U}^{\mu}$,

$$\mathbf{U}_i^{\mu} = \left[U_i^{i} \vdots U_i^{(1)} \vdots ... \vdots U_i^{(\mu)} \right]^{T} \in \mathfrak{R}^{\mu+1}, \; i = 1, 2, ..., r, \tag{1.29}$$

$$\mathbf{U}^{\mu} = \left[\mathbf{U}_1^{(\mu)} \vdots \mathbf{U}_2^{(\mu)} \vdots ... \vdots \mathbf{U}_r^{(\mu)} \right]^{T} \in \mathfrak{R}^{(\mu+1)r}. \tag{1.30}$$

The instantaneous values of the control variables U_i and of the control vector
$\mathbf{U}$ *at an instant $t \in \mathfrak{T}$ are $U_i\left(t\right)$ and $\mathbf{U}\left(t\right)$, respectively.*
   *A system that creates, generates, the control for the given system is **the***
***controller** $\mathcal{C}$ for the given system. Its output vector $\mathbf{Y}_C$ is the control vector*
$\mathbf{U}$, $\mathbf{Y}_C = \mathbf{U}$.

The physical nature of a control variable can be anyone.

Note 19 *Rejection or compensation?*

In this book we accept to use the term "compensation (for disturbance action)" rather than the term "rejection (the disturbance action)" for the reasons explained in [41, Remark 134, p. 62] and [45, Remark 234, pp. 169, 170].

Definition 20 *Control system*

The system composed of a plant and of its controller is the control system CS of the plant.

Note 21 *Plant input and output vectors*

The plant $\mathcal{P}$ input vectors are in general the disturbance vector $\mathbf{D}$ and the control vector $\mathbf{U}$, so that the plant input vector $\mathbf{I}_P$ has two subvectors:

$$\mathbf{I}_P = \begin{bmatrix} \mathbf{D}^T & \mathbf{U}^T \end{bmatrix}^T \in \mathfrak{R}^{d+r}. \tag{1.31}$$

The plant $\mathcal{P}$ output vector $\mathbf{Y}_p$ is in general denoted by $\mathbf{Y}$,

$$\mathbf{Y}_p = \mathbf{Y} \in \mathfrak{R}^N. \tag{1.32}$$

Note 22 *Controller input and output vectors*

The controller $\mathcal{C}$ input vectors are in general the disturbance vector $\mathbf{D}$, the plant output vector $\mathbf{Y}$, and the plant desired output vector $\mathbf{Y}_d$, so that the controller input vector in general

$$\mathbf{I}_C = \begin{bmatrix} \mathbf{D}^T & \mathbf{Y}^T & \mathbf{Y}_d^T \end{bmatrix}^T \in \mathfrak{R}^{d+2N}, \tag{1.33}$$

The feedback controller $\mathcal{C}_f$ input vectors are in general the plant output vector $\mathbf{Y}$ and the plant desired output vector $\mathbf{Y}_d$ so that the feedback controller input vector

$$\mathbf{I}_{Cf} = \begin{bmatrix} \mathbf{Y}^T & \mathbf{Y}_d^T \end{bmatrix}^T \in \mathfrak{R}^{2N}, \tag{1.34}$$

Usually we treat mathematically *the output error vector* $\mathbf{e}$,

$$\mathbf{e} = \mathbf{Y}_d - \mathbf{Y}, \tag{1.35}$$

as the feedback controller input vector,

$$\mathbf{I}_{Cf} = \mathbf{e}, \tag{1.36}$$

although the controller receives the signals on $\mathbf{Y}_d$ and $\mathbf{Y}$, determines their difference, i.e., the output error vector $\mathbf{e} = \mathbf{Y}_d - \mathbf{Y}$, and creates the error signal $\xi_\mathbf{e}$ usually proportional to $\mathbf{e}$, $\xi_\mathbf{e} = k_\mathbf{e}\mathbf{e}$.

The controller $\mathcal{C}$ output vector $\mathbf{Y}_C$ in general and the feeback controller $\mathcal{C}_f$ output vector $\mathbf{Y}_{Cf}$ in particular are the control vector $\mathbf{U}$,

$$\mathbf{Y}_C = \mathbf{Y}_{Cf} = \mathbf{U} \in \mathfrak{R}^r. \tag{1.37}$$

Note 23 *Control system input and output vectors*

The control system $\mathcal{CS}$ input vectors and the closed loop, i.e., feedback, control system $\mathcal{CS}_f$ input vectors are in general the disturbance vector $\mathbf{D}$ and the plant desired output vector $\mathbf{Y}_d$ so that the control system input vector

$$\mathbf{I}_{CS} = \mathbf{I}_{CSf} = \begin{bmatrix} \mathbf{D}^T & \mathbf{Y}_d^T \end{bmatrix}^T \in \mathfrak{R}^{d+2N}. \tag{1.38}$$

The control system $\mathcal{CS}$ output vector $\mathbf{Y}_C$ and the closed loop, i.e., feedback, control system $\mathcal{CS}_f$ output vector $\mathbf{Y}_{Cf}$ are the same and are the plant output vector $\mathbf{Y}_P$,

$$\mathbf{Y}_{CS} = \mathbf{Y}_{Cf} = \mathbf{Y}_P \in \mathfrak{R}^N. \tag{1.39}$$

Chapter 2

IO systems

2.1 *IO* system mathematical model

2.1.1 Time domain

This section deals with physical dynamical systems in general and control systems in particular, which are mathematically described directly in the form of a *time*-invariant linear vector ***Input-Output*** (*IO*) differential equation of the classical form (2.1) ,

$$\sum_{k=0}^{k=\nu} A_k \mathbf{Y}^{(k)}(t) = \sum_{k=0}^{k=\eta} D_k \mathbf{D}^{(k)}(t) + \sum_{k=0}^{k=\mu} B_k \mathbf{U}^{(k)}(t) = \sum_{k=0}^{k=\xi} H_k \mathbf{I}^{(k)}(t), \ \forall t \in \mathfrak{T}_0,$$

$$\nu \geq 1, \ \xi = \max\left(\eta, \mu\right), \ \mathbf{Y}^{(k)}(t) = \frac{d^k \mathbf{Y}(t)}{dt^k}, \ 0 \leq \eta \leq \nu, \ 0 \leq \mu \leq \nu,$$

$$A_k \in \mathfrak{R}^{N \times N}, \ D_k \in \mathfrak{R}^{N \times d}, \ B_k \in \mathfrak{R}^{N \times r}, \ k = 0, 1, .., \nu, \ det A_\nu \neq 0,$$

$$\eta < \nu \Longrightarrow D_i = O_{N,d}, \ i = \eta + 1, \ \eta + 2, ..., \nu.$$

$$\mu < \nu \Longrightarrow B_i = O_{N,r}, \ i = \mu + 1, \ \mu + 2, ..., \nu. \tag{2.1}$$

Note 24 *System, plant, and control system*
  If and only if there is $k \in \{0, 1, ..., \mu\}$ such that $B_k \neq O_{N,r}$ then the IO system (2.1) becomes the IO plant (2.1) (Definition 9, Section 1.4). Otherwise, the IO system (2.1) represents the IO control system, $\xi = \eta$ and $H_k \equiv \left[D_k \vdots O_{N,r} \right].$

The disturbance vector $\mathbf{D}$ (1.12) (Section 1.4) and the control vector $\mathbf{U}$ (1.28) (Section 1.5) compose the system input vector $\mathbf{I}$ (2.4) (Section 1.4):

$$\mathbf{I} = \mathbf{I}_{IO} = \left[\begin{array}{c} \mathbf{D} \\ \mathbf{U} \end{array} \right] \in \mathfrak{R}^{d+r}, \ M = d + r. \tag{2.2}$$

We accept the following:

Condition 25 *The matrix A_ν of the IO system (2.1) is nonsingular, i.e., it obeys*

$$det A_\nu \neq 0. \tag{2.3}$$

Note 26 ***Throughout this book we accept the validity of Condition 25.***

Note 27 *The condition on the nonsingularity of the matrix A_ν imposed in Condition 25 guarantees*

$$\exists s \in \mathbb{C} \Longrightarrow \det \left(\sum_{k=0}^{k=\nu} A_k s^k \right) \neq 0,$$

and permits the solvability of the Laplace transform of (2.1) for $\mathbf{Y}(s)$ [40].

Besides, the condition $det A_\nu \neq 0$ is a sufficient condition, but not necessary condition, for all the output variables of the system (2.1) to have the same order ν of their highest derivatives.

$\mathfrak{R}^k$ is the k-dimensional real vector space, $k \in \{1, 2, ...\}$, (Section 1.4). $\mathbb{C}^k$ denotes the k-dimensional complex vector space (Section 1.4). $O_{M \times N}$ is the zero matrix in $\mathfrak{R}^{M \times N}$, and O_N is the zero matrix in $\mathfrak{R}^{N \times N}$, $O_N = O_{N \times N}$. The vector $\mathbf{0}_k \in \mathfrak{R}^k$ is the zero vector in $\mathfrak{R}^k$ and $\mathbf{1}_k \in \mathfrak{R}^k$ is the unit vector in $\mathfrak{R}^k$,, (Section 1.2).

The total *input vector*

$$\mathbf{I} = [I_1 \ I_2 \ ... \ I_M]^T \in \mathfrak{R}^M, \tag{2.4}$$

its subvectors

$$\begin{array}{rcl} \mathbf{D} & = & [D_1 \ D_2 \ ... \ D_d]^T \in \mathfrak{R}^d, \\ \mathbf{U} & = & [U_1 \ U_2 \ ... \ U_r]^T \in \mathfrak{R}^r, \end{array} \tag{2.5} \tag{2.6}$$

(Definition 10), and the total *output vector*

$$\mathbf{Y} = [Y_1 \ Y_2 \ ... \ Y_N]^T \in \mathfrak{R}^N, \tag{2.7}$$

(Definition 12) (Section 1.4). The values I_i, D_j, U_k, and Y_l are the total values of the input and the output variables, respectively. *The total value* of a variable signifies that its value is measured with respect to its total zero, if it has the total zero value, and if it does not have the total zero value then an appropriate value is accepted to play the role of the total zero value.

The form of the system mathematical model (2.1) is too complex and makes the system study unreasonably cumbersome. We simplify it by applying the elegant and simple compact notation for the extended matrices proposed in [35] and in brief explained in Note 6 (Section 1.2). At first we introduce the extended matrices $A^{(\nu)}$, $B^{(\mu)}$, and $D^{(\eta)}$,

$$A^{(\nu)} = \left[A_0 \vdots A_1 \vdots ... \vdots A_\nu \right] \in \mathfrak{R}^{N \times (\nu+1)N},$$

$$B^{(\mu)} = \left[B_0 \vdots B_1 \vdots ... \vdots B_\mu \right] \in \mathfrak{R}^{N \times (\mu+1)r},$$

$$D^{(\eta)} = \left[D_0 \vdots D_1 \vdots ... \vdots D_\eta \right] \in \mathfrak{R}^{N \times (\eta+1)d}, \tag{2.8}$$

and then the very simple extended vectors $\mathbf{D}^\eta(t)$, $\mathbf{I}^\xi(t)$, $\mathbf{U}^\mu(t)$, and $\mathbf{Y}^\nu(t)$:

$$\mathbf{D}^\eta(t) = \left[\mathbf{D}^T(t) \vdots \mathbf{D}^{(1)^T}(t) \vdots ... \vdots \mathbf{D}^{(\eta)^T}(t) \right]^T \in \mathfrak{R}^{(\eta+1)d}, \tag{2.9}$$

$$\mathbf{I}^\xi(t) = \left[\mathbf{I}^T(t) \vdots \mathbf{I}^{(1)^T}(t) \vdots ... \vdots \mathbf{I}^{(\xi)^T}(t) \right]^T \in \mathfrak{R}^{(\xi+1)M} \tag{2.10}$$

$$\mathbf{U}^\mu(t) = \left[\mathbf{U}^T(t) \vdots \mathbf{U}^{(1)^T}(t) \vdots ... \vdots \mathbf{U}^{(\mu)^T}(t) \right]^T \in \mathfrak{R}^{(\mu+1)r} \tag{2.11}$$

$$\mathbf{Y}^\nu(t) = \left[\mathbf{Y}^T(t) \vdots \mathbf{Y}^{(1)^T}(t) \vdots ... \vdots \mathbf{Y}^{(\nu)^T}(t) \right]^T \in \mathfrak{R}^{(\nu+1)N}, \tag{2.12}$$

They induce the corresponding initial vectors $\mathbf{D}_0^{\eta-1} = \mathbf{D}^{\eta-1}(0)$, $\mathbf{I}_0^{\xi-1} = \mathbf{I}^{\xi-1}(0)$, $\mathbf{U}_0^{\mu-1} = \mathbf{U}^{\mu-1}(0)$, and $\mathbf{Y}_0^{\nu-1} = \mathbf{Y}^{\nu-1}(0)$.

We repeat that the upper index ν in the parentheses in $A^{(\nu)}$ makes $A^{(\nu)}$ essentially different from the ν-*th* power A^ν of A,

$$A^{(\nu)} = \left[A_0 \vdots A_1 \vdots ... \vdots A_\nu \right] \neq A^\nu = \underbrace{AA....A}_{\nu \ times}. \tag{2.13}$$

Notice also that for the extended vector $\mathbf{Y}^v$ the superscript v is not in the parentheses in order to distinguish it from the v-th derivative $d^v \mathbf{Y}(t)/dt^v$ of $\mathbf{Y}(t)$,

$$\mathbf{Y}^v(t) = \left[\mathbf{Y}^T(t) \,\vdots\, \mathbf{Y}^{(1)^T}(t) \,\vdots\, ... \,\vdots\, \mathbf{Y}^{(\nu)^T}(t)\right]^T \neq \mathbf{Y}^{(\nu)}(t) = \frac{d^\nu \mathbf{Y}(t)}{dt^\nu}. \quad (2.14)$$

The application of the above compact notation (2.8)-(2.12) to the IO vector differential equation (2.1) transforms it into the following simple, elegant, and compact form:

$$A^{(\nu)}\mathbf{Y}^\nu(t) = D^{(\eta)}\mathbf{D}^\eta(t) + B^{(\mu)}\mathbf{U}^\mu(t) = H^{(\mu)}\mathbf{I}^\mu(t), \,\forall t \in \mathfrak{T}_0,$$

$$H^{(\mu)} = \left[D^{(\mu)} \,\vdots\, B^{(\mu)}\right], \,\mathbf{I}^\mu(t) = \left[\left(\mathbf{D}^{(\mu)}\right)^T \,\vdots\, \left(\mathbf{U}^{(\mu)}\right)^T\right]^T. \quad (2.15)$$

Note 28 *The state vector* $\mathbf{S}_{IO}$ *of the IO system (2.15) is defined in (1.18) (Section 1.4) by:*

$$\mathbf{S}_{IO} = \mathbf{Y}^{\nu-1} = \left[\mathbf{Y}^T \,\vdots\, \mathbf{Y}^{(1)^T} \,\vdots\, ... \,\vdots\, \mathbf{Y}^{(\nu-1)^T}\right]^T \in \mathfrak{R}^n, \, n = \nu N, \quad (2.16)$$

This new vector notation $\mathbf{Y}^{\nu-1}$ *has permitted us to define the state of the IO system (2.15) by preserving the physical sense. It enabled us to establish in [40] the direct link between the definitions of the Lyapunov and of BI stability properties with the corresponding conditions for them in the complex domain. It enables us to discover in what follows the complex domain criteria for observability, controllability, and trackability directly from their definitions. Such criteria possess the complete physical meaning.*

The state variables and the state vector $\mathbf{S}_{IO} = \mathbf{Y}^{\nu-1}$, *Equation (2.16), have the well-known form called* **phase form***, i.e.,* $\mathbf{S}_{IO} = \mathbf{Y}^{\nu-1}$ *is the phase (state) vector.*

2.1.2   Complex domain

The following complex matrix functions [35], [36], [40], the first one of which is $S_i^{(k)}(.) : \mathbb{C} \longrightarrow \mathbb{C}^{\,i(k+1)\times i}$, essentially simplify the system study via the complex domain,

$$S_i^{(k)}(s) = \left[s^0 I_i \,\vdots\, s^1 I_i \,\vdots\, s^2 I_i \,\vdots\, ... \,\vdots\, s^k I_i\right]^T \in \mathbb{C}^{\,i(k+1)\times i},$$

$$(k, i) \in \{(\mu, M), \,(\nu, N)\}. \quad (2.17)$$

The matrix I_i is the *i-th* order identity matrix, $I_i \in \mathfrak{R}^{i \times i}$. Another complex function is $Z_k^{(\varsigma-1)}(.) : \mathbb{C} \to \mathbb{C}^{(\varsigma+1)k \times \varsigma k}$,

$$
Z_k^{(\varsigma-1)}(s) = \begin{bmatrix} O_k & O_k & O_k & \dots & O_k \\ s^0 I_k & O_k & O_k & \dots & O_k \\ \dots & \dots & \dots & \dots & \dots \\ s^{\varsigma-1} I_k & s^{\varsigma-2} I_k & s^{\varsigma-3} I_k & \dots & s^0 I_k \end{bmatrix}, \ \varsigma \geq 1,
$$

$$
\varsigma = 1 \implies Z_k^{(1-1)}(s) = Z_k^{(0)}(s) = s^0 I_k = I_k,
$$

$$
Z_k^{(\varsigma-1)}(s) \in \mathbb{C}^{(\varsigma+1)k \times \varsigma k}, \ (\varsigma, k) \in \{(\mu, M), \ (\nu, N)\}, \tag{2.18}
$$

where the final entry of $Z_k^{(\varsigma-1)}(s)$ is always $s^0 I_k$.

Note 29 *[35], [36], [40] If $\varsigma = 0$ then the matrix $Z_k^{(\varsigma-1)}(s) = Z_k^{(-1)}(s)$ is not defined and should be completely omitted rather than to be replaced by the zero matrix. Because the matrix $Z_k^{(\varsigma-1)}(s)$ is not defined for $\varsigma \leq 0$, i.e., it does not exist for $\varsigma \leq 0$. Derivatives exist only for natural numbers, i.e., $\mathbf{Y}^{(\varsigma)}(t)$ can exist only for $\varsigma \geq 1$. Matrix function $Z_k^{(\varsigma-1)}(.)$ is related to the Laplace transform of derivatives only.*

It is well known that the Laplace transform of

$$
\sum_{k=0}^{k=\nu} A_k \mathbf{Y}^{(k)}(t) \tag{2.19}
$$

contains the Laplace transform $\mathbf{Y}(s)$ of $\mathbf{Y}(t)$ multiplied by a matrix polynomial in s and a double sum containing the products of powers of the complex variable s and initial values of $\mathbf{Y}(t)$ and of its derivatives up to the order of $\nu - 1$, all multiplied by the corresponding system matrices. The references [35], [36], and [40] contain the proof that the simple, compact, and elegant form of the Laplace transform of (2.19) reads:

$$
\mathcal{L} \left\{ \sum_{k=0}^{k=\nu} A_k \mathbf{Y}^{(k)}(t) \right\} = A^{(\nu)} S_N^{(\nu)}(s) \mathbf{Y}(s) - A^{(\nu)} Z_N^{(\nu-1)}(s) \mathbf{Y}_0^{\nu-1}, \tag{2.20}
$$

where the matrices $S_N^{(\nu)}(s)$ and $Z_N^{(\nu-1)}(s)$ are defined in (2.17) and in (2.18), respectively. Analogously,

$$
\mathcal{L} \left\{ \sum_{k=0}^{k=\eta} D_k \mathbf{D}^{(k)}(t) \right\} = D^{(\eta)} S_d^{(\eta)}(s) \mathbf{D}(s) - D^{(\eta)} Z_d^{(\eta-1)}(s) \mathbf{D}^{\eta-1}(0), \tag{2.21}
$$

$$\mathcal{L}\left\{\sum_{k=0}^{k=\mu} B_k \mathbf{U}^{(k)}(t)\right\} = B^{(\mu)} S_r^{(\mu)}(s)\mathbf{U}(s) - B^{(\mu)} Z_r^{(\mu-1)}(s)\mathbf{U}^{\mu-1}(0). \quad (2.22)$$

Equations (2.20), (2.21) and (2.22) determine the simple compact form of the Laplace transform of (2.1), hence of (2.15):

$$A^{(\nu)} S_N^{(\nu)}(s)\mathbf{Y}(s) - A^{(\nu)} Z_N^{(\nu-1)}(s)\mathbf{Y}^{\nu-1}(0) =$$
$$= D^{(\eta)} S_d^{(\eta)}(s)\mathbf{D}(s) - D^{(\eta)} Z_d^{(\eta-1)}(s)\mathbf{D}^{\eta-1}(0)+$$
$$+B^{(\mu)} S_r^{(\mu)}(s)\mathbf{U}(s) - B^{(\mu)} Z_r^{(\mu-1)}(s)\mathbf{U}^{\mu-1}(0). \quad (2.23)$$

This equation determines $\mathbf{Y}(s)$:

$$\mathbf{Y}(s) = \left(A^{(\nu)} S_N^{(\nu)}(s)\right)^{-1} \bullet$$

$$\bullet \left[D^{(\eta)} S_d^{(\eta)}(s)\vdots B^{(\mu)} S_r^{(\mu)}(s)\vdots - D^{(\eta)} Z_d^{(\eta-1)}(s)\vdots - B^{(\mu)} Z_r^{(\mu-1)}(s)\vdots A^{(\nu)} Z_N^{(\nu-1)}(s)\right]$$

$$\bullet \begin{bmatrix} \mathbf{D}(s) \\ \mathbf{U}(s) \\ \mathbf{D}^{\eta-1}(0) \\ \mathbf{U}^{\mu-1}(0) \\ \mathbf{Y}^{\nu-1}(0) \end{bmatrix} = F_{IO}(s)\,\mathbf{V}_{IO}(s), \quad (2.24)$$

since the inverse of $A^{(\nu)} S_N^{(\nu)}(s)$ exists due to Condition 25. The plant full transfer function matrix results from (2.24)

$$F_{IO}(s) = \left(A^{(\nu)} S_N^{(\nu)}(s)\right)^{-1} \bullet$$

$$\bullet \left[D^{(\eta)} S_d^{(\eta)}(s)\vdots B^{(\mu)} S_r^{(\mu)}(s)\vdots - D^{(\eta)} Z_d^{(\eta-1)}(s)\vdots - B^{(\mu)} Z_r^{(\mu-1)}(s)\vdots A^{(\nu)} Z_N^{(\nu-1)}(s)\right]$$
$$(2.25)$$

The inverse Laplace transform of $F_{IO}(s)$ is *the IO* system *full fundamental matrix* $\Psi_{IO}(t)$, $\Psi_{IO}(t) = \mathcal{L}^{-1}\{F_{IO}(s)\}$, and the inverse Laplace transform of

$$\left(A^{(\nu)} S_N^{(\nu)}(s)\right)^{-1} = \frac{adj\left(A^{(\nu)} S_N^{(\nu)}(s)\right)}{p_{IO}(s)} = p_{IO}^{-1}(s)\,adj\left(A^{(\nu)} S_N^{(\nu)}(s)\right) \quad (2.26)$$

is *the IO* system *fundamental matrix* $\Phi_{IO}(t)$ [40]:

$$\Phi_{IO}(t) = \mathcal{L}^{-1}\left\{\left(A^{(\nu)} S_N^{(\nu)}(s)\right)^{-1}\right\}. \quad (2.27)$$

Equation (2.25) discovers that the polynomial $p_{IO}(s)$,

$$p_{IO}(s) = \det\left(A^{(\nu)}S_N^{(\nu)}(s)\right),\tag{2.28}$$

is the characteristic polynomial of the IO system (2.15) and the denominator polynomial of all its transfer function matrices due to Equation (2.25) as shown also in what follows. Equation (2.25) induces also the matrix polynomial $L_{IO}(s)$ defined by

$$L_{IO}(s) = adj\left(A^{(\nu)}S_N^{(\nu)}(s)\right)B^{(\mu)}S_r^{(\mu)}(s), \quad L_{IO}(s) \in \mathbb{C}^{N\times r}.\tag{2.29}$$

It is the numerator matrix polynomial of the plant transfer function matrix $G_{IOU}(s)$ relative to the control vector $\mathbf{U}$:

$$G_{IOU}(s) = \left(A^{(\nu)}S_N^{(\nu)}(s)\right)^{-1}B^{(\mu)}S_r^{(\mu)}(s) =$$
$$= \frac{adj\left(A^{(\nu)}S_N^{(\nu)}(s)\right)B^{(\mu)}S_r^{(\mu)}(s)}{p_{IO}(s)} = \frac{L_{IO}(s)}{p_{IO}(s)},\tag{2.30}$$

Equation (2.24) determines also all other specific transfer function matrices of the IO system (2.15):

- With respect to the disturbance $\mathbf{D}$:

$$G_{IOD}(s) = \frac{adj\left(A^{(\nu)}S_N^{(\nu)}(s)\right)D^{(\eta)}S_d^{(\eta)}(s)}{p_{IO}(s)},\tag{2.31}$$

- With respect to the initial conditions $\mathbf{D}^{\eta-1}(0)$ of the disturbance $\mathbf{D}$:

$$G_{IOD_0}(s) = -\frac{adj\left(A^{(\nu)}S_N^{(\nu)}(s)\right)D^{(\eta)}Z_d^{(\eta-1)}(s)}{p_{IO}(s)},\tag{2.32}$$

- With respect to the initial conditions $\mathbf{U}^{\mu-1}(0)$ of the control $\mathbf{U}$:

$$G_{IOU_0}(s) = -\frac{adj\left(A^{(\nu)}S_N^{(\nu)}(s)\right)B^{(\mu)}Z_r^{(\mu-1)}(s)}{p_{IO}(s)},\tag{2.33}$$

- With respect to the initial conditions $\mathbf{Y}^{\nu-1}(0)$ of the output $\mathbf{Y}$:

$$G_{IOY_0}(s) = \frac{adj\left(A^{(\nu)}S_N^{(\nu)}(s)\right)A^{(\nu)}Z_N^{(\nu-1)}(s)}{p_{IO}(s)}.\tag{2.34}$$

The Laplace transform $\mathbf{V}_{IO}(s)$ of the IO system action vector $\mathbf{V}_{IO}(t)$ reads

$$\mathbf{V}_{IO}(s) = \begin{bmatrix} \mathbf{D}(s) \\ \mathbf{U}(s) \\ \mathbf{D}^{\eta-1}(0) \\ \mathbf{U}^{\mu-1}(0) \\ \mathbf{Y}^{\nu-1}(0) \end{bmatrix} = \begin{bmatrix} \mathbf{I}_{IO}(s) \\ \mathbf{C}_{IO0} \end{bmatrix}. \tag{2.35}$$

The Laplace transform $\mathbf{I}_{IO}(s)$ of the IO system input vector $\mathbf{I}_{IO}(t)$ reads:

$$\mathbf{I}_{IO}(s) = \begin{bmatrix} \mathbf{D}(s) \\ \mathbf{U}(s) \end{bmatrix}. \tag{2.36}$$

The vector $\mathbf{C}_{IO0}$ of all IO system initial conditions has the following form:

$$\mathbf{C}_{IO0} = \begin{bmatrix} \mathbf{D}^{\eta-1}(0) \\ \mathbf{U}^{\mu-1}(0) \\ \mathbf{Y}^{\nu-1}(0) \end{bmatrix}. \tag{2.37}$$

The IO system output response $\mathbf{Y}\left(t; \mathbf{Y}_0^{\nu-1}; \mathbf{D}; \mathbf{U}\right)$ is the inverse Laplace transform of (2.24):

$$\mathbf{Y}\left(t; \mathbf{Y}_0^{\nu-1}; \mathbf{D}; \mathbf{U}\right) = \mathcal{L}^{-1}\left\{F_{IO}(s)\,\mathbf{V}_{IO}(s)\right\}. \tag{2.38}$$

Example 30 *Appendix B.1 contains an IO system Example 227.*

What follows shows an important physical meaning of the system full transfer function matrix $F(s)$. For the definition, types and properties of the Dirac unit impulse $\delta(.)$, see [40].

Definition 31 *[40, Definition 181, p. 171, Note 182, p. 172]   A matrix function $\boldsymbol{\Psi}_{IO}(.) : \mathfrak{T} \longrightarrow \mathfrak{R}^{N\times[(\mu+1)M+\nu N]}$ is the **full fundamental matrix function** of the IO system (2.15) if and only if it obeys both (i) and (ii) for an arbitrary input vector function $\mathbf{I}(.)$, Equation (2.2), and for arbitrary initial conditions $\mathbf{I}_{0-}^{\mu-1}$ and $\mathbf{Y}_{0-}^{\nu-1}$,*
    (i)

$$\mathbf{Y}(t; \mathbf{Y}_{0-}^{\nu-1}; \mathbf{I}) = \int_{0-}^{t} \left\{ \boldsymbol{\Psi}_{IO}(\tau) \begin{bmatrix} \mathbf{I}(t-\tau) \\ \delta(t-\tau)\mathbf{I}_{0-}^{\mu-1} \\ \delta(t-\tau)\mathbf{Y}_{0-}^{\nu-1} \end{bmatrix} \right\} d\tau =$$

$$= \int_{0-}^{t} \left\{ \boldsymbol{\Psi}_{IO}(t-\tau) \begin{bmatrix} \mathbf{I}(\tau) \\ \delta(\tau)\mathbf{I}_{0-}^{\mu-1} \\ \delta(\tau)\mathbf{Y}_{0-}^{\nu-1} \end{bmatrix} \right\} d\tau, \tag{2.39}$$

equivalently

$$\mathbf{Y}(t; \mathbf{Y}_{0-}^{\nu-1}; \mathbf{I}) = \int_{0-}^{t} \Gamma_{IO}(\tau)\mathbf{I}(t-\tau)d\tau + \Gamma_{IOi_0}(t)\mathbf{I}_{0-}^{\mu-1} + \Gamma_{IOy_0}(t)\mathbf{Y}_{0-}^{\nu-1},$$

$$\int_{0-}^{t} [\Gamma_{IO}(\tau)\mathbf{I}(t-\tau)d\tau] = \int_{0-}^{t} [\Gamma_{IO}(t-\tau)\mathbf{I}(\tau)d\tau],$$

$$\mathbf{I}(t-\tau) = \mathbf{I}(t,\tau), \ \Gamma_{IO}(t-\tau) = \Gamma_{IO}(t,\tau), \tag{2.40}$$

and

$$\Psi_{IO}(t) = \left[\Gamma_{IO}(t) \vdots \Gamma_{IOi_0}(t) \vdots \Gamma_{IOy_0}(t)\right],$$

$$\Gamma_{IO}(t) \in \mathfrak{R}^{N\times M}, \ \Gamma_{IOi_0}(t) \in \mathfrak{R}^{N\times\mu M}, \Gamma_{IOy_0}(t) \in \mathfrak{R}^{N\times\nu N}, \tag{2.41}$$

(ii)

$$\Gamma_{IOi_0}(0^-) = \left[\Gamma_{IOi_01} \vdots O_{N,(\mu-1)M}\right] \ where$$

$$\Gamma_{IOi_01}(0^-)\mathbf{i}_{0-} = -\int_{0-}^{0-} [\Gamma_{IO}(\tau)\mathbf{i}(t-\tau)d\tau],$$

$$\Gamma_{IOy_0}(0^-) \equiv \left[\mathbf{I}_N \vdots O_{N,(\nu-1)N}\right]. \tag{2.42}$$

Note 32 *[40, Equation (10.4), p. 172] The second Equation (2.39) under (i) of Definition 31 results from its first equation and from the properties of* $\delta(.)$:

$$\mathbf{Y}(t; \mathbf{Y}_{0-}^{\nu-1}; \mathbf{I}) =$$

$$= \int_{0-}^{t} \Gamma_{IO}(\tau)\mathbf{I}(t,\tau)d\tau + \Gamma_{IOi_0}(t)\mathbf{I}_{0-}^{\mu-1} + \Gamma_{IOy_0}(t)\mathbf{Y}_{0-}^{\nu-1}, \ t \in \mathfrak{T}_0. \tag{2.43}$$

The matrix $\Gamma_{IO}(t)$ *is the* **output fundamental matrix** *of the IO system (2.15),*

Theorem 33 *[40, Theorem 183, pp. 172, 173] (i) The full fundamental matrix function* $\Psi_{IO}(.)$ *of the IO system (2.15) is the inverse of the left Laplace transform of the system full transfer function matrix* $F_{IO}(s)$,

$$\Psi_{IO}(t) = \mathcal{L}^{-1}\{F_{IO}(s)\}. \tag{2.44}$$

(ii) The full transfer function matrix $F_{IO}(s)$ of the IO system (2.15) is the left Laplace transform of the system full fundamental matrix $\Psi_{IO}(t)$,

$$F_{IO}(s) = \mathcal{L}^{-}\{\Psi_{IO}(t)\}. \tag{2.45}$$

(iii) The submatrices $\Gamma_{IO}(t)$, $\Gamma_{IOi_0}(t)$ and $\Gamma_{IOy_0}(t)$ are the inverse Laplace transforms of $G_{IO}(s)$, $G_{IOi_0}(s)$ and $G_{IOy_0}(s)$, respectively,

$$\Gamma_{IO}(t) = \mathcal{L}^{-1}\{G_{IO}(s)\} = \mathcal{L}^{-1}\left\{\Phi_{IO}(s)\left[B^{(\mu)}S_M^{(\mu)}(s)\right]\right\},$$

$$\Gamma_{IOi_0}(t) = \mathcal{L}^{-1}\{G_{IOi_0}(s)\} = -\mathcal{L}^{-1}\left\{\Phi_{IO}(s)\left[B^{(\mu)}Z_M^{(\mu-1)}(s)\right]\right\},$$

$$\Gamma_{IOy_0}(t) = \mathcal{L}^{-1}\{G_{IOy_0}(s)\} = \mathcal{L}^{-1}\left\{\Phi_{IO}(s)\left[A^{(\nu)}Z_N^{(\nu-1)}(s)\right]\right\}, \tag{2.46}$$

*where $\Phi_{IO}(s)$ is the left Laplace transform of the IO **system fundamental matrix function** $\Phi_{IO}(.) : \mathfrak{T} \longrightarrow \mathfrak{R}^{N \times N}$,,*

$$\Phi_{IO}(s) = \mathcal{L}^{-}\{\Phi_{IO}(t)\}, \ \ \Phi_{IO}(t) = \mathcal{L}^{-1}\{\Phi_{IO}(s)\}, \tag{2.47}$$

and

$$\Phi_{IO}(s) = \left(A^{(\nu)}S_N^{(\nu)}(s)\right)^{-1}, \ \ \Phi_{IO}(t) = \mathcal{L}^{-1}\left\{\left(A^{(\nu)}S_N^{(\nu)}(s)\right)^{-1}\right\}. \tag{2.48}$$

(iv) The IO system full fundamental matrix $\Psi_{IO}(t)$ and its fundamental matrix $\Phi_{IO}(t)$ are linked as follows:

$$\Psi_{IO}(t) = \mathcal{L}^{-1}\left\{\Phi_{IO}(s)\left[B^{(\mu)}S_M^{(\mu)}(s) \vdots - B^{(\mu)}Z_M^{(\mu-1)}(s) \vdots A^{(\nu)}Z_N^{(\nu-1)}(s)\right]\right\},$$

$$\Psi_{IO}(s) = \Phi_{IO}(s)\left[B^{(\mu)}S_M^{(\mu)}(s) \vdots - B^{(\mu)}Z_M^{(\mu-1)}(s) \vdots A^{(\nu)}Z_N^{(\nu-1)}(s)\right]. \tag{2.49}$$

2.2   *IO* plant desired regime

We accept the following definition of a desired regime by following [36], [40]:

Definition 34 *Desired regime*

*A system (plant) is in **a desired** (called also: **nominal** or **nonperturbed**) **regime on** $\mathfrak{T}_0$ (for short: in **a desired regime**) if and only if it realizes its desired (output) response $\mathbf{Y}_d(t)$ all the time on $\mathfrak{T}_0$,*

$$\mathbf{Y}(t) = \mathbf{Y}_d(t), \ \forall t \in \mathfrak{T}_0. \tag{2.50}$$

*The terms **nominal** and **nonperturbed** are meaningful in general, i.e., for any system, e.g., for plants, controllers and control systems; while the term **desired** has the full sense only for plants.*

Proposition 35 *[40] In order for the plant to be in a desired (nominal, nonperturbed) regime, i.e.,*

$$\mathbf{Y}(t) = \mathbf{Y}_d(t), \ \forall t \in \mathfrak{T}_0,$$

it is necessary that the initial real output vector is equal to the initial desired output vector,

$$\mathbf{Y}_0 = \mathbf{Y}_{d0}.$$

The system cannot be in a nominal regime (on $\mathfrak{T}_0$) if its initial real output vector is different from the initial desired output vector:

$$\mathbf{Y}_0 \neq \mathbf{Y}_{d0} \Longrightarrow \exists \sigma \in \mathfrak{T}_0 \Longrightarrow \mathbf{Y}(\sigma) \neq \mathbf{Y}_d(\sigma).$$

The real initial output vector $\mathbf{Y}(0) = \mathbf{Y}_0$ is most often different from the desired initial output vector $\mathbf{Y}_d(0) = \mathbf{Y}_{d0}$. The system is most often in a *nondesired (non-nominal, perturbed, disturbed)* regime.

Definition 36 *Nominal control $\mathbf{U}_N(.)$] relative to $[\mathbf{D}(.), \mathbf{Y}_d(.)]$ of the IO plant (2.15)*
   *A control vector function $\mathbf{U}^*(.)$ of the IO plant (2.15) is **nominal relative to** $[\mathbf{D}(.), \mathbf{Y}_d(.)]$, which is denoted by $\mathbf{U}_N(.)$, if and only if*

$$\mathbf{U}(.)] = \mathbf{U}^*(.)$$

ensures that the corresponding real response $\mathbf{Y}(.) = \mathbf{Y}^*(.)$ to the input action of $\mathbf{D}(.)$ on the plant obeys $\mathbf{Y}^*(t) = \mathbf{Y}_d(t)$ all the time as soon as all the internal and the output system initial conditions are desired (nominal, nonperturbed).

This definition and (2.15) imply the following theorem:

Theorem 37 *[36, Theorem 50, pp. 46-48], [40, Theorem 56, pp. 49-52] In order for a control vector function $\mathbf{U}^*(.)$ to be nominal for the IO plant (2.15) relative to $[\mathbf{D}(.), \mathbf{Y}_d(.)]$: $\mathbf{U}^*(.) = \mathbf{U}_N(.)$, it is necessary and sufficient that 1) and 2) hold:*
   1) rank $rankB^{(\mu)}S_r^{(\mu)}(s) = N \leq r$, i.e., $rankB^{(\mu)} = N \leq r$, and
   2) any one of the following equations is valid:

$$B^{(\mu)}\mathbf{U}^{*^\mu}(t) = -D^{(\eta)}\mathbf{D}^\eta(t) + A^{(\nu)}\mathbf{Y}_d^\nu(t), \ \forall t \in \mathfrak{T}_0, \qquad (2.51)$$

or, equivalently in the complex domain:

$$\mathbf{U}^*(s) = \left(B^{(\mu)} S_r^{(\mu)}(s) \right)^T \left[\left(B^{(\mu)} S_r^{(\mu)}(s) \right) \left(B^{(\mu)} S_r^{(\mu)}(s) \right)^T \right]^{-1} \bullet$$

$$\bullet \left\langle \begin{array}{c} B^{(\mu)} Z_r^{(\mu-1)}(s) \mathbf{U}^{*^{\mu-1}}(0) + A^{(\nu)} \left[S_N^{(\nu)}(s) \mathbf{Y}_d(s) - Z_N^{(\nu-1)}(s) \mathbf{Y}_d^{\nu-1}(0) \right] - \\ -D^{(\eta)} \left[S_d^{(\eta)}(s) \mathbf{D}(s) - Z_d^{(\eta-1)}(s) \mathbf{D}^{\eta-1}(0) \right] \end{array} \right\rangle .$$

$$(2.52)$$

This theorem holds for all *IO* plants (2.15).

Condition 38 *The desired output response* $\mathbf{Y}_d(t)$ *of the IO system (2.15) is realizable, i.e.,* $N \le r$. *Both it and the nominal input* $\mathbf{I}_N(.)$ *are known.*

The compact form of the *IO* plant (2.15) in terms of the deviations follows from (2.53), (2.55), (9.3),

$$\mathbf{d} = \mathbf{D} - \mathbf{D_N}, \tag{2.53}$$

$$\mathbf{i} = \mathbf{I} - \mathbf{I}_N \tag{2.54}$$

$$\mathbf{u} = \mathbf{U} - \mathbf{U}_N, \tag{2.55}$$

$$\mathbf{y} = \mathbf{Y} - \mathbf{Y}_d, \tag{2.56}$$

and (2.15):

$$A^{(\nu)} \mathbf{y}^\nu(t) = D^{(\eta)} \mathbf{d}^\eta(t) + B^{(\mu)} \mathbf{u}^\mu(t) = H^{(\mu)} \mathbf{i}^\mu(t), \ \forall t \in \mathfrak{T}_0. \tag{2.57}$$

Note 39 *Equation (2.57) is the IO system model determined in terms of the deviations of all variables. It has exactly the same form, the same order, and the same matrices as the system model expressed in total values of the variables (2.15). They possess the same characteristics and properties by noting once more that* $\mathbf{y} = \mathbf{0}_N$ *represents* $\mathbf{Y} = \mathbf{Y}_d$. *For example, they have the same transfer function matrices, and the stability properties of* $\mathbf{y}^{\gamma-1} = \mathbf{0}_{\nu N}$ *of (2.57) are simultaneously the same stability properties of* $\mathbf{Y}_d^{\gamma-1}(t)$ *of (2.15).*

2.3   Exercises

Exercise 40 *1. Select an IO physical plant.*
   2. Determine its time domain IO mathematical model.

Chapter 3

ISO systems

3.1 *ISO* system mathematical model

3.1.1 Time domain

The dynamical systems theory and the control theory have been mainly developed for the linear **Input-State-Output** (*ISO*) **(dynamical, control) systems**. Their mathematical models contain *the state vector differential equation* (3.1) and *the output algebraic vector equation* (3.2),

$$\frac{d\mathbf{X}(t)}{dt} = A\mathbf{X}(t) + D\mathbf{D}(t) + B\mathbf{U}(t) = A\mathbf{X}(t) + P\mathbf{I}(t), \ \forall t \in \mathfrak{T}_0,$$

$$A \in \mathfrak{R}^{n \times n}, D \in \mathfrak{R}^{n \times d}, B \in \mathfrak{R}^{n \times r}, P = \begin{bmatrix} D \vdots B \end{bmatrix} \in \mathfrak{R}^{n \times (d+r)}, \qquad (3.1)$$

$$\mathbf{Y}(t) = C\mathbf{X}(t) + V\mathbf{D}(t) + U\mathbf{U}(t) = C\mathbf{X}(t) + Q\mathbf{I}(t), \ \forall t \in \mathfrak{T}_0,$$

$$C \in \mathfrak{R}^{N \times n}, V \in \mathfrak{R}^{N \times d}, U \in \mathfrak{R}^{N \times r}, Q = \begin{bmatrix} V \vdots U \end{bmatrix} \in \mathfrak{R}^{N \times (d+r)}. \qquad (3.2)$$

The *ISO* mathematical model (3.1), (3.2) is well-known also as the *state-space system (description)*.

Note 41 *System, plant, and control system*
If and only if $B \neq O_{n,r}$ then the ISO system (3.1), (3.2) becomes the ISO plant (3.1), (3.2) (Definition 9, Section 1.4). Otherwise, the ISO system (3.1), (3.2) represents the ISO control system.

The state vector $\mathbf{S}_{ISO}$ of the *ISO* system (3.1), (3.2) is the vector $\mathbf{X}$ (1.19).

31

The fundamental matrix function

$$\Phi_{ISO}(.,t_0) \equiv \Phi(.,t_0) : \mathfrak{T} \longrightarrow \mathfrak{R}^{n\times n} \tag{3.3}$$

of the system (3.1), (3.2),

$$\Phi(t,t_0) = e^{At}\left(e^{At_0}\right)^{-1} = e^{A(t-t_0)} \in \mathfrak{R}^{n\times n}, \tag{3.4}$$

has the following well-known properties:

$$det\Phi(t,t_0) \neq 0, \ \forall t \in \mathfrak{T}_0, \ \forall t_0 \in \mathfrak{T}, \tag{3.5}$$

$$\Phi(t,t_0)\,\Phi(t_0,t) = e^{A(t-t_0)}e^{A(t_0-t)} \equiv e^{A0} = I_n \Longrightarrow$$
$$\Phi(t_0,t) = \Phi^{-1}(t,t_0), \tag{3.6}$$

$$\Phi^{(1)}(t,t_0) = A\Phi(t,t_0) = \Phi(t,t_0)\,A. \tag{3.7}$$

By applying the classical method to solve the state equation (3.1) by its integration we determine its solution:

$$\mathbf{X}(t;t_0;\mathbf{X}_0;\mathbf{I}) = \Phi(t,t_0)\,\mathbf{X}_0 + \int_{t_0}^{t}\Phi(t,\tau)\,\mathbf{PI}(\tau)\,d\tau =$$

$$= \Phi(t,t_0)\left[\mathbf{X}_0 + \int_{t_0}^{t}\Phi(t_0,\tau)\,\mathbf{PI}(\tau)\,d\tau\right], \ \forall t \in \mathfrak{T}_0. \tag{3.8}$$

This and the system output Equation (3.2) determine the system response:

$$\mathbf{Y}(t;t_0;\mathbf{X}_0;\mathbf{I}) = C\Phi(t,t_0)\,\mathbf{X}_0 + \int_{t_0}^{t}C\Phi(t,\tau)\,\mathbf{PI}(\tau)\,d\tau + Q\mathbf{I}(t) =$$

$$= C\Phi(t,t_0)\left[\mathbf{X}_0 + \int_{t_0}^{t}\Phi(t_0,\tau)\,\mathbf{PI}(\tau)\,d\tau\right] + Q\mathbf{I}(t), \ \forall t \in \mathfrak{T}_0, \tag{3.9}$$

Note 42 *The IO system (2.1), Section 2, can be formally mathematically transformed into the equivalent ISO system (3.1), (3.2) (for such transformation in the general case of the IO system (2.15) see Appendix C.1 and for more details: [40, Appendix C.1, pp. 417-420]). The obtained state variables are without any physical meaning if $\mu > 0$ in the IO system (2.15). Also, the ISO system (3.1), (3.2) can be transformed into the IO system (2.15) (for the transformation in the general case see [40, Appendix C.2, p. 421]).*

3.1.2 Complex domain

We recall (1.3) (Section 1.2) that I is the identity matrix of the dimension n: $I_n = I$.

The application of the Laplace transform to the ISO system (3.1), (3.2) gives its complex domain description:

$$s\mathbf{X}(s) - \mathbf{X}_0 = A\mathbf{X}(s) + D\mathbf{D}(s) + B\mathbf{U}(s), \ \mathbf{Y}(s) = C\mathbf{X}(s) + V\mathbf{D}(s) + U\mathbf{U}(s).$$

We determine first $\mathbf{X}(s)$ from the first equation, and then replace the solution into the second equation to get the well-known result for $\mathbf{Y}(s)$:

$$\mathbf{X}(s) = (sI - A)^{-1} [D\mathbf{D}(s) + B\mathbf{U}(s) + \mathbf{X}_0], \qquad (3.10)$$

$$\mathbf{Y}(s) = C (sI - A)^{-1} [D\mathbf{D}(s) + B\mathbf{U}(s) + \mathbf{X}_0] + V\mathbf{D}(s) + U\mathbf{U}(s), \quad (3.11)$$

which we can set into the following forms:

$$\mathbf{X}(s) = (sI - A)^{-1} \left[D \vdots B \vdots I \right] \begin{bmatrix} \mathbf{D}(s) \\ \mathbf{U}(s) \\ \mathbf{X}_0 \end{bmatrix} = F_{ISOIS}(s) \mathbf{V}_{ISO}(s), \quad (3.12)$$

$$\mathbf{Y}(s) = \left[C (sI - A)^{-1} D + V \vdots C (sI - A)^{-1} B + U \vdots C (sI - A)^{-1} \right] \cdot$$

$$\cdot \begin{bmatrix} \mathbf{D}(s) \\ \mathbf{U}(s) \\ \mathbf{X}_0 \end{bmatrix} = F_{ISO}(s) \mathbf{V}_{ISO}(s), \qquad (3.13)$$

where:
- $F_{ISO}(s)$,

$$F_{ISO}(s) =$$

$$= \left[C (sI - A)^{-1} D + V \vdots C (sI - A)^{-1} B + U \vdots C (sI - A)^{-1} \right], \quad (3.14)$$

is the ISO plant (3.1), (3.2) *input to output (IO) full transfer function matrix*, the inverse Laplace transform of which is the plant *IO full fundamental matrix* $\Psi_{ISO}(t)$ [40],

$$\Psi_{ISO}(t) = \mathcal{L}^{-1} \{F_{ISO}(s)\}, \qquad (3.15)$$

- $p_{ISO}(s)$,

$$p_{ISO}(s) = \det (sI - A), \qquad (3.16)$$

is the characteristic polynomial of the *ISO* plant (3.1), (3.2) and the denominator polynomial of all its transfer function matrices,
- $G_{ISOD}(s)$,

$$G_{ISOD}(s) = C(sI - A)^{-1}D + V =$$
$$= p_{ISO}^{-1}(s)[Cadj(sI - A)D + p_{ISO}(s)V], \tag{3.17}$$

is the *ISO* plant (3.1), (3.2) *transfer function matrix relative to the disturbance* **D**,
- $G_{ISOU}(s)$,

$$G_{ISOU}(s) = C(sI - A)^{-1}B + U =$$
$$= p_{ISO}^{-1}(s)[Cadj(sI - A)B + p_{ISO}(s)U] = \frac{L_{ISO}(s)}{p_{ISO}(s)}, \tag{3.18}$$

is the *ISO* plant (3.1), (3.2) *transfer function matrix relative to the control* **U**,
- $G_{ISOX_0}(s)$,

$$G_{ISOX_0}(s) = C(sI - A)^{-1} = p_{ISO}^{-1}(s)Cadj(sI - A), \tag{3.19}$$

is the *ISO* plant (3.1), (3.2) *transfer function matrix relative to the initial state* $\mathbf{X}_0$,
- $\mathbf{V}_{ISO}(s)$ and $\mathbf{C}_{ISO0}$,

$$\mathbf{V}_{ISO}(s) = \begin{bmatrix} \mathbf{I}_{ISO}(s) \\ \mathbf{C}_{ISO0} \end{bmatrix}, \ \mathbf{I}_{ISO}(s) = \begin{bmatrix} \mathbf{D}(s) \\ \mathbf{U}(s) \end{bmatrix}, \ \mathbf{C}_{ISO0} = \mathbf{X}_0, \tag{3.20}$$

are the Laplace transform of the action vector $\mathbf{V}_{ISOP}(t)$ and the vector $\mathbf{C}_{ISO0}$ of all plant initial conditions, respectively.

Example 43 *Let the ISO system (3.1), (3.2) be defined by*

$$\underbrace{\begin{bmatrix} \frac{dX_1}{dt} \\ \frac{dX_2}{dt} \\ \frac{dX_3}{dt} \end{bmatrix}}_{\mathbf{X}^{(1)}} = \underbrace{\begin{bmatrix} -2 & 0 & 1 \\ 3 & 2 & 2 \\ 0 & 4 & -3 \end{bmatrix}}_{A} \underbrace{\begin{bmatrix} X_1 \\ X_2 \\ X_3 \end{bmatrix}}_{\mathbf{X}} + \underbrace{\begin{bmatrix} 7 & 5 \\ 10 & -6 \\ -4 & 3 \end{bmatrix}}_{B} \underbrace{\begin{bmatrix} U_1 \\ U_2 \end{bmatrix}}_{\mathbf{U}}, \tag{3.21}$$

$$\underbrace{\begin{bmatrix} Y_1 \\ Y_2 \end{bmatrix}}_{\mathbf{Y}} = \underbrace{\begin{bmatrix} -1 & 1 & 4 \\ 0 & 2 & 3 \end{bmatrix}}_{C} \underbrace{\begin{bmatrix} X_1 \\ X_2 \\ X_3 \end{bmatrix}}_{\mathbf{X}} + \underbrace{\begin{bmatrix} 6 & 7 \\ 5 & 8 \end{bmatrix}}_{U} \underbrace{\begin{bmatrix} U_1 \\ U_2 \end{bmatrix}}_{\mathbf{U}}.. \tag{3.22}$$

The resolvent matrix

$$(sI_3 - A)^{-1} = \begin{bmatrix} s+2 & 0 & -1 \\ -3 & s-2 & -2 \\ 0 & -4 & s+3 \end{bmatrix}^{-1} =$$

$$= \begin{bmatrix} \frac{s^2+s-14}{s^3+3s^2-12s-40} & \frac{4}{s^3+3s^2-12s-40} & \frac{s-2}{s^3+3s^2-12s-40} \\ \frac{3s+9}{s^3+3s^2-12s-40} & \frac{s^2+5s+6}{s^3+3s^2-12s-40} & \frac{2s+7}{s^3+3s^2-12s-40} \\ 12\frac{s+2}{(s+2)(s^3+3s^2-12s-40)} & \frac{4s+8}{s^3+3s^2-12s-40} & \frac{s^2-4}{s^3+3s^2-12s-40} \end{bmatrix} \qquad (3.23)$$

The equations (3.12) for $D = O_{3,d}$ yields:

$$F_{ISOIS}(s) =$$

$$= \begin{bmatrix} \frac{s^2+s-14}{s^3+3s^2-12s-40} & \frac{4}{s^3+3s^2-12s-40} & \frac{s-2}{s^3+3s^2-12s-40} \\ \frac{3s+9}{s^3+3s^2-12s-40} & \frac{s^2+5s+6}{s^3+3s^2-12s-40} & \frac{2s+7}{s^3+3s^2-12s-40} \\ 12\frac{s+2}{(s+2)(s^3+3s^2-12s-40)} & \frac{4s+8}{s^3+3s^2-12s-40} & \frac{s^2-4}{s^3+3s^2-12s-40} \end{bmatrix} \bullet$$

$$\bullet \begin{bmatrix} 7 & 5 & 1 & 0 & 0 \\ 10 & -6 & 0 & 1 & 0 \\ -4 & 3 & 0 & 0 & 1 \end{bmatrix} = \begin{bmatrix} G_{ISOISU}(s) \vdots G_{ISOXoIS}(s) \end{bmatrix}, \qquad (3.24)$$

where

$$G_{ISOISU}(s) = \begin{bmatrix} \frac{s^2+s-14}{s^3+3s^2-12s-40} & \frac{4}{s^3+3s^2-12s-40} & \frac{s-2}{s^3+3s^2-12s-40} \\ \frac{3s+9}{s^3+3s^2-12s-40} & \frac{s^2+5s+6}{s^3+3s^2-12s-40} & \frac{2s+7}{s^3+3s^2-12s-40} \\ 12\frac{s+2}{(s+2)(s^3+3s^2-12s-40)} & \frac{4s+8}{s^3+3s^2-12s-40} & \frac{s^2-4}{s^3+3s^2-12s-40} \end{bmatrix} \bullet$$

$$\bullet \begin{bmatrix} 7 & 5 \\ 10 & -6 \\ -4 & 3 \end{bmatrix} \Longrightarrow$$

$$G_{ISOISU}(s) = \begin{bmatrix} \frac{7s^2+3s-50}{s^3+3s^2-12s-40} & \frac{5s^2+8s-100}{s^3+3s^2-12s-40} \\ \frac{10s^2+63s+95}{s^3+3s^2-12s-40} & \frac{-6s^2-9s+30}{s^3+3s^2-12s-40} \\ \frac{-4s^2+124s+264}{(s+2)(s^3+3s^2-12s-40)} & \frac{3s^2+36s+60}{s^3+3s^2-12s-40} \end{bmatrix}, \qquad (3.25)$$

$$G_{ISOXoIS}(s) = \begin{bmatrix} \frac{s^2+s-14}{s^3+3s^2-12s-40} & \frac{4}{s^3+3s^2-12s-40} & \frac{s-2}{s^3+3s^2-12s-40} \\ \frac{3s+9}{s^3+3s^2-12s-40} & \frac{s^2+5s+6}{s^3+3s^2-12s-40} & \frac{2s+7}{s^3+3s^2-12s-40} \\ 12\frac{s+2}{(s+2)(s^3+3s^2-12s-40)} & \frac{4s+8}{s^3+3s^2-12s-40} & \frac{s^2-4}{s^3+3s^2-12s-40} \end{bmatrix}.$$

$$(3.26)$$

Finally,

$$F_{ISOIS}(s) =$$

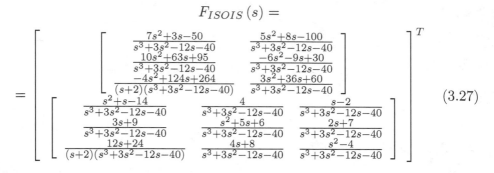

$$= \left[\begin{array}{c} \left[\begin{array}{cc} \frac{7s^2+3s-50}{s^3+3s^2-12s-40} & \frac{5s^2+8s-100}{s^3+3s^2-12s-40} \\ \frac{10s^2+63s+95}{s^3+3s^2-12s-40} & \frac{-6s^2-9s+30}{s^3+3s^2-12s-40} \\ \frac{-4s^2+124s+264}{(s+2)(s^3+3s^2-12s-40)} & \frac{3s^2+36s+60}{s^3+3s^2-12s-40} \end{array} \right] \\ \left[\begin{array}{ccc} \frac{s^2+s-14}{s^3+3s^2-12s-40} & \frac{4}{s^3+3s^2-12s-40} & \frac{s-2}{s^3+3s^2-12s-40} \\ \frac{3s+9}{s^3+3s^2-12s-40} & \frac{s^2+5s+6}{s^3+3s^2-12s-40} & \frac{2s+7}{s^3+3s^2-12s-40} \\ \frac{12s+24}{(s+2)(s^3+3s^2-12s-40)} & \frac{4s+8}{s^3+3s^2-12s-40} & \frac{s^2-4}{s^3+3s^2-12s-40} \end{array} \right] \end{array} \right]^T \quad (3.27)$$

Equation (3.14), (Subsection 3.1.2), becomes for $D = O_{3,d}$ and $V = O_{2,d}$,

$$F_{ISO}(s) = \Big[G_{ISO}(s) \,\vdots\, G_{ISOX_0}(s) \Big] =$$

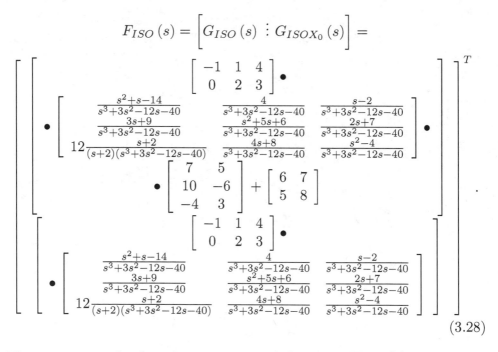

$$(3.28)$$

The system transfer function matrices and the vector $\mathbf{V}_{ISO}(s)$ are, in view *of (3.18)-(3.20), (Subsection 3.1.2):*

$$G_{ISOU}(s) =$$

$$\left\{ \begin{bmatrix} -1 & 1 & 4 \\ 0 & 2 & 3 \end{bmatrix} \bullet \right.$$

$$\bullet \begin{bmatrix} \dfrac{s^2+s-14}{s^3+3s^2-12s-40} & \dfrac{4}{s^3+3s^2-12s-40} & \dfrac{s-2}{s^3+3s^2-12s-40} \\ \dfrac{3s+9}{s^3+3s^2-12s-40} & \dfrac{s^2+5s+6}{s^3+3s^2-12s-40} & \dfrac{2s+7}{s^3+3s^2-12s-40} \\ 12\dfrac{s+2}{(s+2)(s^3+3s^2-12s-40)} & \dfrac{4s+8}{s^3+3s^2-12s-40} & \dfrac{s^2-4}{s^3+3s^2-12s-40} \end{bmatrix} \bullet$$

$$\left. \bullet \begin{bmatrix} 7 & 5 \\ 10 & -6 \\ -4 & 3 \end{bmatrix} + \begin{bmatrix} 6 & 7 \\ 5 & 8 \end{bmatrix} \right\} \implies$$

$$G_{ISO}(s) = \begin{bmatrix} \dfrac{\left(6s^3+2s^2+476s+1121\right)(s+2)}{(s^3+3s^2-12s-40)(s+2)} & \dfrac{\left(7s^3+19s^2+52s+105\right)(s+2)}{(s^3+3s^2-12s-40)(s+2)} \\ \dfrac{\left(5s^3+23s^2+638s+512\right)(s+2)}{(s^3+3s^2-12s-40)(s+2)} & \dfrac{\left(8s^3+21s^2-6s-80\right)(s+2)}{(s^3+3s^2-12s-40)(s+2)} \end{bmatrix}, \quad (3.29)$$

$$G_{ISOX_0}(s) =$$

$$= \left\{ \begin{bmatrix} -1 & 1 & 4 \\ 0 & 2 & 3 \end{bmatrix} \bullet \right.$$

$$\left. \bullet \begin{bmatrix} \dfrac{s^2+s-14}{s^3+3s^2-12s-40} & \dfrac{4}{s^3+3s^2-12s-40} & \dfrac{s-2}{s^3+3s^2-12s-40} \\ \dfrac{3s+9}{s^3+3s^2-12s-40} & \dfrac{s^2+5s+6}{s^3+3s^2-12s-40} & \dfrac{2s+7}{s^3+3s^2-12s-40} \\ 12\dfrac{s+2}{(s+2)(s^3+3s^2-12s-40)} & \dfrac{4s+8}{s^3+3s^2-12s-40} & \dfrac{s^2-4}{s^3+3s^2-12s-40} \end{bmatrix} \right\} \implies$$

$$G_{ISOX_0}(s) = \begin{bmatrix} \dfrac{\left(-s^2+50s+119\right)(s+2)}{(s+2)(s^3+3s^2-12s-40)} & \dfrac{s^2+21s+34}{s^3+3s^2-12s-40} & \dfrac{4s^2+3s-7}{s^3+3s^2-12s-40} \\ \dfrac{(42s+90)(s+2)}{(s+2)(s^3+3s^2-12s-40)} & \dfrac{2s^2+22s+36}{s^3+3s^2-12s-40} & \dfrac{3s^2+4s+2}{s^3+3s^2-12s-40} \end{bmatrix}. $$

$$(3.30)$$

Finally,

$$F_{ISO}(s) = \begin{bmatrix} G_{ISO}(s) & \vdots & G_{ISOX_0}(s) \end{bmatrix} = \begin{bmatrix} G_{ISO}^T(s) \\ G_{ISOX_0}^T \end{bmatrix}^T =$$

$$\begin{bmatrix} \begin{bmatrix} \dfrac{\left(6s^3+2s^2+476s+1121\right)(s+2)}{(s^3+3s^2-12s-40)(s+2)} & \dfrac{\left(7s^3+19s^2+52s+105\right)(s+2)}{(s^3+3s^2-12s-40)(s+2)} \\ \dfrac{\left(5s^3+23s^2+638s+512\right)(s+2)}{(s^3+3s^2-12s-40)(s+2)} & \dfrac{\left(8s^3+21s^2-6s-80\right)(s+2)}{(s^3+3s^2-12s-40)(s+2)} \end{bmatrix}^T \\ \begin{bmatrix} \dfrac{\left(-s^2+50s+119\right)(s+2)}{(s+2)(s^3+3s^2-12s-40)} & \dfrac{s^2+21s+34}{s^3+3s^2-12s-40} & \dfrac{4s^2+3s-7}{s^3+3s^2-12s-40} \\ \dfrac{(42s+90)(s+2)}{(s+2)(s^3+3s^2-12s-40)} & \dfrac{2s^2+22s+36}{s^3+3s^2-12s-40} & \dfrac{3s^2+4s+2}{s^3+3s^2-12s-40} \end{bmatrix}^T \end{bmatrix}^T \quad (3.31)$$

together with

$$\mathbf{V}_{ISO}(s) = \begin{bmatrix} \mathbf{U}(s) \\ \mathbf{C}_{ISO0} \end{bmatrix}, \quad \mathbf{U}(s) = \begin{bmatrix} U_1(s) \\ U_2(s) \end{bmatrix}, \quad \mathbf{C}_{ISO0} = \mathbf{X}_0 = \begin{bmatrix} X_{10} \\ X_{20} \\ X_{30} \end{bmatrix}.$$

$$(3.32)$$

3.2 *ISO* plant desired regime

The following definition clarifies the meaning of the nominal input control vector $\mathbf{U}_N$ and of the nominal state vector $\mathbf{X}_N$ with respect to a chosen or given disturbance vector function $\mathbf{D}(.)$ and the desired output function $\mathbf{Y}_d(.)$ of the *ISO* plant (3.1), (3.2).

Definition 44 *A functional vector control-state pair* $[\mathbf{U}*(.), \mathbf{X}*(.)]$ *is **nominal** for the ISO plant (3.1), (3.2) **relative to the functional pair***

$$[\mathbf{D}(.), \mathbf{Y}_d(.)],$$

which is denoted by $[\mathbf{U}_N(.), \mathbf{X}_N(.)]$, *if and only if* $[\mathbf{U}(.), \mathbf{X}(.)] = [\mathbf{U}*(.), \mathbf{X}*(.)]$ *ensures that the corresponding real response* $\mathbf{Y}(.) = \mathbf{Y}*(.)$ *of the plant obeys* $\mathbf{Y}*(t) = \mathbf{Y}_d(t)$ *all the time,*

$$[\mathbf{U}*(.), \mathbf{X}*(.)] = [\mathbf{U}_N(.), \mathbf{X}_N(.)] \iff \langle \mathbf{Y}*(t) = \mathbf{Y}_d(t), \ \forall t \in \mathfrak{T}_0 \rangle.$$

The nominal motion $\mathbf{X}_N(.; \mathbf{X}_{N0}; \mathbf{D}; \mathbf{U}_N)$, $\mathbf{X}_N(0; \mathbf{X}_{N0}; \mathbf{D}; \mathbf{U}_N) \equiv \mathbf{X}_{N0}$, *is the desired motion* $\mathbf{X}_d(.; \mathbf{X}_{d0}; \mathbf{D}; \mathbf{U}_N)$ *of the ISO plant (3.1), (3.2) **relative to the functional vector pair** $[\mathbf{D}(.), \mathbf{Y}_d(.)]$, for short: **the desired motion of the system**,*

$$\mathbf{X}_d(.; \mathbf{X}_{d0}; \mathbf{D}; \mathbf{U}_N) \equiv \mathbf{X}_N(.; \mathbf{X}_{N0}; \mathbf{D}; \mathbf{U}_N),$$
$$\mathbf{X}_d(0; \mathbf{X}_{d0}; \mathbf{D}; \mathbf{U}_N) \equiv \mathbf{X}_{d0} \equiv \mathbf{X}_{N0}. \quad (3.33)$$

Notice that the full system matrix [40, Section 11.2, pp. 192-199]

$$\begin{bmatrix} -B & sI - A \\ U & C \end{bmatrix} \in \mathbb{C}^{(n+N) \times (r+n)} \quad (3.34)$$

is a rectangular matrix in general.

Definition 44 and (3.1), (3.2) imply the following theorem:

Theorem 45 *In order for the vector control-state pair* $[\mathbf{U}^*(.), \mathbf{X}^*(.)]$ *to be nominal for the ISO plant (3.1), (3.2) relative to the functional vector pair* $[\mathbf{D}(.), \mathbf{Y}_d(.)]$, $[\mathbf{U}^*(.), \mathbf{X}^*(.)] = [\mathbf{U}_N(.), \mathbf{X}_N(.)]$, *it is necessary and sufficient that it obeys the following equations:*

$$-B\mathbf{U}^*(t) + \frac{d\mathbf{X}^*(t)}{dt} - A\mathbf{X}^*(t) = DD(t), \ \forall t \in \mathfrak{T}_0, \qquad (3.35)$$

$$U\mathbf{U}^*(t) + C\mathbf{X}^*(t) = \mathbf{Y}_d(t) - V\mathbf{D}(t), \ \forall t \in \mathfrak{T}_0, \qquad (3.36)$$

or equivalently in the complex domain,

$$\begin{bmatrix} -B & sI - A \\ U & C \end{bmatrix} \begin{bmatrix} \mathbf{U}^*(s) \\ \mathbf{X}^*(s) \end{bmatrix} = \begin{bmatrix} \mathbf{X}_0^* + DD(s) \\ \mathbf{Y}_d(s) - VD(s) \end{bmatrix}. \qquad (3.37)$$

Let us consider the existence of the solutions of the equations (3.35), (3.36), or equivalently of (3.37). There are $(n + r)$ unknown variables and $(N + n)$ equations. The unknown variables are the entries of $\mathbf{U}^*(s) \in \mathbb{C}^r$ and of $\mathbf{X}^*(s) \in \mathbb{C}^n$. There are $(n + r)$ unknown variables and $(N + n)$ equations. The unknown variables are the entries of $\mathbf{U}^*(s) \in \mathbb{C}^r$ and of $\mathbf{X}^*(s) \in \mathbb{C}^n$.

Claim 46 *In order to exist a nominal functional vector control-state pair*

$$[\mathbf{U}_N(.), \mathbf{X}_N(.)]$$

for the ISO plant (3.1), (3.2) relative to its desired response $\mathbf{Y}_d(.)$ *it is necessary and sufficient that* $N \leq r$. *Then, the functional vector control-state pair* $[\mathbf{U}_N(.), \mathbf{X}_N(.)]$ *is nominal relative to the desired response* $\mathbf{Y}_d(.)$ *of the plant (3.1), (3.2) in view of Theorem 45.*

Proof. The dimension of the matrix (3.34) is $(n + N) \times (r + n)$. It is well-known (e.g., [2, p. 115]) that for Equation (3.37) to have a solution it is necessary and sufficient that the rank of the matrix (3.34) is equal to $n + N$, which is possible if and only if $n + N \leq r + n$, i.e., if and only if $N \leq r$. ∎

Claim 46 resolves completely the problem of the existence of a nominal functional vector control-state pair $[\mathbf{U}_N(.), \mathbf{X}_N(.)]$ for the *ISO* plant (3.1), (3.2) relative to the functional vector pair $[\mathbf{D}(.), \mathbf{Y}_d(.)]$.

Condition 47 *The desired output response of the ISO plant (3.1), (3.2) is realizable, i.e.,* $N \leq r$. *The nominal control-state pair* $[\mathbf{U}_N(.), \mathbf{X}_N(.)]$ *is known.*

The *ISO* plant description in terms of the deviations (3.38),

$$\mathbf{x} = \mathbf{X} - \mathbf{X}_N = \mathbf{X} - \mathbf{X}_d, \tag{3.38}$$

(9.3), and (2.53), (2.55) (Section 2.2) reads:

$$\frac{d\mathbf{x}(t)}{dt} = A\mathbf{x}(t) + D\mathbf{d}(t) + B\mathbf{u}(t), \ \forall t \in \mathfrak{T}_0, \tag{3.39}$$

$$\mathbf{y}(t) = C\mathbf{x}(t) + V\mathbf{d}(t) + U\mathbf{u}(t), \ \forall t \in \mathfrak{T}_0, \tag{3.40}$$

3.3   Exercises

Exercise 48 *1. Select a physical ISO plant.*
2. Determine its time domain ISO mathematical model.

Chapter 4

EISO systems

4.1 *EISO* system mathematical model

4.1.1 Time domain

A slightly more general class than the *ISO* systems (3.1), (3.2) is the family of the **Extended Input-State-Output systems** (*EISO* **systems**) described in terms of the total coordinates by

$$\frac{d\mathbf{X}(t)}{dt} = A\mathbf{X}(t) + D^{(\mu)}\mathbf{D}^\mu(t) + \mathrm{B}^{(\mu)}\mathbf{U}^\mu(t) = A\mathbf{X}(t) + P^{(\mu)}\mathbf{I}^\mu(t), \ \forall t \in \mathfrak{T}_0,$$

$$A \in \mathfrak{R}^{n \times n}, \ D^{(\mu)} \in \mathfrak{R}^{n \times (\mu+1)d}, \ \mathrm{B}^{(\mu)} \in \mathfrak{R}^{n \times (\mu+1)r}, \ P^{(\mu)} \in \mathfrak{R}^{n \times (\mu+1)M}, \quad (4.1)$$

$$\mathbf{Y}(t) = C\mathbf{X}(t) + V\mathbf{D}(t) + U\mathbf{U}(t) = C\mathbf{X}(t) + Q\mathbf{I}(t), \ \forall t \in \mathfrak{T}_0.$$

$$C \in \mathfrak{R}^{N \times n}, \ V \in \mathfrak{R}^{N \times d}, \ U \in \mathfrak{R}^{N \times r}, \ Q \in \mathfrak{R}^{N \times M}. \quad (4.2)$$

The overall input mathematical data of the *EISO* system are the input vector $\mathbf{I}$ and the matrix $P^{(\mu)}$ related to the extended input vector $\mathbf{I}^\mu$:

$$\mathbf{I} = \mathbf{I}_{EISO} = \begin{bmatrix} \mathbf{D}^T & \mathbf{U}^T \end{bmatrix}^T \in \mathfrak{R}^{d+r}, \ M = d+r,$$

$$P^{(\mu)} = \begin{bmatrix} P_0 \ \vdots \ P_1 \ \vdots \ \dots \ \vdots \ P_\mu \end{bmatrix} \in \mathfrak{R}^{n \times (\mu+1)M},$$

$$P^{(\mu)} = \begin{bmatrix} D^{(\mu)} \ \vdots \ \mathrm{B}^{(\mu)} \end{bmatrix} \in \mathfrak{R}^{n \times (\mu+1)(d+r)}, \ Q = \begin{bmatrix} V \ \vdots \ U \end{bmatrix} \in \mathfrak{R}^{N \times (d+r)},$$

$$\mathrm{B}^{(\mu)} = \begin{bmatrix} \mathrm{B}_0 \vdots \mathrm{B}_1 \vdots \dots \vdots \mathrm{B}_\mu \end{bmatrix} \in \mathfrak{R}^{n \times (\mu+1)r}, \ D^{(\mu)} = \begin{bmatrix} D_0 \vdots D_1 \vdots \dots \vdots D_\mu \end{bmatrix} \in \mathfrak{R}^{n \times (\mu+1)d}.$$

$$(4.3)$$

If $\mu = 0$ then the *EISO* system (4.1), (4.2) becomes the *ISO* system (3.1), (3.2) (Section 3.1). In order to present one physical origin of the *EISO* system (4.1), (4.2) we discover it in the physical IO systems (2.15), (Section 2.1), as shown in the following:

Theorem 49 *The EISO form (4.1), (4.2) of the IO system (2.15)*

The IO system (2.15) can be transformed into the EISO form (4.1), (4.2) by preserving the physical meaning of all variables, where we should distinguish the case $\nu > 1$ from the case $\nu = 1$ in the IO system (2.15), for which

$$\nu N = n, \tag{4.4}$$

with the following choice of the system physical substate vectors $\mathbf{X}_i$ of the system physical state vector $\mathbf{X}$:

$$
\left\{
\begin{array}{c}
\nu > 1 \Longrightarrow \mathbf{X}_i = \mathbf{Y}^{(i-1)} \in \mathfrak{R}^N,\ \forall i = 1, 2, ..., \nu,\ i.e., \\[2mm]
\mathbf{X} = \left[\mathbf{X}_1^T \vdots \mathbf{X}_2^T \vdots ... \vdots \mathbf{X}_\nu^T \right]^T = \left[\mathbf{Y}^T \vdots \mathbf{Y}^{(1)T} \vdots ... \vdots \mathbf{Y}^{(\nu-1)^T} \right]^T = \\[2mm]
= \mathbf{Y}^{\nu-1} \in \mathfrak{R}^n
\end{array}
\right\}
$$

$$\nu = 1 \Longrightarrow \mathbf{X}_1 = \mathbf{X} = \mathbf{Y} \in \mathfrak{R}^N, \tag{4.5}$$

and with the following matrices of the EISO form (4.1), (4.2) in terms of the matrices A_k, $k = 0, 1, ..., \nu$, of the IO system (2.15):

$$\nu > 1 \Longrightarrow A =$$

$$
\left[
\begin{array}{ccccc}
O_N & I_N & ... & O_N & O_N \\
O_N & O_N & ... & O_N & O_N \\
O_N & O_N & ... & I_N & O_N \\
... & ... & ... & ... & ... \\
O_N & O_N & ... & O_N & I_N \\
-A_\nu^{-1}A_0 & -A_\nu^{-1}A_1 & ... & -A_\nu^{-1}A_{\nu-2} & -A_\nu^{-1}A_{\nu-1}
\end{array}
\right] \in \mathfrak{R}^{\nu N \times \nu N} \ ,
$$

$$\nu = 1 \Longrightarrow A = -A_1^{-1}A_0 \in \mathfrak{R}^{N \times N},$$

$$\tag{4.6}$$

$$
P^{(\mu)} =
\left\{
\begin{array}{c}
\left[\begin{array}{c} O_{(\nu-1)N,(\mu+1)M} \\ A_\nu^{-1}H^{(\mu)} \end{array} \right] \in \mathfrak{R}^{n \times (\mu+1)M},\ \nu > 1, \\[3mm]
A_1^{-1}H^{(\mu)} \in \mathfrak{R}^{N \times (\mu+1)M},\ \nu = 1,
\end{array}
\right\}, \tag{4.7}
$$

or, equivalently,

$$
P^{(\mu)} =
\left\{
\begin{array}{c}
\left[\begin{array}{c} O_{(\nu-1)N,N} \\ I_N \end{array} \right] A_\nu^{-1}H^{(\mu)},\ \nu > 1, \\[3mm]
I_N A_1^{-1}H^{(\mu)},\ \nu = 1,
\end{array}
\right\}, \tag{4.8}
$$

$$
P_{inv} =
\left\{
\begin{array}{c}
\left[\begin{array}{c} O_{(\nu-1)N,N} \\ I_N \end{array} \right] \in \mathfrak{R}^{n \times N},\ \nu > 1, \\[3mm]
I_N \in \mathfrak{R}^{N \times N},\ \nu = 1,
\end{array}
\right\}, \tag{4.9}
$$

i.e.,

$$P^{(\mu)} = P_{inv}A_\nu^{-1}H^{(\mu)} \in \Re^{n\times(\mu+1)M} \tag{4.10}$$

$$C = [I_N \quad O_N \quad O_N \quad O_N \quad ... \quad O_N] \in \Re^{N\times n}, \quad Q = O_{N,M} \in \Re^{N\times M}. \tag{4.11}$$

Proof. of Theorem 49, Section 4.1.
Let Equation (4.4) hold and let $\mathbf{X}$ be defined by (4.5) so that

$$\nu > 1 \Longrightarrow$$
$$\mathbf{X}_1(t) = \mathbf{Y}(t),$$
$$\mathbf{X}_2(t) = \mathbf{Y}^{(1)}(t) = \mathbf{X}_1^{(1)}(t)$$
$$\mathbf{X}_3(t) = \mathbf{Y}^{(2)}(t) = \mathbf{X}_2^{(1)}(t)$$
$$\cdot \quad \cdot \quad \cdot \quad \cdot$$
$$\mathbf{X}_{\nu-1}(t) = \mathbf{Y}^{(\nu-2)}(t) = \mathbf{X}_{\nu-2}^{(1)}(t)$$
$$\mathbf{X}_\nu(t) = \mathbf{Y}^{(\nu-1)}(t) = \mathbf{X}_{\nu-1}^{(1)}(t)$$
$$\nu = 1 \Longrightarrow$$
$$\mathbf{X}(t) = \mathbf{X}_1(t) = \mathbf{Y}(t). \tag{4.12}$$

We can solve the preceding equations for the derivatives:

$$\nu > 1 \Longrightarrow$$
$$\mathbf{X}_1^{(1)}(t) = \mathbf{X}_2(t),$$
$$\mathbf{X}_2^{(1)}(t) = \mathbf{X}_3(t),$$
$$\cdot \quad \cdot \quad \cdot \quad \cdot$$
$$\mathbf{X}_{\nu-2}^{(1)}(t) = \mathbf{X}_{\nu-1}(t)$$
$$\mathbf{X}_{\nu-1}^{(1)}(t) = \mathbf{X}_\nu(t)$$
$$\nu = 1 \Longrightarrow$$
$$\mathbf{X}_1^{(1)}(t) = \mathbf{Y}^{(1)}(t). \tag{4.13}$$

Equations (4.12) and (4.13) transform *IO* Equation (2.1) into

$$\nu > 1 \Longrightarrow A_\nu \mathbf{X}_\nu^{(1)}(t) + \sum_{k=0}^{k=\nu-1} A_k \mathbf{X}_{k+1}(t) = H^{(\mu)}\mathbf{I}^\mu(t),$$

$$\nu = 1 \Longrightarrow A_1 \mathbf{X}_1^{(1)}(t) + A_0 \mathbf{X}_1(t) = H^{(\mu)}\mathbf{I}^{\mu:}(t) = H^{(1)}\mathbf{I}^{1:}(t),$$

which implies the following due to $det A_\nu \neq 0$ determined in (2.1):

$$\nu > 1 \Longrightarrow$$

$$\mathbf{X}_\nu^{(1)}(t) = -A_\nu^{-1} \left[\sum_{k=0}^{k=\nu-1} A_k \mathbf{X}_{k+1}(t) - H^{(\mu)} \mathbf{I}^{\mu:}(t) \right],$$

$$\nu = 1 \Longrightarrow A_1 \mathbf{X}_1^{(1)}(t) + A_0 \mathbf{X}_1(t) = H^{(\mu)} \mathbf{I}^{\mu:}(t),$$

i.e.,

$$\nu > 1 \Longrightarrow$$

$$\mathbf{X}_\nu^{(1)}(t) = \left\{ \begin{array}{l} -A_\nu^{-1} A_0 \mathbf{X}_1(t) - A_\nu^{-1} A_1 \mathbf{X}_2(t) - \ldots \\ \ldots - A_\nu^{-1} A_{\nu-1} \mathbf{X}_\nu(t) + A_\nu^{-1} H^{(\beta)} \mathbf{I}^\mu(t) \end{array} \right\},$$

$$\nu = 1 \Longrightarrow$$

$$\mathbf{X}_1^{(1)}(t) = -A_1^{-1} A_0 \mathbf{X}_1(t) + A_1^{-1} H^{(\mu)} \mathbf{I}^{\mu:}(t). \tag{4.14}$$

This and (4.13) yield

$$\nu > 1 \Longrightarrow$$

$$\mathbf{X}_1^{(1)}(t) = \mathbf{X}_2(t),$$

$$\mathbf{X}_2^{(1)}(t) = \mathbf{X}_3(t),$$

$$\ldots$$

$$\mathbf{X}_{\nu-2}^{(1)}(t) = \mathbf{X}_{\nu-1}(t)$$

$$\mathbf{X}_{\nu-1}^{(1)}(t) = \mathbf{X}_\nu(t)$$

$$\mathbf{X}_\nu^{(1)}(t) = \left\{ \begin{array}{l} -A_\nu^{-1} A_0 \mathbf{X}_1(t) - A_\nu^{-1} A_1 \mathbf{X}_2(t) - \ldots \\ \ldots - A_\nu^{-1} A_{\nu-1} \mathbf{X}_\nu(t) + A_\nu^{-1} H^{(\mu)} \mathbf{I}^{\mu:}(t) \end{array} \right\},$$

$$\nu = 1 \Longrightarrow$$

$$\mathbf{X}_1^{(1)}(t) = -A_1^{-1} A_0 \mathbf{X}_1(t) + A_1^{-1} H^{(\mu)} \mathbf{I}^{\mu:}(t), \tag{4.15}$$

or,

$$\nu > 1 \Longrightarrow \begin{bmatrix} \mathbf{X}_1^{(1)}(t) \\ \mathbf{X}_2^{(1)}(t) \\ \ldots \\ \mathbf{X}_{\nu-2}^{(1)}(t) \\ \mathbf{X}_{\nu-1}^{(1)}(t) \\ \mathbf{X}_\nu^{(1)}(t) \end{bmatrix} =$$

$$= \begin{bmatrix} \text{block } diag\left\{I_{(\nu-1)N} \quad -A_\nu^{-1}\right\} \bullet \\ \begin{bmatrix} O_N & I_N & O_N & O_N & \cdots & O_N \\ O_N & O_N & I_N & O_N & \cdots & O_N \\ O_N & O_N & O_N & I_N & \cdots & O_N \\ \cdots & \cdots & \cdots & \cdots & \cdots & \cdots \\ O_N & O_N & O_N & O_N & \cdots & I_N \\ A_0 & A_1 & A_2 & A_3 & \cdots & A_{\nu-1} \end{bmatrix} \end{bmatrix} \underbrace{\begin{bmatrix} \mathbf{X}_1(t) \\ \mathbf{X}_2(t) \\ \cdots \\ \mathbf{X}_{\nu-2}(t) \\ \mathbf{X}_{\nu-1}(t) \\ \mathbf{X}_\nu(t) \end{bmatrix}}_{\mathbf{X}} +$$

$$\underbrace{A}$$

$$+ \underbrace{\begin{bmatrix} O_{N,(\mu+1)M} \\ O_{N,(\mu+1)M} \\ O_{N,(\mu+1)M} \\ \cdots \\ O_{N,(\mu+1)M} \\ A_\nu^{-1}H^{(\mu)} \end{bmatrix}}_{P} \mathbf{I}^\mu(t),$$

$$\nu = 1 \Longrightarrow \mathbf{X}_1^{(1)}(t) = -A_1^{-1}A_0\mathbf{X}_1(t) + A_1^{-1}H^{(\mu)}\mathbf{I}^{\mu:}(t),$$

which imply (4.6)-(4.11). ■

Note 50 *The substate vectors $\mathbf{X}_i$ and the state vector $\mathbf{X}$ composed of them and all defined by (4.5) have the full physical meaning. They are the system output vector $\mathbf{Y}$ and its derivatives. The EISO system (4.1), (4.2) determined by (4.4)-(4.11) retains the full physical sense as the original IO system (2.15). They have the same properties.*

The EISO form (4.1), (4.2) of the original IO system (2.15) differs from the well-known ISO form (3.1), (3.2), i.e., (C.2), (C.3) (Appendix C.1), of the IO system (2.15) for the preservation of the derivatives of the input vector in the state equation (4.1), which has not been accepted so far: Equation (3.1). The physical nature of the IO system (2.15) introduces the derivatives of the input vector in the state equation. The formal mathematical transformation given by Equations (C.4)-(C.10) (Section C.1) ignores the explicit action of the input vector derivatives on the physical state of the IO system (2.15).

The existing formal mathematical transformation of the IO system (2.15) into the ISO form (3.1), (3.2) loses the physical sense if $\mu > 0$ so that the chosen state variables and the state vector are physically meaningless.

This book develops the state theory for the IO systems (2.15) by exploiting their EISO form (4.1), (4.2), (4.4)-(4.11) in order to preserve the full physical sense of the original IO system (2.15). A useful tool to achieve this is the new simple compact calculus based on the compact notation

$$\left[\mathbf{Y}^T \vdots \mathbf{Y}^{(1)T} \vdots ... \vdots \mathbf{Y}^{(\nu-1)^T} \right]^T = \mathbf{Y}^{\nu-1},$$

which enabled us to define the physical (and mathematical) state vector of the IO systems (2.15) in the form $\mathbf{X} = \mathbf{Y}^{\nu-1}$.

Note 51 *The matrix* $P_{inv}^{(\mu)}$ *(4.7), (4.8), (4.10) is the invariant submatrix of the matrix* $P^{(\mu)}$. *It is invariant relative to both all matrices* A_i, $i = 0, 1, 2, ..., \nu$, *and all submatrices* H_k *of* $H^{(\mu)}$, $k = 0, 1, 2, ..., \mu$, *of the original IO system (2.15). In other words, the matrix* P_{inv} *is independent of both all matrices* A_i, $i = 0, 1, 2, ..., \nu$, *and all matrices* H_k, $k = 0, 1, 2, ..., \mu$.

Note 52 *Let* $\nu > 1$. *Then* $O_{(\nu-1)N,M}$ *is* $(\nu - 1) N \times M$ *zero matrix.*

If and only if $\nu = 1$, *then the matrix* $O_{(\nu-1)N,M}$ *becomes formally* $O_{0,M}$ *that does not exist. Then it should be simply omitted.*

Conclusion 53 *For the existence of the* $(\nu - 1) N \times M$ *zero matrix* $O_{(\nu-1)N,M}$ *to exist it is necessary and sufficient that the natural number* ν *obeys* $\nu > 1$:

$$\exists O_{(\nu-1)N,M} \in \mathfrak{R}^{(\nu-1)N \times M} \iff \nu \in \{2, 3, ..., n, ...\}. \tag{4.16}$$

By referring to the well-known form of the solution of the *ISO* systems (3.1), (3.2) we easily show that the solution of the *EISO* system (4.1), (4.2) is determined by

$$\mathbf{X}(t; t_0; \mathbf{I}^{\mu}) = e^{A(t-t_0)}\mathbf{X}_0 + \int_{t_0}^{t} e^{A(t-\tau)} P^{(\mu)}\mathbf{I}^{\mu}(\tau) \, d\tau, \tag{4.17}$$

$$= e^{A(t-t_0)} \left[\mathbf{X}_0 + \int_{t_0}^{t} e^{A(t_0-\tau)} P^{(\mu)}\mathbf{I}^{\mu}(\tau) \, d\tau \right], \ \forall t \in \mathfrak{T}_0, \tag{4.18}$$

or equivalently by

$$\mathbf{X}(t; t_0; \mathbf{I}^{\mu}) = \Phi(t, t_0)\mathbf{X}_0 + \int_{t_0}^{t} \Phi(t, \tau) P^{(\mu)}\mathbf{I}^{\mu}(\tau) \, d\tau = \tag{4.19}$$

$$= \Phi(t, t_0) \left[\mathbf{X}_0 + \int_{t_0}^{t} \Phi(t_0, \tau) P^{(\mu)}\mathbf{I}^{\mu}(\tau) \, d\tau \right], \ \forall t \in \mathfrak{T}_0, \tag{4.20}$$

for $\Phi(t, t_0)$ (3.3) (Section 3.1), i.e.,

$$\Phi(t, t_0) = \Phi(t, 0)\Phi^{-1}(t_0, 0) = e^{At} \left(e^{At_0} \right)^{-1} = e^{A(t-t_0)} \in \mathfrak{R}^{n \times n}. \tag{4.21}$$

These equations and Equation (4.2) determine the *EISO* system response to the initial state vector $\mathbf{X}_0$ and to the extended input vector function $\mathbf{I}^\mu(.)$:

$$\mathbf{Y}(t;t_0;\mathbf{X}_0;\mathbf{I}^\mu) = C\Phi(t,t_0)\mathbf{X}_0 + \int_{t_0}^t C\Phi(t,\tau)P^{(\mu)}\mathbf{I}^\mu(\tau)\,d\tau + Q\mathbf{I}(t) = \tag{4.22}$$

$$= C\Phi(t,t_0)\left[\mathbf{X}_0 + \int_{t_0}^t \Phi(t_0,\tau)P^{(\mu)}\mathbf{I}^\mu(\tau)\,d\tau\right] + Q\mathbf{I}(t),\ \forall t \in \mathfrak{T}_0. \tag{4.23}$$

4.1.2 Complex domain

The Laplace transform of the *EISO* system (4.1), (4.2) relative to $D(t)$ and $U(t)$ reads:

$$s\mathbf{X}(s) - \mathbf{X}_0 = A\mathbf{X}(s) + \left\{ \begin{array}{l} D^{(\mu)}S_d^{(\mu)}(s)\mathbf{D}(s) - D^{(\mu)}Z_d^{(\mu-1)}(s)\mathbf{D}_0^{\mu-1}+ \\ +\mathrm{B}^{(\mu)}S_r^{(\mu)}(s)\mathbf{U}(s) - \mathrm{B}^{(\mu)}Z_r^{(\mu-1)}(s)\mathbf{U}_0^{\mu-1} \end{array} \right\}, \tag{4.24}$$

$$\mathbf{Y}(s) = C\mathbf{X}(s) + V\mathbf{D}^{(\mu)}(s) + V\mathbf{U}(s). \tag{4.25}$$

These equations lead to:

$$\mathbf{X}(s) = (sI - A)^{-1}\left[\begin{array}{cc} D^{(\mu)}S_d^{(\mu)}(s) \vdots \mathrm{B}^{(\mu)}S_r^{(\mu)}(s) \vdots - \\ D^{(\mu)}Z_d^{(\mu-1)}(s) \vdots - \mathrm{B}^{(\mu)}Z_r^{(\mu-1)}(s) \vdots I \end{array} \right]$$

$$\bullet \left[\mathbf{D}^T(s) \vdots \mathbf{U}^T(s) \vdots \mathbf{D}^{(\mu)}{}_0^{\mu-1^T} \vdots \mathbf{U}_0^{\mu-1^T} \vdots \mathbf{X}_0^T\right]^T =$$

$$= F_{EISOIS}(s)\mathbf{V}_{EISO}(s), \tag{4.26}$$

$$\mathbf{Y}(s) = \left(\begin{bmatrix} C(sI-A)^{-1}D^{(\mu)}S_d^{(\mu)}(s) + V \vdots \\ \vdots C(sI-A)^{-1}\mathrm{B}^{(\mu)}S_r^{(\mu)}(s) + U \vdots \\ \vdots - C(sI-A)^{-1}D^{(\mu)}Z_d^{(\mu-1)}(s) \vdots \\ \vdots - C(sI-A)^{-1}\mathrm{B}^{(\mu)}Z_r^{(\mu-1)}(s) \vdots \\ \vdots C(sI-A)^{-1} \end{bmatrix} \bullet \right.$$

$$\left. \bullet \left[\mathbf{D}^T(s) \vdots \mathbf{U}^T(s) \vdots \mathbf{D}_0^{\mu-1^T} \vdots \mathbf{U}_0^{\mu-1^T} \vdots \mathbf{X}_0^T\right]^T \right)$$

$$= F_{EISO}(s)\mathbf{V}_{EISO}(s), \tag{4.27}$$

where
- $F_{EISOIS}(s)$,

$$F_{EISOIS}(s) = (sI - A)^{-1} \bullet$$

$$\bullet \left[D^{(\mu)} S_d^{(\mu)}(s) \,\vdots\, \mathrm{B}^{(\mu)} S_r^{(\mu)}(s) \,\vdots\, - D^{(\mu)} Z_d^{(\mu-1)}(s) \,\vdots\, - \mathrm{B}^{(\mu)} Z_r^{(\mu-1)}(s) \,\vdots\, I \right],$$
(4.28)

is the *EISO* system (4.1), (4.2) *input to state (IS) full transfer function matrix*, the inverse Laplace transform of which is the plant *IS full fundamental matrix* $\Psi_{EISOIS}(t)$,

$$\Psi_{EISOIS}(t) = \mathcal{L}^{-1}\{F_{EISOIS}(s)\},$$
(4.29)

- $F_{EISO}(s)$

$$F_{EISO}(s) = \begin{bmatrix} C(sI-A)^{-1} D^{(\mu)} S_d^{(\mu)}(s) + V \,\vdots \\ \vdots\, C(sI-A)^{-1} \mathrm{B}^{(\mu)} S_r^{(\mu)}(s) + U \,\vdots \\ \vdots\, - C(sI-A)^{-1} D^{(\mu)} Z_d^{(\mu-1)}(s) \,\vdots \\ \vdots\, - C(sI-A)^{-1} \mathrm{B}^{(\mu)} Z_r^{(\mu-1)}(s) \,\vdots \\ \vdots\, C(sI-A)^{-1} \end{bmatrix},$$
(4.30)

is the *EISO* plant (4.1), (4.2) *input to output (IO) full transfer function matrix* $F_{EISO}(s)$ *relative to the input pair* $[D^{(\mu)}(t), \mathbf{U}(t)]$ *and the initial vectors* $D^{(\mu)}{}_0^{\mu-1}$, $\mathbf{U}_0^{\mu-1}$ *and* $\mathbf{X}_0$.

The inverse Laplace transform of $F_{EISO}(s)$ is the plant *IO full fundamental matrix* $\Psi_{EISO}(t)$ [40],

$$\Psi_{EISO}(t) = \mathcal{L}^{-1}\{F_{EISO}(s)\},$$
(4.31)

- $p_{EISO}(s)$, for short $p(s)$,

$$p(s) = p_{EISO}(s) = \det(sI - A),$$
(4.32)

is the characteristic polynomial of the *EISO* plant (4.1), (4.2) and the denominator polynomial of all its transfer function matrices,

- $G_{EISOD}(s)$, for short $G_{D^{(\mu)}}(s)$,

$$G_{D^{(\mu)}}(s) = G_{EISOD}(s) = C(sI-A)^{-1} D^{(\mu)} S_d^{(\mu)}(s) + V =$$

$$= p^{-1}(s) \left[C \, adj(sI-A)^{-1} D^{(\mu)} S_d^{(\mu)}(s) + p(s) V \right],$$
(4.33)

is the *EISO* plant (4.1), (4.2) *transfer function matrix relating the output to the disturbance* $D^{(\mu)}$,

   - $G_{EISOU}(s)$, for short $G_U(s)$,

$$G_U(s) = G_{EISOU}(s) = C(sI - A)^{-1} B^{(\mu)} S_r^{(\mu)}(s) + U =$$
$$= p^{-1}(s)\left[Cadj(sI - A)B^{(\mu)} S_r^{(\mu)}(s) + p(s)U\right] \in \mathbb{C}^{N \times r}, \qquad (4.34)$$

is the *EISO* plant (4.1), (4.2) *transfer function matrix relating the output to the control* $\mathbf{U}$,

   - $G_{EISOD_0}(s)$, for short $G_{D_0}(s)$,

$$G_{D_0}(s) = G_{EISOD_0}(s) = -C(sI - A)^{-1} D^{(\mu)} Z_d^{(\mu-1)}(s) =$$
$$= -p^{-1}(s)\left[Cadj(sI - A)D^{(\mu)} Z_d^{(\mu-1)}(s)\right], \qquad (4.35)$$

is the *EISO* plant (4.1), (4.2) *transfer function matrix relating the output to the initial extended disturbance vector* $D^{(\mu)}{}_0^{\mu-1}$,

   - $G_{EISOU_0}(s)$, for short $G_{U_0}(s)$,

$$G_{U_0}(s) = G_{EISOU_0}(s) = -C(sI - A)^{-1} B^{(\mu)} Z_r^{(\mu-1)}(s) =$$
$$= -p^{-1}(s)\left[Cadj(sI - A)B^{(\mu)} Z_r^{(\mu-1)}(s)\right], \qquad (4.36)$$

is the *EISO* plant (4.1), (4.2) *transfer function matrix relating the output to the extended initial control* $\mathbf{U}_0^{\mu-1}$,

   - $G_{EISOX_0}(s)$, for short $G_{X_0}(s)$,

$$G_{EISOX_0}(s) = G_{X_0}(s) = C(sI - A)^{-1} =$$
$$= p^{-1}(s)\,Cadj(sI - A), \qquad (4.37)$$

is the *EISO* plant (4.1), (4.2) *transfer function matrix relating the output to the initial state* $\mathbf{X}_0$.

   - $\mathbf{V}_{EISO}(s)$ and $\mathbf{C}_{EISO0}$,

$$\mathbf{V}_{EISO}(s) = \left[\begin{array}{c} \mathbf{I}_{EISO}(s) \\ \mathbf{C}_{EISO0} \end{array}\right], \quad \mathbf{I}(s) = \mathbf{I}_{EISO}(s) = \left[\begin{array}{c} \mathbf{D}(s) \\ \mathbf{U}(s) \end{array}\right],$$

$$\mathbf{C}_{EISO0} = \left[\begin{array}{ccc} \left(\mathbf{D}_0^{\mu-1}\right)^T & \left(\mathbf{U}_0^{\mu-1}\right)^T & \mathbf{X}_0^T \end{array}\right]^T, \qquad (4.38)$$

are the Laplace transform of the action vector $\mathbf{V}_{EISO}(t)$ and the vector $\mathbf{C}_{EISO0}$ of all initial conditions, respectively.

If we consider the whole extended vectors $D^\mu(t)$ and $\mathbf{U}^\mu(t)$ as the system input vectors then other forms of the system transfer function matrices result. To show that let

$$V^{(\mu)} = \left[V \,\vdots\, O_{N,d} \,\vdots\, \cdots \,\vdots\, O_{N,d} \right] \in \mathfrak{R}^{N \times (\mu+1)d}, \qquad (4.39)$$

$$U^{(\mu)} = \left[U \,\vdots\, O_{N,r} \,\vdots\, \cdots \,\vdots\, O_{N,r} \right] \in \mathfrak{R}^{N \times (\mu+1)r}. \qquad (4.40)$$

The Laplace transform of the *EISO* system (4.1), (4.2) relative to the extended vectors $D^{(\mu)\mu}(t)$ and $\mathbf{U}^\mu(t)$ reads:

$$s\mathbf{X}(s) - \mathbf{X}_0 = A\mathbf{X}(s) + D^{(\mu)}\mathcal{L}\left\{\mathbf{D}^\mu(t)\right\} + \mathbf{B}^{(\mu)}\mathcal{L}\left\{\mathbf{U}^\mu(t)\right\},$$

$$\mathbf{Y}(s) = C\mathbf{X}(s) + V^{(\mu)}\mathcal{L}\left\{\mathbf{D}^\mu(t)\right\} + U^{(\mu)}\mathcal{L}\left\{\mathbf{U}^\mu(t)\right\},$$

so that the Laplace transform $\mathbf{Y}(s)$ of the system output vector $\mathbf{Y}(t)$ can be set also in the following form:

$$\mathbf{Y}(s) = \left(\begin{bmatrix} \left(C\left(sI - A\right)^{-1} D^{(\mu)} + V^{(\mu)} \right) \,\vdots\, \\ \vdots\, \left(C\left(sI - A\right)^{-1} \mathbf{B}^{(\mu)} + U^{(\mu)} \right) \,\vdots\, \\ \vdots\, C\left(sI - A\right)^{-1} \end{bmatrix} \bullet \right) =$$

$$\bullet \left[\mathcal{L}\left\{\mathbf{D}^\mu(t)\right\}^T \,\vdots\, \mathcal{L}\left\{\mathbf{U}^\mu(t)\right\}^T \,\vdots\, \mathbf{X}_0^T \right]^T$$

$$= F_{EISOU^\mu}(s)\, \mathbf{V}_{EISOU^\mu}(s), \qquad (4.41)$$

where:
- $F_{EISOU^\mu}(s)$,

$$F_{EISO^\mu}(s) = \begin{bmatrix} C\left(sI - A\right)^{-1} D^{(\mu)} + V^{(\mu)} \,\vdots\, \\ \vdots\, C\left(sI - A\right)^{-1} \mathbf{B}^{(\mu)} + U^{(\mu)} \,\vdots\, \\ \vdots\, C\left(sI - A\right)^{-1} \end{bmatrix}, \qquad (4.42)$$

is the *EISO* plant (4.1), (4.2) *input to output (IO) full transfer function matrix* $F_{EISO^\mu}(s)$ *relative to the extended input pair* $[D^{(\mu)\mu}(t),\ \mathbf{U}^\mu(t)]$ and the initial state vector $\mathbf{X}_0$,
- $G_{EISOD^\mu}(s)$, for short $G_{D^{(\mu)}\mu}(s)$,

$$G_{D^{(\mu)}\mu}(s) = G_{EISOD^\mu}(s) = C\left(sI - A\right)^{-1} D^{(\mu)} + V^{(\mu)} =$$

$$= p^{-1}(s) \underbrace{\left[Cadj\left(sI - A\right)^{-1} D^{(\mu)} + p(s) V^{(\mu)} \right]}_{N_{D^{(\mu)}}(s)} \in \mathbb{C}^{N \times d}, \qquad (4.43)$$

is the *EISO* plant (4.1), (4.2) *transfer function matrix relating the output to the extended disturbance* $D^{(\mu)\mu}$,
 - and $G_{EISOU^\mu}(s)$, for short $G_{U^\mu}(s)$,

$$G_{U^\mu}(s) = G_{EISOU^\mu}(s) = C(sI - A)^{-1} B^{(\mu)} + U^{(\mu)} = \qquad (4.44)$$

$$= p^{-1}(s) \left[C \, adj\,(sI - A) \, B^{(\mu)} + p(s) U^{(\mu)} \right] \in \mathbb{C}^{N \times (\mu+1)r} \implies \qquad (4.45)$$

$$G_U(s) = G_{U^\mu}(s) \, S_r^{(\mu)}(s), \qquad (4.46)$$

is the *EISO* plant (4.1), (4.2) *transfer function matrix* $G_{EISOU^\mu}(s)$, *for short* $G_{U^\mu}(s)$, *relating the output to the extended control* $\mathbf{U}^\mu$.

Theorem 54 *Properties of the EISO system (4.1), (4.2), (4.4)-(4.11)*

   The EISO system (4.1), (4.2), (4.4)-(4.11) possesses the following properties:

I)   *a) If* $\nu = 1$ *and* $0 \leq \mu \leq 1$, *the matrix* $\left[(sI - A) \, \vdots \, P_{inv} \right]$ *has the full rank n for every complex number s and for every matrix A (4.6) including every eigenvalue* $s_i(A)$ *of the matrix A (4.6), and for every matrix A (4.6):*

$$rank \left[(sI - A) \, \vdots \, P_{inv} \right] = n, \ \forall (s, A) \in \mathbb{C} \times \mathfrak{R}^{n \times n}. \qquad (4.47)$$

The rank n of the matrix $\left[(sI - A) \, \vdots \, P^{(\mu)} \right]$ *is invariant and full relative to every* $(s, A) \in \mathbb{C} \times \mathfrak{R}^{n \times n}$, *A given by (4.6).*

   b) If $\nu = 1$ *and* $0 \leq \mu \leq 1$ *then for the matrix* $\left[(sI - A) \, \vdots \, P^{(\mu)} \right]$ *to have the full rank n for every complex number s including every eigenvalue* $s_i(A)$ *of the matrix A, and for every matrix A , it is necessary and sufficient that the matrix* $H^{(1)}$ *has the full rank* $n = N$:

$$rank\, H^{(1)} = n = N. \qquad (4.48)$$

II)   *a) If* $\nu > 1$ *and* $0 \leq \mu < \infty$, *then the matrix* $\left[(sI - A) \, \vdots \, P_{inv} \right]$ *has the full rank n for every complex number s including every eigenvalue* $s_i(A)$ *of the matrix A and for every matrix* $A \in \mathfrak{R}^{n \times n}$:

$$rank \left[(sI - A) \, \vdots \, P_{inv} \right] = n, \ \forall (s, A) \in \mathbb{C} \times \mathfrak{R}^{n \times n}. \qquad (4.49)$$

The rank of the matrix $\left[(sI - A) \vdots P_{inv}\right]$ *is invariant and full relative to every* $(s, A) \in \mathbb{C} \times \mathfrak{R}^{n \times n}$.

b) *If* $\nu > 1$ *and* $0 \leq \mu < \infty$, *then for the matrix* $\left[(sI - A) \vdots P^{(\mu)}\right]$ *to have the full rank* n *for every complex number* s *including every eigenvalue* $s_i(A)$ *of the matrix* A *and for every matrix* $A \in \mathfrak{R}^{n \times n}$:

$$rank \left[(sI - A) \vdots P^{(\mu)}\right] = n, \ \forall (s, A) \in \mathbb{C} \times \mathfrak{R}^{n \times n}. \qquad (4.50)$$

it is necessary and sufficient that the extended matrix $H^{(\mu)}$ *has the full rank* N,

$$rank H^{(\mu)} = N. \qquad (4.51)$$

Proof. The matrix A of the *EISO* system (4.1), (4.2), (4.4)-(4.11) is determined by (4.6) .

I)   a) Let $\nu = 1$ and $0 \leq \mu \leq 1$ due to (2.1). Then $N = n$ and the matrix $\left[(sI_n - A) \vdots P_{inv}\right]$ has the following form due to (4.9):

$$rank \left[(sI_n - A) \vdots P_{inv}\right] = rank \left[sI_n + A_1^{-1} A_0 \vdots I_n\right] =$$
$$= rank I_n = n \Longrightarrow \forall s = s_i(A) \in \mathbb{C}, \forall i = 1, 2, ..., n; \ i.e.,$$
$$\forall s \in \mathbb{C}, \ \forall A_k \in \mathfrak{R}^{n \times n}, \ k = 0, 1.$$

This proves the statement under I-a).

b) Let $\nu = 1$ and $0 \leq \mu \leq 1$ due to (2.1). Let (4.48) be valid. For the statement under I-b) we use the matrix $\left[(sI_n - A) \vdots P^{(\mu)}\right]$ that has the following form due to (4.8):

$$\left[(sI_n - A) \vdots P^{(\mu)}\right] = \left[sI_n + A_1^{-1} A_0 \vdots A_1^{-1} H^{(\mu)}\right] \Longrightarrow$$
$$rank \left[(sI_n - A) \vdots P^{(\mu)}\right] = rank \left[sI_n + A_1^{-1} A_0 \vdots A_1^{-1} H^{(\mu)}\right].$$

Necessity. Let $A_0 = O_n$ and $s = 0$ in $\left[(sI_n - A) \vdots P^{(\mu)}\right]$:

$$\left[(0I_n - O_n) \vdots P^{(\mu)}\right] = \left[O_n \vdots A_1^{-1} H^{(\mu)}\right] \Longrightarrow$$

Let the matrix

$$\left[(sI_n - A) \;\vdots\; P^{(\mu)} \right] = \left[O_n \;\vdots\; A_1^{-1} H^{(\mu)} \right]$$

have the full rank n :

$$n = rank \left[(sI_n - A) \;\vdots\; P^{(\mu)} \right] = rank \left[O_n \;\vdots\; A_1^{-1} H^{(\mu)} \right] =$$

$$= rank A_1^{-1} H^{(\mu)} = rank H^{(\mu)} \ due \ to \ det A_1^{-1} \neq 0.$$

The rank of $H^{(\mu)}$ equals n. Equation (4.48) holds, which proves its necessity.

Sufficiency. Let $rank H^{(\mu)} = N = n$ due to (4.48). This and $det A_1^{-1} \neq 0$ yield

$$N = n = rank H^{(\mu)} = rank A_1^{-1} H^{(\mu)} =$$

$$= rank \left[sI_n + A_1^{-1} A_0 \;\vdots\; A_1^{-1} H^{(\mu)} \right] = rank \left[(sI_n - A) \;\vdots\; P^{(\mu)} \right],$$

$$0 \leq \mu \leq 1, \ \forall s \in \mathbb{C}, \ \forall A_k \in \mathfrak{R}^{n \times n}, \ k = 0, 1,$$

This proves that for the rank of the matrix $\left[(sI_n - A) \;\vdots\; P^{(\mu)} \right]$ to be full, i.e., to be equal to n, it is sufficient that the rank of the matrix $H^{(1)}$ is full, i.e., equal to n. Hence,

$$rank H^{(1)} = n \implies rank \left[(sI_n - A) \;\vdots\; P^{(\mu)} \right] \equiv n.$$

This proves the statement under I-b).

II) Let $\nu > 1$ and $\mu \geq 0$, $\mu < \infty$.

a) The matrix $\left[(sI_n - A) \;\vdots\; P_{inv} \right]$ has the following form due to (4.9):

$$\nu > 1, \ \mu \geq 0 \implies \left[(sI_n - A) \;\vdots\; P_{inv} \right] =$$

$$= \left[sI_n - A \;\vdots\; \begin{matrix} O_{(\nu-1)N,N} \\ I_N \end{matrix} \right] =$$

$$= \begin{bmatrix} sI_N & -I_N & O_N & \cdots & O_N & O_N & O_N \\ O_N & sI_N & -I_N & \cdots & O_N & O_N & O_N \\ O_N & O_N & sI_N & \cdots & O_N & O_N & O_N \\ \cdots & \cdots & \cdots & \cdots & \cdots & \cdots & \cdots \\ O_N & O_N & O_N & \cdots & sI_N & -I_N & O_N \\ A_\nu^{-1} A_0 & A_\nu^{-1} A_1 & A_\nu^{-1} A_2 & \cdots & A_\nu^{-1} A_{\nu-2} & sI_N + A_\nu^{-1} A_{\nu-1} & I_N \end{bmatrix}.$$

This implies

$$rank \left[(sI_n - A) \vdots P_{inv} \right] =$$

$$rank \left[sI_n - A \vdots \begin{bmatrix} O_{(\nu-1)N,N} \\ I_N \end{bmatrix} \right] =$$

$$= rank \begin{bmatrix} sI_N & -I_N & \dots & O_N & O_N & O_N \\ O_N & sI_N & \dots & O_N & O_N & O_N \\ O_N & O_N & \dots & O_N & O_N & O_N \\ \dots & \dots & \dots & \dots & \dots & \dots \\ O_N & O_N & \dots & sI_N & -I_N & O_N \\ A_\nu^{-1}A_0 & A_\nu^{-1}A_1 & \dots & A_\nu^{-1}A_{\nu-2} & sI_N + A_\nu^{-1}A_{\nu-1} & I_N \end{bmatrix} =$$

$$= rank \begin{bmatrix} -I_N & O_N & \dots & O_N & O_N & O_N \\ sI_N & -I_N & \dots & O_N & O_N & O_N \\ O_N & sI_N & \dots & O_N & O_N & O_N \\ \dots & \dots & \dots & \dots & \dots & \dots \\ O_N & O_N & \dots & sI_N & -I_N & O_N \\ A_\nu^{-1}A_1 & A_\nu^{-1}A_2 & \dots & A_\nu^{-1}A_{\nu-2} & -sI_N + A_\nu^{-1}A_{\nu-1} & I_N \end{bmatrix} =$$

$$= \nu N = n, \ \forall (s, A) \in \mathbb{C} \times \mathfrak{R}^{n \times n}.$$

This proves the invariance of the matrix $\left[(sI_n - A) \vdots P_{inv} \right]$ relative to every $(s, A) \in \mathbb{C} \times \mathfrak{R}^{n \times n}$. The first statement under II) is true.

　　b) Let $\nu > 1$ and $0 \leq \mu < \infty$.

　　Necessity. Let $A_k = O_N, \ \forall k = 0, 1, ..., \nu - 1$ and $s = 0$ in $\left[(sI_n - A) \vdots P^{(\mu)} \right]$ and let $rank \left[(sI_n - A) \vdots P^{(\mu)} \right] = n$:

$$\left[(sI_n - A) \vdots P^{(\mu)} \right] = \begin{bmatrix} O_N & -I_N & \dots & O_N & O_N & O_N \\ O_N & O_N & \dots & O_N & O_N & O_N \\ O_N & O_N & \dots & O_N & O_N & O_N \\ \dots & \dots & \dots & \dots & \dots & \dots \\ O_N & O_N & \dots & O_N & -I_N & O_N \\ O_N & O_N & \dots & O_N & O_N & A_\nu^{-1}H^{(\mu)} \end{bmatrix} \implies$$

$$n = \nu N = rank \left[(sI_n - A) \vdots P^{(\mu)} \right] =$$

$$= rank \begin{bmatrix} O_N & -I_N & ... & O_N & O_N & O_N \\ O_N & O_N & ... & O_N & O_N & O_N \\ O_N & O_N & ... & O_N & O_N & O_N \\ ... & ... & ... & ... & ... & ... \\ O_N & O_N & ... & O_N & -I_N & O_N \\ O_N & O_N & ... & O_N & O_N & A_\nu^{-1} H^{(\mu)} \end{bmatrix} =$$

$$= rank \begin{bmatrix} -I_N & ... & O_N & O_N & O_N \\ O_N & ... & O_N & O_N & O_N \\ O_N & ... & O_N & O_N & O_N \\ ... & ... & ... & ... & ... \\ O_N & ... & O_N & -I_N & O_N \\ O_N & ... & O_N & O_N & A_\nu^{-1} H^{(\mu)} \end{bmatrix} =$$

$$= (\nu - 1) N + rank A_\nu^{-1} H^{(\mu)} \Longrightarrow$$

$$N = rank A_\nu^{-1} H^{(\mu)} = rank H^{(\mu)} \text{ due to } det A_\nu^{-1}.$$

This proves the validity of the condition (4.51), i.e., its necessity.

Sufficiency. Let the condition (4.51) hold. The matrix

$$\left[(sI_n - A) \vdots P^{(\mu)} \right] = \left[sI_n - A \vdots \begin{array}{c} O_{(\nu-1)N,(\mu+1)M} \\ A_\nu^{-1} H^{(\mu)} \end{array} \right]$$

has the following form in view of (4.7)-(4.9):

$$\left[sI_n - A \vdots \begin{array}{c} O_{(\nu-1)N,(\mu+1)M} \\ A_\nu^{-1} H^{(\mu)} \end{array} \right] =$$

$$= \begin{bmatrix} sI_N & -I_N & O_N & ... & O_N & O_N & O_{N,(\mu+1)M} \\ O_N & sI_N & -I_N & ... & O_N & O_N & O_{N,(\mu+1)M} \\ O_N & O_N & sI_N & ... & O_N & O_N & O_{N,(\mu+1)M} \\ ... & ... & ... & ... & ... & ... & ... \\ O_N & O_N & O_N & ... & sI_N & -I_N & O_{N,(\mu+1)M} \\ A_\nu^{-1} A_0 & A_\nu^{-1} A_1 & A_\nu^{-1} A_2 & ... & A_{\nu-2} & sI_N + A_\nu^{-1} A_{\nu-1} & A_\nu^{-1} H^{(\mu)} \end{bmatrix}.$$

Having in mind that for the matrix

$$\left[(sI_n - A) \vdots \begin{array}{c} O_{(\nu-1)N,(\mu+1)M} \\ A_\nu^{-1} H^{(\mu)} \end{array} \right]$$

to have the full rank n for every eigenvalue $s_i(A)$ of the matrix $s_i(A)$ it is sufficient that its following submatrix has the full rank n:

$$
\begin{bmatrix}
-I_N & O_N & \dots & O_N & O_N & O_{N,(\mu+1)M} \\
sI_N & -I_N & \dots & O_N & O_N & O_{N,(\mu+1)M} \\
O_N & sI_N & \dots & O_N & O_N & O_{N,(\mu+1)M} \\
\dots & \dots & \dots & \dots & \dots & \dots \\
O_N & O_N & \dots & sI_N & -I_N & O_{N,(\mu+1)M} \\
A_\nu^{-1}A_1 & A_\nu^{-1}A_2 & \dots & A_\nu^{-1}A_{\nu-2} & sI_N + A_\nu^{-1}A_{\nu-1} & A_\nu^{-1}H^{(\mu)}
\end{bmatrix}
$$

which is true because the matrix $H^{(\mu)}$ has the rank N due to $rank H^{(\mu)} = N$ (4.51) and implies $N = rank H^{(\mu)} = rank A_\nu^{-1}H^{(\mu)}$ due to $det A_\nu^{-1} \neq 0$:

$$
rank \left[sI_n - A \ \vdots \ \begin{bmatrix} O_{(\nu-1)N,(\mu+1)M} \\ A_\nu^{-1}H^{(\mu)} \end{bmatrix} \right] =
$$

$$
= rank
\begin{bmatrix}
-I_N & O_N & .. & O_N & O_N & O_{N,(\mu+1)M} \\
sI_N & -I_N & .. & O_N & O_N & O_{N,(\mu+1)M} \\
O_N & sI_N & .. & O_N & O_N & O_{N,(\mu+1)M} \\
\dots & \dots & .. & \dots & \dots & \dots \\
O_N & O_N & .. & sI_N & -I_N & O_{N,(\mu+1)M} \\
A_\nu^{-1}A_1 & A_\nu^{-1}A_2 & . & A_\nu^{-1}A_{\nu-2} & \begin{matrix} sI_N+ \\ +A_\nu^{-1}A_{\nu-1} \end{matrix} & A_\nu^{-1}H^{(\mu)}
\end{bmatrix}
=
$$

$$
= rank
\begin{bmatrix}
-I_N & O_N & \dots & O_N & O_N \\
sI_N & -I_N & \dots & O_N & O_N \\
O_N & sI_N & \dots & O_N & O_N \\
\dots & \dots & \dots & \dots & \dots \\
O_N & O_N & \dots & sI_N & -I_N
\end{bmatrix}
+ rank A_\nu^{-1}H^{(\mu)} =
$$

$$
= (\nu - 1)N + rank A_\nu^{-1}H^{(\mu)} = (\nu - 1)N + N = \nu N = n,
$$
$$
\forall s_i(A) \in \mathbb{C}, \ i.e., \ \forall s \in \mathbb{C}, \ \forall A \in \mathfrak{R}^n, \tag{4.52}
$$

This proves the second statement under II) and completes the proof. ∎

Comment 55 *If $det A_\nu \neq 0$ then the statements of this theorem largely simplify the verification of the rank of the complex valued matrices*

$$
\left[(sI - A) \ \vdots \ P_{inv} \right] \ and \ \left[(sI - A) \ \vdots \ P^{(\mu)} \right].
$$

While the rank of the matrix $\left[(sI - A) \vdots P_{inv}\right]$ *is invariantly equal to* n,

the rank of the matrix $\left[(sI - A) \vdots P^{(\mu)}\right]$ *is not invariant.*

However, if the rank of the matrix $H^{(\mu)}$ *is full,* $\operatorname{rank}H^{(\mu)} = N$, *then the*

rank of the matrix $\left[(sI - A) \vdots P^{(\mu)}\right]$ *is independent of both* A *and* $s \in \mathbb{C}$,

i.e., it is also invariant.

Example 56 *Let the EISO system (4.1), (4.2) be defined by*

$$
\underbrace{\begin{bmatrix} \frac{dX_1}{dt} \\ \frac{dX_2}{dt} \\ \frac{dX_3}{dt} \end{bmatrix}}_{\mathbf{X}^{(1)}} = \underbrace{\begin{bmatrix} -2 & 0 & 1 \\ 3 & 2 & 2 \\ 0 & 4 & -3 \end{bmatrix}}_{A} \underbrace{\begin{bmatrix} X_1 \\ X_2 \\ X_3 \end{bmatrix}}_{\mathbf{X}} +
$$

$$
+ \underbrace{\begin{bmatrix} 7 & 5 \\ 10 & -6 \\ -4 & 3 \end{bmatrix}}_{B_0} \underbrace{\begin{bmatrix} U_1 \\ U_2 \end{bmatrix}}_{\mathbf{U}} + \underbrace{\begin{bmatrix} 0 & 1 \\ 1 & -1 \\ -2 & 2 \end{bmatrix}}_{B_1} \underbrace{\begin{bmatrix} U_1^{(1)} \\ U_2^{(1)} \end{bmatrix}}_{\mathbf{I}^{(1)}}, \tag{4.53}
$$

$$
\underbrace{\begin{bmatrix} Y_1 \\ Y_2 \end{bmatrix}}_{\mathbf{Y}} = \underbrace{\begin{bmatrix} -1 & 1 & 4 \\ 0 & 2 & 3 \end{bmatrix}}_{C} \underbrace{\begin{bmatrix} X_1 \\ X_2 \\ X_3 \end{bmatrix}}_{\mathbf{X}} + \underbrace{\begin{bmatrix} 6 & 7 \\ 5 & 8 \end{bmatrix}}_{U} \underbrace{\begin{bmatrix} U_1 \\ U_2 \end{bmatrix}}_{\mathbf{U}}.. \tag{4.54}
$$

In this example

$$
n = 3, \ M = 2, \ \mu = 1, \ r = 2.
$$

$$
S_2^{(1)}(s) = \left[s^0 I_2 \vdots s^1 I_2\right]^T, \ Z_2^{(1-1)}(s) = \begin{bmatrix} O_2 \\ I_2 \end{bmatrix}
$$

The extended matrix $P^{(\mu)(1)}$ *and* $\mathbf{I}^1$ *follow from (4.53):*

$$
B^{(1)} = \begin{bmatrix} 7 & 5 & 0 & 1 \\ 10 & -6 & 1 & -1 \\ -4 & 3 & -2 & 2 \end{bmatrix}, \ \mathbf{U}^1 = \begin{bmatrix} \mathbf{U} \\ \mathbf{U}^{(1)} \end{bmatrix} = \begin{bmatrix} U_1 \\ U_2 \\ U_1^{(1)} \\ U_2^{(2)} \end{bmatrix}.
$$

The resolvent matrix $(sI_3 - A)^{-1}$:

$$(sI_3 - A)^{-1} = \begin{bmatrix} s+2 & 0 & -1 \\ -3 & s-2 & -2 \\ 0 & -4 & s+3 \end{bmatrix}^{-1} =$$

$$= \begin{bmatrix} \frac{s^2+s-14}{s^3+3s^2-12s-40} & \frac{4}{s^3+3s^2-12s-40} & \frac{s-2}{s^3+3s^2-12s-40} \\ \frac{3s+9}{s^3+3s^2-12s-40} & \frac{s^2+5s+6}{s^3+3s^2-12s-40} & \frac{2s+7}{s^3+3s^2-12s-40} \\ \frac{12}{(s^3+3s^2-12s-40)} & \frac{4s+8}{s^3+3s^2-12s-40} & \frac{s^2-4}{s^3+3s^2-12s-40} \end{bmatrix} \qquad (4.55)$$

Equation (4.30) yields:

$$F_{EISOIS}(s) = (sI - A)^{-1} \left[\mathrm{B}^{(1)} S_2^{(1)}(s) \vdots - \mathrm{B}^{(1)} Z_2^{(0)}(s) \vdots I_3 \right],$$

for

$$\mathrm{B}^{(1)} S_2^{(1)}(s) = \begin{bmatrix} 7 & 5 & 0 & 1 \\ 10 & -6 & 1 & -1 \\ -4 & 3 & -2 & 2 \end{bmatrix} \begin{bmatrix} 1 & 0 \\ 0 & 1 \\ s & 0 \\ 0 & s \end{bmatrix} = \begin{bmatrix} 7 & 5+s \\ 10+s & -6-s \\ -4-2s & 3+2s \end{bmatrix},$$

$$\mathrm{B}^{(1)} Z_2^{(0)}(s) = \begin{bmatrix} 7 & 5 & 0 & 1 \\ 10 & -6 & 1 & -1 \\ -4 & 3 & -2 & 2 \end{bmatrix} \begin{bmatrix} 0 & 0 \\ 0 & 0 \\ 1 & 0 \\ 0 & 1 \end{bmatrix} = \begin{bmatrix} 0 & 1 \\ 1 & -1 \\ -2 & 2 \end{bmatrix},$$

$$F_{ISOIS}(s) = \begin{bmatrix} \frac{s^2+s-14}{s^3+3s^2-12s-40} & \frac{4}{s^3+3s^2-12s-40} & \frac{s-2}{s^3+3s^2-12s-40} \\ \frac{3s+9}{s^3+3s^2-12s-40} & \frac{s^2+5s+6}{s^3+3s^2-12s-40} & \frac{2s+7}{s^3+3s^2-12s-40} \\ \frac{12}{(s^3+3s^2-12s-40)} & \frac{4s+8}{s^3+3s^2-12s-40} & \frac{s^2-4}{s^3+3s^2-12s-40} \end{bmatrix} \bullet$$

$$\bullet \begin{bmatrix} 7 & 5+s & 0 & 1 & 1 & 0 & 0 \\ 10+s & -6-s & 1 & -1 & 0 & 1 & 0 \\ -4-2s & 3+2s & -2 & 2 & 0 & 0 & 1 \end{bmatrix} =$$

$$= \left[G_{EISOISU}(s) \vdots G_{EISOISU_0}(s) \vdots G_{EISOISX_0}(s) \right], \qquad (4.56)$$

where

$$G_{EISOISU}(s) = \begin{bmatrix} \frac{s^2+s-14}{s^3+3s^2-12s-40} & \frac{4}{s^3+3s^2-12s-40} & \frac{s-2}{s^3+3s^2-12s-40} \\ \frac{3s+9}{s^3+3s^2-12s-40} & \frac{s^2+5s+6}{s^3+3s^2-12s-40} & \frac{2s+7}{s^3+3s^2-12s-40} \\ \frac{12}{(s^3+3s^2-12s-40)} & \frac{4s+8}{s^3+3s^2-12s-40} & \frac{s^2-4}{s^3+3s^2-12s-40} \end{bmatrix} \bullet$$

$$\bullet \begin{bmatrix} 7 & 5+s \\ 10+s & -6-s \\ -4-2s & 3+2s \end{bmatrix} \Longrightarrow$$

$$G_{EISOISU}(s) = \begin{bmatrix} \frac{75s^2+11s-42}{s^3+3s^2-12s-40} & \frac{s^3+8s^2-8s-82}{s^3+3s^2-12s-40} \\ \frac{s^3+15s^2+77s+123}{s^3+3s^2-12s-40} & \frac{-s^3-7s^2-10s+79}{s^3+3s^2-12s-40} \\ \frac{-2s^3+56s+180}{s^3+3s^2-12s-40} & \frac{2s^3-s^2-28s}{s^3+3s^2-12s-40} \end{bmatrix}, \tag{4.57}$$

$$G_{EISOISU_0}(s) = \begin{bmatrix} \frac{s^2+s-14}{s^3+3s^2-12s-40} & \frac{4}{s^3+3s^2-12s-40} & \frac{s-2}{s^3+3s^2-12s-40} \\ \frac{3s+9}{s^3+3s^2-12s-40} & \frac{s^2+5s+6}{s^3+3s^2-12s-40} & \frac{2s+7}{s^3+3s^2-12s-40} \\ \frac{12}{s^3+3s^2-12s-40} & \frac{4s+8}{s^3+3s^2-12s-40} & \frac{s^2-4}{s^3+3s^2-12s-40} \end{bmatrix} \bullet$$

$$\bullet \begin{bmatrix} 0 & 1 \\ 1 & -1 \\ -2 & 2 \end{bmatrix} \Longrightarrow$$

$$G_{EISOISU_0}(s) = \begin{bmatrix} \frac{-2s}{s^3+3s^2-12s-40} & \frac{s^2+3s-22}{s^3+3s^2-12s-40} \\ \frac{s^2+s-8}{s^3+3s^2-12s-40} & \frac{-s^2+2s+17}{s^3+3s^2-12s-40} \\ \frac{-2s^2+4s+16}{s^3+3s^2-12s-40} & \frac{2s^2-4s-4}{s^3+3s^2-12s-40} \end{bmatrix}, \tag{4.58}$$

$$G_{EISOISX_0}(s) = \begin{bmatrix} \frac{s^2+s-14}{s^3+3s^2-12s-40} & \frac{4}{s^3+3s^2-12s-40} & \frac{s-2}{s^3+3s^2-12s-40} \\ \frac{3s+9}{s^3+3s^2-12s-40} & \frac{s^2+5s+6}{s^3+3s^2-12s-40} & \frac{2s+7}{s^3+3s^2-12s-40} \\ \frac{12}{s^3+3s^2-12s-40} & \frac{4s+8}{s^3+3s^2-12s-40} & \frac{s^2-4}{s^3+3s^2-12s-40} \end{bmatrix} \bullet$$

$$\bullet \begin{bmatrix} 1 & 0 & 0 \\ 0 & 1 & 0 \\ 0 & 0 & 1 \end{bmatrix} \Longrightarrow$$

$$G_{EISOISX_0}(s) = \begin{bmatrix} \frac{s^2+s-14}{s^3+3s^2-12s-40} & \frac{4}{s^3+3s^2-12s-40} & \frac{s-2}{s^3+3s^2-12s-40} \\ \frac{3s+9}{s^3+3s^2-12s-40} & \frac{s^2+5s+6}{s^3+3s^2-12s-40} & \frac{2s+7}{s^3+3s^2-12s-40} \\ \frac{12}{s^3+3s^2-12s-40} & \frac{4s+8}{s^3+3s^2-12s-40} & \frac{s^2-4}{s^3+3s^2-12s-40} \end{bmatrix}. \tag{4.59}$$

Finally,

$$F_{EISOIS}(s) =$$

$$= \begin{bmatrix} \begin{bmatrix} \frac{75s^2+11s-42}{s^3+3s^2-12s-40} & \frac{s^3+8s^2-8s-82}{s^3+3s^2-12s-40} \\ \frac{s^3+15s^2+77s+123}{s^3+3s^2-12s-40} & \frac{-s^3-7s^2-10s+79}{s^3+3s^2-12s-40} \\ \frac{-2s^3+56s+180}{s^3+3s^2-12s-40} & \frac{2s^3-s^2-28s}{s^3+3s^2-12s-40} \end{bmatrix}^T \\ \begin{bmatrix} \frac{-2s}{s^3+3s^2-12s-40} & \frac{s^2+3s-22}{s^3+3s^2-12s-40} \\ \frac{s^2+s-8}{s^3+3s^2-12s-40} & \frac{-s^2+2s+17}{s^3+3s^2-12s-40} \\ \frac{-2s^2+4s+16}{s^3+3s^2-12s-40} & \frac{12-4ss-8}{s^3+3s^2-12s-40} \end{bmatrix}^T \\ \begin{bmatrix} \frac{s^2+s-14}{s^3+3s^2-12s-40} & \frac{4}{s^3+3s^2-12s-40} & \frac{s-2}{s^3+3s^2-12s-40} \\ \frac{3s+9}{s^3+3s^2-12s-40} & \frac{s^2+5s+6}{s^3+3s^2-12s-40} & \frac{2s+7}{s^3+3s^2-12s-40} \\ \frac{12}{s^3+3s^2-12s-40} & \frac{4s+8}{s^3+3s^2-12s-40} & \frac{s^2-4}{s^3+3s^2-12s-40} \end{bmatrix}^T \end{bmatrix}^T \quad (4.60)$$

Equation (4.27) yields:

$$F_{EISO}(s) = \begin{bmatrix} G_{EISOU}(s) & \vdots & G_{EISOU_0}(s) & \vdots & G_{EISOX_0}(s) \end{bmatrix} =$$

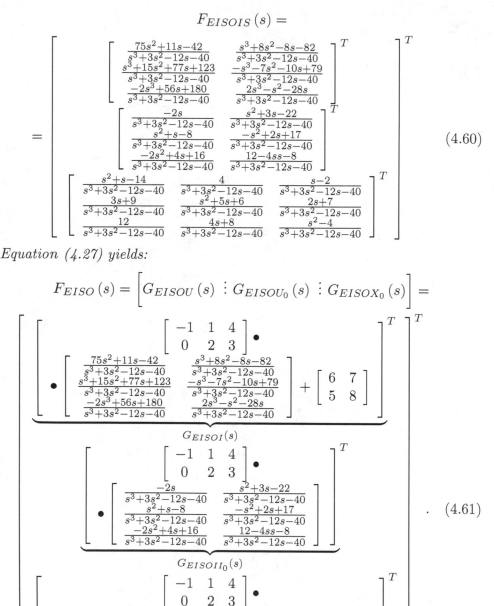

$$. \quad (4.61)$$

The system transfer function matrices and the vector $\mathbf{V}_{ISO}(s)$ are, in view of (4.34)-(3.20), (Subsection 3.1.2):

$$G_{EISOU}(s) =$$

$$= \left[\bullet \left[\begin{array}{cc} \dfrac{75s^2+11s-42}{s^3+3s^2-12s-40} & \dfrac{s^3+8s^2-8s-82}{s^3+3s^2-12s-40} \\[2mm] \dfrac{s^3+15s^2+77s+123}{s^3+3s^2-12s-40} & \dfrac{-s^3-7s^2-10s+79}{s^3+3s^2-12s-40} \\[2mm] \dfrac{-2s^3-28s-72}{(s+2)(s^3+3s^2-12s-40)} & \dfrac{2s^3+11s^2+60s-48}{s^3+3s^2-12s-40} \end{array} \right] \left[\begin{array}{cc} -1 & 1 & 4 \\ 0 & 2 & 3 \end{array} \right] \bullet + \left[\begin{array}{cc} 6 & 7 \\ 5 & 8 \end{array} \right] \right] \Longrightarrow$$

$$G_{EISOU}(s) = \left[\begin{array}{cc} \dfrac{-7s^3-60s^2+284s+42+885}{s^3+3s^2-12s-40} + 6 & \dfrac{6s^3-19s^2-114s+161}{s^3+3s^2-12s-40} + 7 \\[2mm] \dfrac{8s^3+27s^2+70s+246}{s^3+3s^2-12s-40} + 5 & \dfrac{4s^3-17s^2-104s+158^2}{s^3+3s^2-12s-40} + 8 \end{array} \right], \tag{4.62}$$

$$G_{EISOU_0}(s) = \left[\bullet \left[\begin{array}{cc} \dfrac{-2s}{s^3+3s^2-12s-40} & \dfrac{s^2+3s-22}{s^3+3s^2-12s-40} \\[2mm] \dfrac{s^2+s-8}{s^3+3s^2-12s-40} & \dfrac{-s^2+2s+17}{s^3+3s^2-12s-40} \\[2mm] \dfrac{-2s^2+4s+16}{s^3+3s^2-12s-40} & \dfrac{2s^2-4s-4}{s^3+3s^2-12s-40} \end{array} \right] \left[\begin{array}{cc} -1 & 1 & 4 \\ 0 & 2 & 3 \end{array} \right] \bullet \right] \Longrightarrow$$

$$G_{EISOU_0}(s) = \left[\begin{array}{cc} \dfrac{-7s^2+19s+56}{s^3+3s^2-12s-40} & \dfrac{6s^2-17s+23}{s^3+3s^2-12s-40} \\[2mm] \dfrac{-4s^2+14s+32}{s^3+3s^2-12s-40} & \dfrac{4s^2-8s+22}{s^3+3s^2-12s-40} \end{array} \right], \tag{4.63}$$

$$G_{EISOX_0}(s) =$$

$$= \left[\bullet \left[\begin{array}{ccc} \dfrac{s^2+s-14}{s^3+3s^2-12s-40} & \dfrac{4}{s^3+3s^2-12s-40} & \dfrac{s-2}{s^3+3s^2-12s-40} \\[2mm] \dfrac{3s+9}{s^3+3s^2-12s-40} & \dfrac{s^2+5s+6}{s^3+3s^2-12s-40} & \dfrac{2s+7}{s^3+3s^2-12s-40} \\[2mm] \dfrac{12}{s^3+3s^2-12s-40} & \dfrac{4s+8}{s^3+3s^2-12s-40} & \dfrac{s^2-4}{s^3+3s^2-12s-40} \end{array} \right] \left[\begin{array}{cc} -1 & 1 & 4 \\ 0 & 2 & 3 \end{array} \right] \bullet \right] \Longrightarrow$$

$$G_{EISOX_0}(s) = \left[\begin{array}{ccc} \dfrac{-s^2+5s+71}{s^3+3s^2-12s-40} & \dfrac{s^2+21s+34}{s^3+3s^2-12s-40} & \dfrac{4s^2+3s-7}{s^3+3s^2-12s-40} \\[2mm] \dfrac{6s+18+36}{s^3+3s^2-12s-40} & \dfrac{2s^2+22s+36}{s^3+3s^2-12s-40} & \dfrac{3s^2+4s+2}{s^3+3s^2-12s-40} \end{array} \right]. \tag{4.64}$$

Altogether,

$$F_{EISO}(s) = \left[G_{EISOU}(s) \vdots G_{EISOU_0}(s) \vdots G_{EISOX_0}(s) \right] =$$

$$= \left[\begin{array}{c} G_{EISOU}^T(s) \\ G_{EISOU_0}^T \\ G_{EISOX_0}^T \end{array} \right]^T =$$

$$\left[\left[\begin{array}{cc} \frac{-7s^3-60s^2+284s+42+885}{s^3+3s^2-12s-40}+6 & \frac{6s^3-19s^2-114s+161}{s^3+3s^2-12s-40}+7 \\ \frac{8s^3+27s^2+70s+246}{s^3+3s^2-12s-40}+5 & \frac{4s^3-17s^2-104s+158^2}{s^3+3s^2-12s-40}+8 \end{array} \right]^T \right.$$
$$\left[\begin{array}{cc} \frac{-7s^2+19s+56}{s^3+3s^2-12s-40} & \frac{6s^2-17s+23}{s^3+3s^2-12s-40} \\ \frac{-4s^2+14s+32}{s^3+3s^2-12s-40} & \frac{4s^2-8s+22}{s^3+3s^2-12s-40} \end{array} \right]^T$$
$$\left. \left[\begin{array}{ccc} \frac{-s^2+5s+71}{s^3+3s^2-12s-40} & \frac{s^2+21s+34}{s^3+3s^2-12s-40} & \frac{4s^2+3s-7}{s^3+3s^2-12s-40} \\ \frac{6s+18+36}{s^3+3s^2-12s-40} & \frac{2s^2+22s+36}{s^3+3s^2-12s-40} & \frac{3s^2+4s+2}{s^3+3s^2-12s-40} \end{array} \right]^T \right]^T \quad (4.65)$$

together with

$$V_{EISOU}(s) = \left[\begin{array}{c} \mathbf{U}(s) \\ \mathbf{C}_{EISOU0} \end{array} \right], \quad \mathbf{U}(s) = \left[\begin{array}{c} U_1(s) \\ U_2(s) \end{array} \right],$$

$$\mathbf{C}_{EISO0} = \left[\begin{array}{c} \mathbf{U}_0^1 \\ \mathbf{X}_0 \end{array} \right] = \left[\begin{array}{c} U_{10} \\ U_{20} \\ X_{10} \\ X_{20} \\ X_{30} \end{array} \right]. \quad (4.66)$$

4.2   EISO plant desired regime

Definition 44, (Section 3.2), slightly changes its formulation as follows.

Definition 57 *A functional vector control-state pair* $[\mathbf{U}*(.), \mathbf{X}*(.)]$ *is **nominal** for the EISO plant (4.1), (4.2) **relative to the functional pair***

$$[\mathbf{D}(.), \mathbf{Y}_d(.)],$$

which is denoted by $[\mathbf{U}_N(.), \mathbf{X}_N(.)]$, *if and only if* $[\mathbf{U}(.), \mathbf{X}(.)] = [\mathbf{U}*(.), \mathbf{X}*(.)]$ *ensures that the corresponding real response* $\mathbf{Y}(.) = \mathbf{Y}*(.)$ *of the plant obeys* $\mathbf{Y}*(t) = \mathbf{Y}_d(t)$ *all the time,*

$$[\mathbf{U}*(.), \mathbf{X}*(.)] = [\mathbf{U}_N(.), \mathbf{X}_N(.)] \Longleftrightarrow \langle \mathbf{Y}*(t) = \mathbf{Y}_d(t), \ \forall t \in \mathfrak{T}_0 \rangle.$$

The nominal motion $\mathbf{X}_N(.;\mathbf{X}_{N0};D^\mu;\mathbf{U}_N^\mu)$, $\mathbf{X}_N(0;\mathbf{X}_{N0};D^\mu;\mathbf{U}_N^\mu) \equiv \mathbf{X}_{N0}$, *is the desired motion* $\mathbf{X}_d(.;\mathbf{X}_{d0};D^\mu;\mathbf{U}_N^\mu)$ *of the EISO plant (4.1), (4.2) relative to the functional vector pair* $[\mathbf{D}(.),\mathbf{Y}_d(.)]$, *for short: the desired motion of the system,*

$$\mathbf{X}_d(.;\mathbf{X}_{d0};\mathbf{D}^\mu;\mathbf{U}_N^\mu) \equiv \mathbf{X}_N(.;\mathbf{X}_{N0};\mathbf{D}^\mu;\mathbf{U}_N^\mu),$$
$$\mathbf{X}_d(0;\mathbf{X}_{d0};\mathbf{D}^\mu;\mathbf{U}_N^\mu) \equiv \mathbf{X}_{d0} \equiv \mathbf{X}_{N0}. \tag{4.67}$$

Definition 57 and the system description (4.1), (4.2) imply the following theorem:

Theorem 58 *In order for the functional vector pair* $[\mathbf{U}^*(.),\mathbf{X}^*(.)]$ *to be nominal for the EISO plant (4.1), (4.2) relative to the functional vector pair* $[\mathbf{D}^{(\mu)}(.),\mathbf{Y}_d(.)]$, $[\mathbf{U}^*(.),\mathbf{X}^*(.)] = [\mathbf{U}_N(.),\mathbf{X}_N(.)]$, *it is necessary and sufficient that it obeys the following equations:*

$$-\mathrm{B}^{(\mu)}\mathbf{U}^{*\mu}(t) + \frac{d\mathbf{X}^*(t)}{dt} - A\mathbf{X}^*(t) = D^{(\mu)}\mathbf{D}^\mu(t), \; \forall t \in \mathfrak{T}_0, \tag{4.68}$$

$$U\mathbf{U}^*(t) + C\mathbf{X}^*(t) = \mathbf{Y}_d(t) - V\mathbf{D}(t), \; \forall t \in \mathfrak{T}_0, \tag{4.69}$$

or equivalently,

$$\begin{bmatrix} -\mathrm{B}^{(\mu)}S_r^{(\mu)}(s) & sI - A \\ U & C \end{bmatrix} \begin{bmatrix} \mathbf{U}^*(s) \\ \mathbf{X}^*(s) \end{bmatrix} =$$

$$= \begin{bmatrix} \mathbf{X}_0^* + \mathrm{B}^{(\mu)}Z_r^{(\mu-1)}(s)\mathbf{U}_0^{\mu-1} + D^{(\mu)} \left\{ \begin{array}{c} S_d^{(\mu)}(s)\mathbf{D}(s)- \\ -Z_d^{(\mu-1)}(s)\mathbf{D}_0^{\mu-1} \end{array} \right\} \\ \mathbf{Y}_d(s) - V\mathbf{D}(s) \end{bmatrix}. \tag{4.70}$$

This theorem opens the problem of the conditions for the existence of the solutions of the equations (4.68), (4.69), or equivalently of (4.70). There are $(r + n)$ unknown variables $\mathbf{U}^*(s) \in \mathbb{C}^r$ and $\mathbf{X}^*(s) \in \mathbb{C}^n$ and $(N + n)$ equations so that the following holds:

Claim 59 *In order to exist a nominal functional vector pair*

$$[\mathbf{U}_N(.),\mathbf{X}_N(.)]$$

for the EISO system (4.1), (4.2) relative to the functional vector pair

$$[\mathbf{D}(.),\mathbf{Y}_d(.)]$$

it is necessary and sufficient that $N \leq r$.

The proof of this Claim is analogous to the proof of Claim 46 (Section 3.2).

Claim 59 provides the full solution to the problem of the existence of a nominal functional vector pair $[\mathbf{U}_N(.), \mathbf{X}_N(.)]$ for the *EISO* plant (4.1), (4.2) relative to the functional vector pair $[\mathbf{D}(.), \mathbf{Y}_d(.)]$.

Condition 60 *The desired output response of the EISO system (4.1), (4.2) is realizable, i.e., $N \leq r$. The nominal control-state pair $[\mathbf{U}_N(.), \mathbf{X}_N(.)]$ is known.*

The *EISO* plant description in terms of the deviations (3.38), (2.53), (9.3) and (2.55) reads:

$$\frac{d\mathbf{x}(t)}{dt} = A\mathbf{x}(t) + D^{(\mu)}\mathbf{d}^{\mu}(t) + \mathrm{B}^{(\mu)}\mathbf{u}^{\mu}(t), \ \forall t \in \mathfrak{T}_0, \qquad (4.71)$$

$$\mathbf{y}(t) = C\mathbf{x}(t) + V\mathbf{d}(t) + U\mathbf{u}(t), \ \forall t \in \mathfrak{T}_0. \qquad (4.72)$$

4.3   Exercises

Exercise 61 *1. Select a physical EISO plant.*
2. Determine its time domain EISO mathematical model.

Chapter 5

HISO systems

5.1 *HISO* system mathematical model

5.1.1 Time domain

The linear **Higher order Input-State-Output** (*HISO*) **(dynamical, control) systems** have not been studied so far. Their mathematical models contain *the α-th* order *linear differential state vector equation* (5.1) and *the linear algebraic output vector equation* (5.2),

$$A^{(\alpha)}\mathbf{R}^\alpha(t) = D^{(\mu)}\mathbf{D}^\mu(t) + B^{(\mu)}\mathbf{U}^\mu(t) = H^{(\mu)}\mathbf{I}^\mu(t), \ \forall t \in \mathfrak{T}_0,$$

$$A^{(\alpha)} \in \mathfrak{R}^{\rho \times (\alpha+1)\rho}, \ \mathbf{R}^\alpha \in \mathfrak{R}^{(\alpha+1)\rho}, \ D^{(\mu)} \in \mathfrak{R}^{\rho \times (\mu+1)d}, \ B^{(\mu)} \in \mathfrak{R}^{\rho \times (\mu+1)r},$$

$$H^{(\mu)} = \left[D^{(\mu)} \vdots B^{(\mu)} \right] \in \mathfrak{R}^{\rho \times (\mu+1)(d+r)}, \ \mathbf{I}^\mu = \left[(\mathbf{D}^\mu)^T \vdots (\mathbf{U}^\mu)^T \right]^T \in \mathfrak{R}^{(\mu+1)(d+r)},$$

$$(5.1)$$

$$\mathbf{Y}(t) = R^{(\alpha)}\mathbf{R}^\alpha(t) + V\mathbf{D}(t) + U\mathbf{U}(t) = R^{(\alpha)}\mathbf{R}^\alpha(t) + Q\mathbf{I}(t), \ \forall t \in \mathfrak{T}_0,$$

$$V \in \mathfrak{R}^{N \times d}, \ U \in \mathfrak{R}^{N \times r}, \ Q = \left[V \vdots U \right] \in \mathfrak{R}^{N \times (d+r)},$$

$$R^{(\alpha)} = \left[R_0 \vdots R_1 \vdots ... \vdots R_{\alpha-1} \vdots O_{N,\rho} \right], \ R_\alpha = O_{N,\rho}. \tag{5.2}$$

The zero matrix value of $R_{y\alpha}$, $R_{y\alpha} = O_{N,\rho}$, ensures that the highest derivative $R^{(\alpha)}$ of the vector $\mathbf{R}$ does not act on the system output vector $\mathbf{Y}$.

The output vector $\mathbf{Y}$ depends only linearly and algebraically on the state vector $\mathbf{S} = \mathbf{R}^{\alpha-1} \in \mathfrak{R}^{\alpha\rho}$ and on the output vector $\mathbf{I}$. It does not depend on the vector $R^{(\alpha)}$ because it does not depend on the whole extended vector $\mathbf{R}^\alpha$ due to $R_{y\alpha} = O_{N,\rho}$.

Note 62 *The state vector* $\mathbf{S}_{HISO}$ *of the HISO system (5.1), (5.2), is defined in (1.22), (1.24) (Section 1.4) by:*

$$\mathbf{S}_{HISO} = \mathbf{R}^{\alpha-1} = \left[\mathbf{R}^T \vdots \mathbf{R}^{(1)^T} \vdots ... \vdots \mathbf{R}^{(\alpha-1)^T} \right]^T \in \mathfrak{R}^n, \ n = \alpha\rho, \quad (5.3)$$

This new vector notation $\mathbf{R}^{\alpha-1}$ *has permitted us to define the state of the HISO system (5.1), (5.2), by preserving the physical sense. It enables us to discover in what follows the complex domain criteria for observability, controllability and trackability directly from their definitions. Such criteria possess the complete physical meaning.*

5.1.2　Complex domain

After the application of the Laplace transform to Equations (5.1) and (5.2) they are transformed into:

$$\mathbf{R}(s) = \left(A^{(\alpha)} S_\rho^{(\alpha)}(s) \right)^{-1} \bullet$$

$$\bullet \left[\begin{array}{c} D^{(\mu)} S_d^{(\mu)}(s) \vdots B^{(\mu)} S_r^{(\mu)}(s) \vdots -D^{(\mu)} Z_d^{(\mu-1)}(s) \vdots A^{(\alpha)} Z_\rho^{(\alpha-1)}(s) \vdots \\ \vdots -B^{(\mu)} Z_r^{(\mu-1)}(s) \end{array} \right] \bullet$$

$$\bullet \left[\mathbf{D}^T(s) \vdots \mathbf{U}^T(s) \vdots \left(\mathbf{D}_0^{\mu-1} \right)^T \vdots \left(\mathbf{R}_0^{\alpha-1} \right)^T \vdots \left(\mathbf{U}_0^{\mu-1} \right)^T \right]^T =$$

$$= F_{HISOIS}(s) \, \mathbf{V}_{HISOIS}(s), \quad (5.4)$$

$$\mathbf{Y}(s) = F_{HISO}(s) \, \mathbf{V}_{HISO}(s) =$$

$$= \left[\begin{array}{c} \left(R^{(\alpha)} S_\rho^{(\alpha)}(s) \left(A^{(\alpha)} S_\rho^{(\alpha)}(s) \right)^{-1} D^{(\mu)} S_d^{(\mu)}(s) + V \right)^T \vdots \\ \vdots \left(R^{(\alpha)} S_\rho^{(\alpha)}(s) \left(A^{(\alpha)} S_\rho^{(\alpha)}(s) \right)^{-1} B^{(\mu)} S_r^{(\mu)}(s) + U \right)^T \vdots \\ \vdots \left(-R^{(\alpha)} S_\rho^{(\alpha)}(s) \left(A^{(\alpha)} S_\rho^{(\alpha)}(s) \right)^{-1} D^{(\mu)} Z_M^{(\mu-1)}(s) \right)^T \vdots \\ \vdots \left(\begin{array}{c} R^{(\alpha)} S_\rho^{(\alpha)}(s) \left(A^{(\alpha)} S_\rho^{(\alpha)}(s) \right)^{-1} A^{(\alpha)} Z_\rho^{(\alpha-1)}(s) - \\ -R^{(\alpha)} Z_\rho^{(\alpha-1)}(s) \end{array} \right)^T \vdots \\ \vdots \left(-R^{(\alpha)} S_\rho^{(\alpha)}(s) \left(A^{(\alpha)} S_\rho^{(\alpha)}(s) \right)^{-1} B^{(\mu)} Z_r^{(\mu-1)}(s) \right)^T \end{array} \right]^T \bullet$$

$$\bullet \left[\mathbf{D}^T(s) \vdots \mathbf{U}^T(s) \vdots \left(\mathbf{D}_0^{\mu-1} \right)^T \vdots \left(\mathbf{R}_0^{\alpha-1} \right)^T \vdots \left(\mathbf{U}_0^{\mu-1} \right)^T \right]^T, \quad (5.5)$$

where:

- $F_{HISOIS}(s)$,

$$F_{HISOIS}(s) = \left(A^{(\alpha)}S_\rho^{(\alpha)}(s)\right)^{-1} \begin{bmatrix} \left[D^{(\mu)}S_d^{(\mu)}(s)\right]^T & \vdots & \\ \vdots & \left[B^{(\mu)}S_r^{(\mu)}(s)\right]^T & \vdots \\ \vdots & \left[-D^{(\mu)}Z_d^{(\mu-1)}(s)\right]^T & \vdots \\ \vdots & \left[A^{(\alpha)}Z_\rho^{(\alpha-1)}(s)\right]^T & \vdots \\ \vdots & \left[-B^{(\mu)}Z_r^{(\mu-1)}(s)\right]^T & \end{bmatrix}^T , \qquad (5.6)$$

is the *HISO* plant (5.1), (5.2) *input to state (IS) full transfer function matrix*, the inverse Laplace transform of which is the plant *IS full fundamental matrix* $\Psi_{HISOIS}(t)$ [40],

$$\Psi_{HISOIS}(t) = \mathcal{L}^{-1}\{F_{HISOIS}(s)\}, \qquad (5.7)$$

and the inverse Laplace transform of $\left(A^{(\alpha)}S_\rho^{(\alpha)}(s)\right)^{-1}$ is *the HISO system fundamental matrix* $\Phi(t)$,,

$$\Phi(t) = \mathcal{L}^{-1}\left\{\left(A^{(\alpha)}S_\rho^{(\alpha)}(s)\right)^{-1}\right\}, \qquad (5.8)$$

- $F_{HISO}(s)$,

$$F_{HISO}(s) =$$

$$\begin{bmatrix} \left(R^{(\alpha)}S_\rho^{(\alpha)}(s)\left(A^{(\alpha)}S_\rho^{(\alpha)}(s)\right)^{-1}D^{(\mu)}S_d^{(\mu)}(s) + V\right)^T & \vdots & \\ \vdots & \left(R^{(\alpha)}S_\rho^{(\alpha)}(s)\left(A^{(\alpha)}S_\rho^{(\alpha)}(s)\right)^{-1}B^{(\mu)}S_r^{(\mu)}(s) + U\right)^T & \vdots \\ \vdots & \left(-R^{(\alpha)}S_\rho^{(\alpha)}(s)\left(A^{(\alpha)}S_\rho^{(\alpha)}(s)\right)^{-1}D^{(\mu)}Z_d^{(\mu-1)}(s)\right)^T & \vdots \\ \vdots & \left(\begin{array}{c}R^{(\alpha)}S_\rho^{(\alpha)}(s)\left(A^{(\alpha)}S_\rho^{(\alpha)}(s)\right)^{-1}A^{(\alpha)}Z_\rho^{(\alpha-1)}(s)- \\ -R^{(\alpha)}Z_\rho^{(\alpha-1)}(s)\end{array}\right)^T & \vdots \\ \vdots & \left(-R^{(\alpha)}S_\rho^{(\alpha)}(s)\left(A^{(\alpha)}S_\rho^{(\alpha)}(s)\right)^{-1}B^{(\mu)}Z_r^{(\mu-1)}(s)\right)^T & \end{bmatrix}^T , \qquad (5.9)$$

is the $HISO$ plant (5.1), (5.2) *input to output (IO) full transfer function matrix*, the inverse Laplace transform of which is the system IO *full fundamental matrix* $\Psi_{HISO}(t)$,

$$\Psi_{HISO}(t) = \mathcal{L}^{-1}\{F_{HISO}(s)\}, \tag{5.10}$$

- $p_{HISO}(s)$,

$$p_{HISO}(s) = \det\left(A^{(\alpha)}S_{\rho}^{(\alpha)}(s)\right), \tag{5.11}$$

is the characteristic polynomial of the $HISO$ plant (5.1), (5.2) and the denominator polynomial of all its transfer function matrices,

- $G_{HISOD}(s)$,

$$G_{HISOD}(s) = R^{(\alpha)}S_{\rho}^{(\alpha)}(s)\left(A^{(\alpha)}S_{\rho}^{(\alpha)}(s)\right)^{-1}D^{(\mu)}S_{d}^{(\mu)}(s) + V =$$

$$= p_{HISO}^{-1}(s)\left[\begin{array}{c} R^{(\alpha)}S_{\rho}^{(\alpha)}(s)adj\left(A^{(\alpha)}S_{\rho}^{(\alpha)}(s)\right)D^{(\mu)}S_{d}^{(\mu)}(s)+ \\ +p_{HISO}(s)\,V \end{array}\right], \tag{5.12}$$

is the $HISO$ plant (5.1), (5.2) *transfer function matrix relative to the disturbance* $\mathbf{D}$,

- $G_{HISOU}(s)$,

$$G_{HISOU}(s) = \left[R^{(\alpha)}S_{\rho}^{(\alpha)}(s)\left(A^{(\alpha)}S_{\rho}^{(\alpha)}(s)\right)^{-1}B^{(\mu)}S_{r}^{(\mu)}(s) + U\right] =$$

$$= p_{HISO}^{-1}(s) \bullet \left[\begin{array}{c} R^{(\alpha)}S_{\rho}^{(\alpha)}(s)adj\left(A^{(\alpha)}S_{\rho}^{(\alpha)}(s)\right)B^{(\mu)}S_{r}^{(\mu)}(s)+ \\ +p_{HISO}(s)\,U \end{array}\right] =$$

$$= p_{HISO}^{-1}(s)\,L_{HISO}(s), \tag{5.13}$$

is the $HISO$ plant (5.1), (5.2) *transfer function matrix relative to the control* $\mathbf{U}$,

- $G_{HISOD_0}(s)$,

$$G_{HISOD_0}(s) = -p_{HISO}^{-1}(s) \bullet$$

$$\bullet \left[R^{(\alpha)}S_{\rho}^{(\alpha)}(s)adj\left(A^{(\alpha)}S_{\rho}^{(\alpha)}(s)\right)D^{(\mu)}Z_{d}^{(\mu-1)}(s)\right] \tag{5.14}$$

is the $HISO$ plant (5.1), (5.2) *transfer function matrix relative to the initial extended disturbance* $\mathbf{D}_{0}^{\mu-1}$,

- $G_{HISOR_0}(s)$,

$$G_{HISOR_0}(s) = p_{HISO}^{-1}(s) \bullet$$

$$\bullet \left(\begin{array}{c} R^{(\alpha)}S_{\rho}^{(\alpha)}(s)adj\left(A^{(\alpha)}S_{\rho}^{(\alpha)}(s)\right)A^{(\alpha)}Z_{\rho}^{(\alpha-1)}(s)- \\ -p_{HISO}(s)\,R^{(\alpha)}Z_{\rho}^{(\alpha-1)}(s) \end{array}\right) \tag{5.15}$$

is the *HISO* plant (5.1), (5.2) *transfer function matrix relative to the initial state vector* $\mathbf{R}_0^{\alpha-1}$,

- $G_{HISOU_0}(s)$,

$$G_{HISOU_0}(s) = -p_{HISO}^{-1}(s) \bullet$$
$$\bullet \left[R^{(\alpha)} S_\rho^{(\alpha)}(s) adj \left(A^{(\alpha)} S_\rho^{(\alpha)}(s) \right) B^{(\mu)} Z_r^{(\mu-1)}(s) \right] \tag{5.16}$$

is the *HISO* plant (5.1), (5.2) *transfer function matrix relative to the extended initial control vector* $\mathbf{U}_0^{\mu-1}$,

- $\mathbf{V}_{HISO}(s)$ and $\mathbf{C}_{HISO0}$,

$$\mathbf{V}_{HISO}(s) = \mathbf{V}_{HISOIS}(s) = \begin{bmatrix} \mathbf{D}^T(s) & \mathbf{U}^T(s) & \mathbf{C}_{HISO0} \end{bmatrix}^T, \tag{5.17}$$

$$\mathbf{C}_{HISO0} = \mathbf{C}_{HISOIS0} = \left[\left(\mathbf{D}_0^{\mu-1} \right)^T \quad \left(\mathbf{R}_0^{\alpha-1} \right)^T \quad \left(\mathbf{U}_0^{\mu-1} \right)^T \right]^T, \tag{5.18}$$

are the Laplace transform of the action vector $\mathbf{V}_{HISOP}(t)$ and the vector $\mathbf{C}_{HISO0}$ of all initial conditions, respectively.

Example 63 *Let the beginning and the end of the HISO system (5.1), (5.2) be defined by*

$$\begin{bmatrix} 1 & 2 \\ 3 & -4 \end{bmatrix} \begin{bmatrix} R_1^{(3)}(t) \\ R_2^{(3)}(t) \end{bmatrix} + \begin{bmatrix} 2 & 3 \\ 1 & 0 \end{bmatrix} \begin{bmatrix} R_1^{(1)}(t) \\ R_2^{(1)}(t) \end{bmatrix} + \begin{bmatrix} 4 & 0 \\ 1 & 1 \end{bmatrix} \begin{bmatrix} R_1(t) \\ R_2(t) \end{bmatrix} =$$
$$= \begin{bmatrix} 5 & 3 \\ 4 & 1 \end{bmatrix} \begin{bmatrix} I_1^{(2)}(t) \\ I_2^{(2)}(t) \end{bmatrix} + \begin{bmatrix} 1 & 0 \\ 0 & 3 \end{bmatrix} \begin{bmatrix} I_1^{(1)}(t) \\ I_2^{(1)}(t) \end{bmatrix} + \begin{bmatrix} 4 & 2 \\ 3 & 6 \end{bmatrix} \begin{bmatrix} I_1(t) \\ I_2(t) \end{bmatrix}, \tag{5.19}$$

$$\begin{bmatrix} Y_1(t) \\ Y_2(t) \end{bmatrix} = \underbrace{\begin{bmatrix} 1 & 0 \\ 0 & 2 \end{bmatrix}}_{R_{y1}} \begin{bmatrix} R_1^{(1)}(t) \\ R_2^{(1)}(t) \end{bmatrix} + \underbrace{\begin{bmatrix} 1 & 2 \\ 2 & 1 \end{bmatrix}}_{Q} \begin{bmatrix} I_1(t) \\ I_2(t) \end{bmatrix} \tag{5.20}$$

In this case

$$\alpha = 3, \ \rho = 2, \ N = 2, \ R_{y0} = R_{y2} = R_{y3} = O_2, \ M = 2, \ \mu = 2 < 3 = \alpha, \tag{5.21}$$

$$R_y^{(\alpha)} = R_y^{(3)} = \begin{bmatrix} R_{y0} \vdots R_{y1} \vdots R_{y2} \vdots R_{y3} \end{bmatrix} = \begin{bmatrix} 0 & 0 & 1 & 0 & 0 & 0 & 0 & 0 \\ 0 & 0 & 0 & 2 & 0 & 0 & 0 & 0 \end{bmatrix}, \tag{5.22}$$

$$R_y^{(3)} S_2^{(3)} = \begin{bmatrix} 0 & 0 & 1 & 0 & 0 & 0 & 0 & 0 \\ 0 & 0 & 0 & 2 & 0 & 0 & 0 & 0 \end{bmatrix} \bullet \begin{bmatrix} s^0 I_2 & \vdots & s^1 I_2 & \vdots & s^2 I_2 & \vdots & s^3 I_2 \end{bmatrix}^T \implies$$

$$R_y^{(3)} S_2^{(3)} = \begin{bmatrix} s & 0 \\ 0 & 2s \end{bmatrix}, \tag{5.23}$$

$$R_y^{(3)} Z_2^{(3-1)} = \begin{bmatrix} 0 & 0 & 1 & 0 & 0 & 0 & 0 & 0 \\ 0 & 0 & 0 & 2 & 0 & 0 & 0 & 0 \end{bmatrix} \bullet \begin{bmatrix} O_2 & O_2 & O_2 \\ s^{3-3} I_2 & O_2 & O_2 \\ s^{3-2} I_k & s^{3-3} I_k & O_2 \\ s^{3-1} I_2 & s^{3-2} I_2 & s^{3-3} I_2 \end{bmatrix} \implies$$

$$R_y^{(3)} Z_2^{(3-1)} = \begin{bmatrix} 1 & 0 & 0 & 0 & 0 & 0 \\ 0 & 2 & 0 & 0 & 0 & 0 \end{bmatrix}. \tag{5.24}$$

The equation (5.19) is the equation (B.1) (Example 227) so that the equation (B.14) (Example 227, Subsection 2.1.2) holds for $\mathbf{Y}$ *replaced by* $\mathbf{R}$, $F_{IO}(s)$ *replaced by* $F_{HISOIS}(s)$, $\mathbf{V}_{IO}^{\mp}(s)$ *replaced by* $\mathbf{V}_{HISOIS}^{\mp}(s)$, *and it is denoted as (5.25):*

$$\mathbf{R}^{\mp}(s) = F_{HISOIS}(s) \, \mathbf{V}_{HISOIS}^{\mp}(s), \tag{5.25}$$

where in this context:

- $F_{HISOIS}(s)$, which is equal to $F_{IO}(s)$ (B.15) (Example 227),

$$F_{HISOIS}(s) = \begin{bmatrix} \frac{4s^3 - 1}{10s^6 + 19s^4 + 17s^3 + 3s^2 + s - 4} & \frac{2s^3 + 3s}{10s^6 + 19s^4 + 17s^3 + 3s^2 + s - 4} \\ \frac{3s^3 + s + 1}{10s^6 + 19s^4 + 17s^3 + 3s^2 + s - 4} & -\frac{s^3 + 2s + 4}{10s^6 + 19s^4 + 17s^3 + 3s^2 + s - 4} \end{bmatrix} \bullet$$

$$\bullet \begin{bmatrix} \begin{bmatrix} 4 + s + 5s^2 & 2 + 3s^2 \\ 3 + 4s^2 & 6 + 3s + s^2 \end{bmatrix} \\ \begin{bmatrix} -1 - 5s & -3s & -5 & -3 \\ -4s & -3 - s & -4 & -1 \end{bmatrix} \\ \begin{bmatrix} 2 + s^2 & 3 + 2s^2 & s & 2s & 1 & 2 \\ 1 + 3s^2 & 3 - 4s^2 & 3s & -4s & 3 & -4 \end{bmatrix} \end{bmatrix}^T$$

$$= \begin{bmatrix} G_{HISOIS}(s) & \vdots & G_{HISOISI_0}(s) & \vdots & G_{HISOISY_0}(s) \end{bmatrix}, \tag{5.26}$$

is the IS full transfer function matrix of the HISO system (5.19), (5.20),

- The system IS transfer function $G_{HISOIS}(s)$ relative to the input vec-

tor **I** :

$$G_{HISOIS}(s) = \begin{bmatrix} \frac{4s^3-1}{10s^6+19s^4+17s^3+3s^2+s-4} & \frac{2s^3+3s}{10s^6+19s^4+17s^3+3s^2+s-4} \\ \frac{3s^3+s+1}{10s^6+19s^4+17s^3+3s^2+s-4} & -\frac{s^3+2s+4}{10s^6+19s^4+17s^3+3s^2+s-4} \end{bmatrix} \bullet$$

$$\bullet \begin{bmatrix} 4+s+5s^2 & 2+3s^2 \\ 3+4s^2 & 6+3s+s^2 \end{bmatrix} =$$

$$= \begin{bmatrix} \frac{28s^5+4s^4+34s^3-5s^2+8s-4}{10s^6+19s^4+17s^3+3s^2+s-4} & \frac{14s^5+6s^4+23s^3+6s^2+18s-2}{10s^6+19s^4+17s^3+3s^2+s-4} \\ \frac{11s^5+3s^4+6s^3-10s^2-s-8}{10s^6+19s^4+17s^3+3s^2+s-4} & \frac{8s^5-3s^4+s^3-7s^2-12s-22}{10s^6+19s^4+17s^3+3s^2+s-4} \end{bmatrix}, \qquad (5.27)$$

- *The system IS transfer function* $G_{HISOISI_0}(s)$ *relative to the extended initial input vector* $\mathbf{I}_{0\mp}^{\mu-1}$:

$$G_{HISOISI_0}(s) = \begin{bmatrix} \frac{4s^3-1}{10s^6+19s^4+17s^3+3s^2+s-4} & \frac{2s^3+3s}{10s^6+19s^4+17s^3+3s^2+s-4} \\ \frac{3s^3+s+1}{10s^6+19s^4+17s^3+3s^2+s-4} & -\frac{s^3+2s+4}{10s^6+19s^4+17s^3+3s^2+s-4} \end{bmatrix} \bullet$$

$$\bullet \begin{bmatrix} -1-5s & -3s & -5 & -3 \\ -4s & -3-s & -4 & -1 \end{bmatrix} \Longrightarrow$$

$$G_{HISOISI_0}(s) =$$

$$= \begin{bmatrix} \frac{1+5s-12s^2-4s^3-28s^4}{10s^6+19s^4+17s^3+3s^2+s-4} & \frac{-1+10s+3s^2-3s^3-3s^4}{10s^6+19s^4+17s^3+3s^2+s-4} \\ \frac{-6s-3s^2-6s^3-14s^4}{10s^6+19s^4+17s^3+3s^2+s-4} & \frac{12+7s-s^2+3s^3-8s^4}{10s^6+19s^4+17s^3+3s^2+s-4} \\ \frac{5-12s-28s^3}{10s^6+19s^4+17s^3+3s^2+s-4} & \frac{11+3s+4s^3}{10s^6+19s^4+17s^3+3s^2+s-4} \\ \frac{3-3s-14s^3}{10s^6+19s^4+17s^3+3s^2+s-4} & \frac{1-s-8s^3}{10s^6+19s^4+17s^3+3s^2+s-4} \end{bmatrix}^T, \qquad (5.28)$$

- *The system IS transfer function* $G_{HISOISR_0}(s)$ *relative to the extended initial output vector* $R_{0\mp}^{\alpha-1}$:

$$G_{HISOISR_0}(s) = \begin{bmatrix} \frac{4s^3-1}{10s^6+19s^4+17s^3+3s^2+s-4} & \frac{2s^3+3s}{10s^6+19s^4+17s^3+3s^2+s-4} \\ \frac{3s^3+s+1}{10s^6+19s^4+17s^3+3s^2+s-4} & -\frac{s^3+2s+4}{10s^6+19s^4+17s^3+3s^2+s-4} \end{bmatrix} \bullet$$

$$\bullet \begin{bmatrix} 2+s^2 & 3+2s^2 & s & 2s & 1 & 2 \\ 1+3s^2 & 3-4s^2 & 3s & -4s & 3 & -4 \end{bmatrix} \Longrightarrow$$

$$G_{HISOISR_0} =$$

$$= \begin{bmatrix} \frac{10s^5+19s^3-s^2+3s-2}{10s^6+19s^4+17s^3+3s^2+s-4} & \frac{-11s^2-2}{10s^6+19s^4+17s^3+3s^2+s-4} \\ \frac{-4s^5+6s^3-2s^2+9s-3}{10s^6+19s^4+17s^3+3s^2+s-4} & \frac{10s^5+16s^3+18s^2-3s-9}{10s^6+19s^4+17s^3+3s^2+s-4} \\ \frac{10s^4+9s^2-s}{10s^6+19s^4+17s^3+3s^2+s-4} & \frac{-5s^2-11s}{10s^6+19s^4+17s^3+3s^2+s-4} \\ \frac{-12s^2-2s}{10s^6+19s^4+17s^3+3s^2+s-4} & \frac{2s^4-6s^2-14s}{10s^6+19s^4+17s^3+3s^2+s-4} \\ \frac{10s^3+9s-1}{10s^6+19s^4+17s^3+3s^2+s-4} & \frac{-5s-11}{10s^6+19s^4+17s^3+3s^2+s-4} \\ \frac{-12s-2}{10s^6+19s^4+17s^3+3s^2+s-4} & \frac{10s^3+10s+18}{10s^6+19s^4+17s^3+3s^2+s-4} \end{bmatrix}^T , \qquad (5.29)$$

- The Laplace transform $\mathbf{V}^{\mp}_{HISOIS}(s)$ of the system action vector $\mathbf{V}_{HISOIS}(t)$ reads

$$\mathbf{V}^{\mp}_{HISOIS}(s) = \begin{bmatrix} \mathbf{I}^{\mp}(s) \\ \mathbf{I}^1_{0\mp} \\ \mathbf{R}^2_{0\mp} \end{bmatrix} = \begin{bmatrix} \mathbf{I}^{\mp}(s) \\ \mathbf{C}_{HISOIS0\mp} \end{bmatrix}, \qquad (5.30)$$

- And the vector $\mathbf{C}_{HISOIS0\mp}$ of all the initial conditions is found to be

$$\mathbf{C}_{HISOIS0\mp} = \begin{bmatrix} \mathbf{I}^1_{0\mp} \\ \mathbf{R}^2_{0\mp} \end{bmatrix}. \qquad (5.31)$$

In order to determine the system IO transfer function matrices we refer to Equation (5.9),

$$F_{HISO}(s) =$$

$$= \begin{bmatrix} \left(R_y^{(\alpha)} S_\rho^{(\alpha)}(s) \left(A^{(\alpha)} S_\rho^{(\alpha)}(s) \right)^{-1} H^{(\mu)} S_M^{(\mu)}(s) + Q \right)^T \\ \left(-R_y^{(\alpha)} S_\rho^{(\alpha)}(s) \left(A^{(\alpha)} S_\rho^{(\alpha)}(s) \right)^{-1} H^{(\mu)} Z_M^{(\mu-1)}(s) \right)^T \\ \left(R_y^{(\alpha)} S_\rho^{(\alpha)}(s) \left(A^{(\alpha)} S_\rho^{(\alpha)}(s) \right)^{-1} A^{(\alpha)} Z_\rho^{(\alpha-1)}(s) - R_y^{(\alpha)} Z_\rho^{(\alpha-1)}(s) \right)^T \end{bmatrix}^T , \qquad (5.32)$$

to the system (5.19), (5.20):

$$F_{HISO}(s) = \begin{bmatrix} G_{HISO}(s) & \vdots & G_{HISOI_0}(s) & \vdots & G_{HISOR_0}(s) \end{bmatrix} =$$

$$= \begin{bmatrix} \left(R_y^{(\alpha)} S_\rho^{(\alpha)}(s) G_{HISOIS}(s) + Q \right)^T \\ \left(R_y^{(\alpha)} S_\rho^{(\alpha)}(s) G_{HISOISI_0}(s) \right)^T \\ \left(R_y^{(\alpha)} S_\rho^{(\alpha)}(s) G_{HISOISR_0}(s) - R_y^{(\alpha)} Z_\rho^{(\alpha-1)}(s) \right)^T \end{bmatrix}^T , \qquad (5.33)$$

where:

*- $F_{HISO}(s)$ is the IO full transfer function matrix of the HISO system
(5.19), (5.20), which together with (5.23), (5.24), determines:*

- $G_{HISOI}(s)$,

$$G_{HISOI}(s) = R_y^{(3)} S_2^{(3)} G_{HISOIS}(s) + Q =$$

$$= Q + \begin{bmatrix} s & 0 \\ 0 & 2s \end{bmatrix} G_{HISOIS}(s) \implies$$

$$G_{HISOI}(s) = \begin{bmatrix} 1 & 2 \\ 2 & 1 \end{bmatrix} + \begin{bmatrix} s & 0 \\ 0 & 2s \end{bmatrix} \bullet$$

$$\bullet \begin{bmatrix} \frac{-4+8s-5s^2+34s^3+4s^4+28s^5}{10s^6+19s^4+17s^3+3s^2+s-4} & \frac{-2+18s+6s^2+23s^3+6s^4+14s^5}{10s^6+19s^4+17s^3+3s^2+s-4} \\ \frac{16+11s+22s^2+28s^3+3s^4+19s^5}{10s^6+19s^4+17s^3+3s^2+s-4} & -\frac{26+26s+13s^2+17s^3+3s^4+10s^5}{10s^6+19s^4+17s^3+3s^2+s-4} \end{bmatrix} \implies$$

$$G_{HISOI}(s) =$$

$$= \begin{bmatrix} \frac{-4-3s+11s^2+12s^3+53s^4+4s^5+38s^6}{-4+s+3s^2+17s^3+19s^4+10s^6} & \frac{-8+24s^2+40s^3+61s^4+6s^5+34s^6}{-4+s+3s^2+17s^3+19s^4+10s^6} \\ \frac{-8+34s+28s^2+78s^3+94s^4+6s^5+58s^6}{-4+s+3s^2+17s^3+19s^4+10s^6} & -\frac{4+51s+49s^2+9s^3+15s^4+6s^5+10s^6}{-4+s+3s^2+17s^3+19s^4+10s^6} \end{bmatrix},$$

$$(5.34)$$

$G_{HISOI}(s)$ *is the IO transfer function matrix of the HISO system (5.19),
(5.20) relative to the input vector* **I**,

- $G_{HISOI_0}(s)$,

$$G_{HISOI_0}(s) = R_y^{(3)} S_2^{(3)} G_{HISOISI_0}(s) =$$

$$= \begin{bmatrix} s & 0 \\ 0 & 2s \end{bmatrix} G_{HISOISI_0}(s) \implies$$

$$G_{HISOI_0}(s) = \begin{bmatrix} s & 0 \\ 0 & 2s \end{bmatrix} \bullet$$

$$\bullet \begin{bmatrix} \frac{1+5s-12s^2-4s^3-28s^4}{10s^6+19s^4+17s^3+3s^2+s-4} & \frac{-1+10s+3s^2-3s^3-3s^4}{10s^6+19s^4+17s^3+3s^2+s-4} \\ \frac{-6s-3s^2-6s^3-14s^4}{10s^6+19s^4+17s^3+3s^2+s-4} & \frac{12+7s-s^2+3s^3-8s^4}{10s^6+19s^4+17s^3+3s^2+s-4} \\ \frac{5-12s-28s^3}{10s^6+19s^4+17s^3+3s^2+s-4} & \frac{11+3s+4s^3}{10s^6+19s^4+17s^3+3s^2+s-4} \\ \frac{3-3s-14s^3}{10s^6+19s^4+17s^3+3s^2+s-4} & \frac{1-s-8s^3}{10s^6+19s^4+17s^3+3s^2+s-4} \end{bmatrix}^T \implies$$

$$G_{HISOI_0}(s) =$$

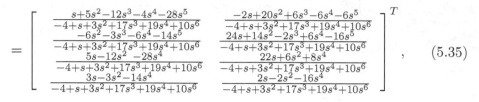

$$
=
\begin{bmatrix}
\dfrac{s+5s^2-12s^3-4s^4-28s^5}{-4+s+3s^2+17s^3+19s^4+10s^6} & \dfrac{-2s+20s^2+6s^3-6s^4-6s^5}{-4+s+3s^2+17s^3+19s^4+10s^6} \\[2ex]
\dfrac{-6s^2-3s^3-6s^4-14s^5}{-4+s+3s^2+17s^3+19s^4+10s^6} & \dfrac{24s+14s^2-2s^3+6s^4-16s^5}{-4+s+3s^2+17s^3+19s^4+10s^6} \\[2ex]
\dfrac{5s-12s^2\ -28s^4}{-4+s+3s^2+17s^3+19s^4+10s^6} & \dfrac{22s+6s^2+8s^4}{-4+s+3s^2+17s^3+19s^4+10s^6} \\[2ex]
\dfrac{3s-3s^2-14s^4}{-4+s+3s^2+17s^3+19s^4+10s^6} & \dfrac{2s-2s^2-16s^4}{-4+s+3s^2+17s^3+19s^4+10s^6}
\end{bmatrix}^{T} , \qquad (5.35)
$$

is the IO transfer function matrix of the HISO system (5.19), (5.20) relative to the extended initial input vector $\mathbf{I}_{0\mp}^{\mu-1}$,

- The IO transfer function matrix $G_{HISOR_0}(s)$ *of the HISO system (5.19), (5.20) relative to the extended initial state vector* $\mathbf{R}_{0\mp}^{\alpha-1}$:

$$G_{HISOR_0}(s) = R_y^{(3)} S_2^{(3)} G_{HISOISR_0}(s) - R_y^{(3)} Z_2^{(3-1)}(s)$$

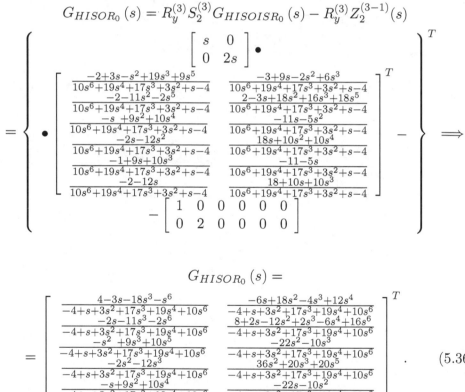

$$G_{HISOR_0}(s) =$$

$$
=
\begin{bmatrix}
\dfrac{4-3s-18s^3-s^6}{-4+s+3s^2+17s^3+19s^4+10s^6} & \dfrac{-6s+18s^2-4s^3+12s^4}{-4+s+3s^2+17s^3+19s^4+10s^6} \\[2ex]
\dfrac{-2s-11s^3-2s^6}{-4+s+3s^2+17s^3+19s^4+10s^6} & \dfrac{8+2s-12s^2+2s^3-6s^4+16s^6}{-4+s+3s^2+17s^3+19s^4+10s^6} \\[2ex]
\dfrac{-s^2\ +9s^3+10s^5}{-4+3s^2+17s^3+19s^4+10s^6} & \dfrac{-22s^2-10s^3}{-4+s+3s^2+17s^3+19s^4+10s^6} \\[2ex]
\dfrac{-2s^2-12s^3}{-4+s+3s^2+17s^3+19s^4+10s^6} & \dfrac{36s^2+20s^3+20s^5}{-4+s+3s^2+17s^3+19s^4+10s^6} \\[2ex]
\dfrac{-s+9s^2+10s^4}{-4+s+3s^2+17s^3+19s^4+10s^6} & \dfrac{-22s-10s^2}{-4+s+3s^2+17s^3+19s^4+10s^6} \\[2ex]
\dfrac{-2s-12s^2}{-4+s+3s^2+17s^3+19s^4+10s^6} & \dfrac{36s+20s^2+30s^4}{-4+s+3s^2+17s^3+19s^4+10s^6}
\end{bmatrix}^{T} . \qquad (5.36)
$$

5.2   The *HISO* plant desired regime

Definition 44, (Section 3.2), takes the special form for the *HISO* plant (5.1), (5.2).

Definition 64 *The functional vector control-state pair* $\left[\mathbf{U}^*(.), \mathbf{R}^{*\alpha-1}(.)\right]$ *is*
nominal for the HISO plant (5.1), (5.2) relative to the functional
vector pair $[\mathbf{D}(.), \mathbf{Y}_d(.)]$, *which is denoted by* $\left[\mathbf{U}_N(.), \mathbf{R}_N^{\alpha-1}(.)\right]$, *if and only*
if

$$\left[\mathbf{U}(.), \mathbf{R}^{\alpha-1}(.)\right] = \left[\mathbf{I}^*(.), \mathbf{R}^{*\alpha-1}(.)\right]$$

ensures that the corresponding real response $\mathbf{Y}(.) = \mathbf{Y}^*(.)$ *of the system*
obeys $\mathbf{Y}^*(t) = \mathbf{Y}_d(t)$ *all the time,*

$$\left[\mathbf{U}^*(.), \mathbf{R}^{*\alpha-1}(.)\right] = \left[\mathbf{U}_N(.), \mathbf{R}_N^{\alpha-1}(.)\right] \iff \langle \mathbf{Y}^*(t) = \mathbf{Y}_d(t), \ \forall t \in \mathfrak{T}_0 \rangle.$$

The system motion $\mathbf{R}_N^{\alpha-1}(.; \mathbf{R}_{N0}^{\alpha-1}; \mathbf{D}; \mathbf{U}_N)$, $\mathbf{R}_N^{\alpha-1}(0; \mathbf{R}_{N0}^{\alpha-1}; \mathbf{D}; \mathbf{U}_N) \equiv \mathbf{R}_{N0}^{\alpha-1}$,
is the desired motion $\mathbf{R}_d^{\alpha-1}(.; \mathbf{R}_{d0}^{\alpha-1}; \mathbf{D}; \mathbf{U}_N)$ *of the HISO plant (5.1),*
(5.2) relative to the functional vector pair $[\mathbf{D}(.), \mathbf{Y}_d(.)]$, *for short:* **the**
system desired motion,

$$\mathbf{R}_d^{\alpha-1}(t; \mathbf{R}_{d0}^{\alpha-1}; \mathbf{D}; \mathbf{U}_N) \equiv \mathbf{R}_N^{\alpha-1}(t; \mathbf{R}_{N0}^{\alpha-1}; \mathbf{D}; \mathbf{U}_N),$$
$$\mathbf{R}_d^{\alpha-1}(0; \mathbf{R}_{d0}^{\alpha-1}; \mathbf{D}; \mathbf{U}_N) \equiv \mathbf{R}_{d0}^{\alpha-1} \equiv \mathbf{R}_{N0}^{\alpha-1}. \tag{5.37}$$

Let

$$\mathbf{v}_1(s) = \left\{ \begin{array}{c} B^{(\mu)}Z_r^{(\mu-1)}(s)\mathbf{U}_0^{\mu-1} - A^{(\alpha)}Z_\rho^{(\alpha-1)}(s)\mathbf{R}_0^{\alpha-1} - \\ -D^{(\nu)}S_d^{(\mu)}(s)\mathbf{D}(s) + D^{(\nu)}Z_d^{(\mu-1)}(s)\mathbf{D}_0^{\mu-1} \end{array} \right\}, \tag{5.38}$$

$$\mathbf{v}_2(s) = \mathbf{Y}_d(s) + R_y^{(\alpha)}Z_\rho^{(\alpha-1)}(s)\mathbf{R}_0^{\alpha-1} - V\mathbf{D}(s). \tag{5.39}$$

Definition 64, the system description (5.1), (5.2), Equations (5.38) and
(5.39) imply:

Theorem 65 *In order for a functional vector pair* $[\mathbf{I}^*(.), \mathbf{R}^*(.)]$ *to be nom-*
inal for the HISO plant (5.1), (5.2) relative to the functional vector pair
$[\mathbf{D}(.), \mathbf{Y}_d(.)]$,
$$\left[\mathbf{U}^*(.), \mathbf{R}^{*\alpha-1}(.)\right] = \left[\mathbf{U}_N(.), \mathbf{R}_N^{\alpha-1}(.)\right],$$
it is necessary and sufficient that it obeys the following equations:

$$B^{(\mu)}\mathbf{U}^{*\mu}(t) - A^{(\alpha)}\mathbf{R}^{*\alpha}(t) = -D^{(\mu)}\mathbf{D}^\mu(t), \ \forall t \in \mathfrak{T}_0, \tag{5.40}$$

$$U\mathbf{U}^*(t) + R_y^{(\alpha)}\mathbf{R}^{*\alpha}(t) = \mathbf{Y}_d(t) - V\mathbf{D}(t), \ \forall t \in \mathfrak{T}_0, \tag{5.41}$$

or equivalently,

$$\begin{bmatrix} B^{(\mu)}S_r^{(\mu)}(s) & -A^{(\alpha)}S_\rho^{(\alpha)}(s) \\ U & R_y^{(\alpha)}S_\rho^{(\alpha)}(s) \end{bmatrix} \begin{bmatrix} \mathbf{U}^*(s) \\ \mathbf{R}^*(s) \end{bmatrix} = \begin{bmatrix} \mathbf{v}_1(s) \\ \mathbf{v}_2(s) \end{bmatrix}. \tag{5.42}$$

Let us consider the existence of the solutions of Equations (5.40), (5.41), i.e., of (5.42). The *HISO* plant (5.1), (5.2) contains $(r+\rho)$ unknown variables and $(N+\rho)$ equations. The unknown variables are the entries of $\mathbf{U}^*(s) \in \mathbb{C}^r$ and of $\mathbf{X}^*(s) \in \mathbb{C}^n$.

Claim 66 *In order to exist a nominal functional vector control-state pair*

$$\left[\mathbf{U}_N(.), \mathbf{R}_N^{\alpha-1}(.)\right]$$

for the HISO plant (5.1), (5.2), relative to the functional vector pair disturbance - desired output $[\mathbf{D}(.), \mathbf{Y}_d(.)]$, *it is necessary and sufficient that* $N \le r$.

The proof of this claim is analogous to the proof of Claim 46 (Section 3.2).

Claim 66 presents the complete solution to the problem of the existence of a nominal functional vector pair $\left[\mathbf{U}_N(.), \mathbf{R}_N^{\alpha-1}(.)\right]$ for the *HISO* plant (5.1), (5.2) relative to the functional vector pair $[\mathbf{D}(.), \mathbf{Y}_d(.)]$.

Condition 67 *The desired output response of the HISO plant (5.1), (5.2) is realizable, i.e.,, $N \le r$. The nominal functional vector pair* $\left[\mathbf{U}_N(.), \mathbf{R}_N^{\alpha-1}(.)\right]$ *is known.*

The *HISO* plant description in terms of the deviations (5.43),

$$\mathbf{r} = \mathbf{R} - \mathbf{R}_N,   \tag{5.43}$$

(2.53), (9.3), and (2.55) reads:

$$A^{(\alpha)}\mathbf{r}^\alpha(t) = D^{(\mu)}\mathbf{d}^\mu(t) + B^{(\mu)}\mathbf{u}^\mu(t), \ \forall t \in \mathfrak{T}_0,   \tag{5.44}$$

$$\mathbf{y}(t) = R^{(\alpha)}\mathbf{R}^\alpha(t) + V\mathbf{d}(t) + U\mathbf{u}(t), \ \forall t \in \mathfrak{T}_0,   \tag{5.45}$$

due to Equations (5.1) and (5.2).

5.3 Exercises

Exercise 68 *1. Select a physical HISO plant.*
2. Determine its time domain HISO mathematical model.

Chapter 6

IIO systems

6.1 *IIO* system mathematical model

6.1.1 Time domain

The general description, in terms of the total vector coordinates, of *time-invariant* continuous-*time* linear ***Input-Internal and Output state systems***, for short *IIO* ***systems***, without a delay, has the following general form:

$$A^{(\alpha)}\mathbf{R}^{\alpha}(t) = D^{(\mu)}\mathbf{D}^{\mu}(t) + B^{(\mu)}\mathbf{U}^{\mu}(t) = H^{(\mu)}\mathbf{I}^{\mu}(t)\,, \ \forall t \in \mathfrak{T}_0, \qquad (6.1)$$

$$E^{(\nu)}\mathbf{Y}^{\nu}(t) = \left\{ \begin{array}{c} R_y^{(\alpha-1)}\mathbf{R}^{\alpha-1}(t) + V^{(\mu)}\mathbf{D}^{\mu}(t) + U^{(\mu)}\mathbf{U}^{\mu}(t) = \\ = R_y^{(\alpha-1)}\mathbf{R}^{\alpha-1}(t) + Q^{(\mu)}\mathbf{I}^{\mu}(t) \end{array} \right\}, \ \forall t \in \mathfrak{T}_0,$$

$$(6.2)$$

where

$$\mathbf{I} = \mathbf{I}_{IIO} = \left[\begin{array}{cc} \mathbf{D}^T & \mathbf{U}^T \end{array} \right]^T \in \mathfrak{R}^{d+r}, \ M = d + r,$$

$$H = \left[D \vdots B \right] \in \mathfrak{R}^{\rho \times (d+r)}, \ Q = \left[V \vdots U \right] \in \mathfrak{R}^{N \times (d+r)},$$

$$R_y^{(\alpha-1)} = \left[R_{y0} \vdots R_{y1} \vdots ... \vdots R_{y,\alpha-1} \right] \in \mathfrak{R}^{N \times \alpha \rho}. \qquad (6.3)$$

Note 69 *If $\nu = 0$ then the IIO system (6.1), (6.2) reduces to the HISO system (5.1), (5.2), Chapter 5.*

We continue to treat the IIO system (6.1), (6.2) with $\nu > 0$.

Condition 70 *The matrices A_α and E_ν obey:*

$$det A_\alpha \neq 0, \ \ which \ implies \ \exists s \in \mathbb{C} \Longrightarrow det \left[\sum_{k=0}^{k=\alpha} s^k A_k \right] \neq 0,$$

$$det E_\nu \neq 0, \ \ which \ implies \ \exists s \in \mathbb{C} \Longrightarrow det \left[\sum_{k=0}^{k=\nu} s^k E_k \right] \neq 0, \qquad (6.4)$$

and

$$\nu \in \{1, 2, ...\} . \qquad (6.5)$$

Note 71 *We accept the validity of Condition 70 in the sequel.*

The left-hand side of Equation (6.1) describes *the internal dynamics of the system*, i.e. *the internal state of the system (Definition 14, Section 1.4)*, and the left-hand side of Equation (6.2) describes *the output dynamics, i.e., the output state* of the system if and only if $\nu > 0$.

Note 72 *The state vector $\mathbf{S}_{IIO}$ of the IIO system (6.1), (6.2) is defined in Equation (1.27) (Section 1.4) by:*

$$\mathbf{S}_{IIO} = \left[\begin{array}{c} \mathbf{R}^{\alpha-1} \\ \mathbf{Y}^{\nu-1} \end{array} \right] = \left[\begin{array}{c} \mathbf{R}^T \ \vdots \ \mathbf{R}^{(1)^T} \ \vdots \ ... \ \vdots \ \mathbf{R}^{(\alpha-1)^T} \\ \mathbf{Y}^T \ \vdots \ \mathbf{Y}^{(1)^T} \ \vdots \ ... \ \vdots \ \mathbf{Y}^{(\nu-1)^T} \end{array} \right]^T \in \mathfrak{R}^n,$$

$$n = \alpha \rho + \nu N, \qquad (6.6)$$

The new vector notation $\mathbf{R}^{\alpha-1}$ and $\mathbf{Y}^{\nu-1}$ has permitted us to define the state of the IIO system (6.1), (6.2) by preserving the physical sense. It enabled us to establish in [40] the direct link between the definitions of the Lyapunov and of BI stability properties with the corresponding conditions for them in the complex domain. It enables us to discover in what follows the complex domain criteria for observability, controllability and trackability directly from their definitions. Such criteria possess the complete physical meaning.

The extended vector $\mathbf{R}^{\alpha-1}$ is the IIO system internal state vector $\mathbf{S}_{IIOI}$. The extended vector $\mathbf{Y}^{\nu-1}$ is the IIO system output state vector $\mathbf{S}_{IIOO}$. They compose the IIO system (full) state vector $\mathbf{S}_{IIO}$,

$$\mathbf{S}_{IIO} = \left[\begin{array}{c} \mathbf{R}^{\alpha-1} \\ \mathbf{Y}^{\nu-1} \end{array} \right] = \left[\begin{array}{c} \mathbf{S}_{IIOI} \\ \mathbf{S}_{IIOO} \end{array} \right] . \qquad (6.7)$$

6.1.2 Complex domain

We transform Equations (6.1), (6.2) by applying the Laplace transform, into

$$\mathbf{R}(s) = F_{IIOIS}(s) \mathbf{V}_{IIOIS}(s), \tag{6.8}$$

$$\mathbf{Y}(s) = F_{IIO}(s) \mathbf{V}_{IIO}(s), \tag{6.9}$$

where:
- $F_{IIOIS}(s)$,

$$F_{IIOIS}(s) = \begin{bmatrix} G^T_{IIOISD}(s) \\ G^T_{IIOISU}(s) \\ G^T_{IIOISD_0}(s) \\ G^T_{IIOISR_0}(s) \\ G^T_{IIOISU_0}(s) \end{bmatrix}^T \tag{6.10}$$

is the IIO system (6.1), (6.2) *input to state (IS) full transfer function matrix*, the inverse Laplace transform of which is the plant *IS full fundamental matrix* $\Psi_{IIOIS}(t)$ [40],

$$\Psi_{IIOIS}(t) = \mathcal{L}^{-1}\{F_{IIOIS}(s)\}, \tag{6.11}$$

and the inverse Laplace transform of $\left(A^{(\alpha)}S_\rho^{(\alpha)}(s)\right)^{-1}$ is *the IIO plant IS fundamental matrix* $\Phi_{IIOIS}(t)$, $\Phi_{IIOIS}(t) = \mathcal{L}^{-1}\left\{\left(A^{(\alpha)}S_\rho^{(\alpha)}(s)\right)^{-1}\right\}$ [40],
- $G_{IIOISD}(s)$,

$$G_{IIOISD}(s) = \left(A^{(\alpha)}S_\rho^{(\alpha)}(s)\right)^{-1} D^{(\mu)}S_d^{(\mu)}(s), \tag{6.12}$$

is the IIO plant (6.1), (6.2) *disturbance to internal state (IS) transfer function matrix*,
- $G_{IIOISU}(s)$,

$$G_{IIOISU}(s) = \left(A^{(\alpha)}S_\rho^{(\alpha)}(s)\right)^{-1} B^{(\mu)}S_r^{(\mu)}(s), \tag{6.13}$$

is the IIO plant (6.1), (6.2) *control to internal state (IS) transfer function matrix*,
- $G_{IIOISD_0}(s)$,

$$G_{IIOISD_0}(s) = -\left(A^{(\alpha)}S_\rho^{(\alpha)}(s)\right)^{-1} D^{(\mu)}Z_d^{(\mu-1)}(s), \tag{6.14}$$

is the IIO plant (6.1), (6.2) *initial disturbance to internal state (IS) transfer function matrix,*

- $G_{IIOISR_0}(s)$,

$$G_{IIOISR_0}(s) = \left(A^{(\alpha)}S_\rho^{(\alpha)}(s)\right)^{-1} A^{(\alpha)} Z_\rho^{(\alpha-1)}(s), \qquad (6.15)$$

is the IIO plant (6.1), (6.2) *initial internal state to internal state (IS) transfer function matrix,*

- $G_{IIOISU_0}(s)$,

$$G_{IIOISU_0}(s) = \left(A^{(\alpha)}S_\rho^{(\alpha)}(s)\right)^{-1} B^{(\mu)} Z_r^{(\mu-1)}(s), \qquad (6.16)$$

is the IIO plant (6.1), (6.2) *initial control to internal state (IS) transfer function matrix,*

$\mathbf{V}_{IIOIS}(s)$,

$$\mathbf{V}_{IIOIS}(s) = \left[\mathbf{D}^T(s) \,\vdots\, \mathbf{U}^T(s) \,\vdots\, \mathbf{C}_{IIOIS0}^T\right]^T, \qquad (6.17)$$

is the Laplace transform of the action vector $\mathbf{V}_{IIOIS}(t)$, and $\mathbf{C}_{IIOIS0}$,

$$\mathbf{C}_{IIOIS0} = \left[\left(\mathbf{D}_0^{\mu-1}\right)^T \,\vdots\, \left(\mathbf{R}_0^{\alpha-1}\right)^T \,\vdots\, \left(\mathbf{U}_0^{\mu-1}\right)^T\right]^T, \qquad (6.18)$$

is the vector of all initial conditions acting on the system internal state,

- $F_{IIO}(s)$,

$$F_{IIO}(s) = \begin{bmatrix} G_{IIOD}^T(s) \\ G_{IIOU}^T(s) \\ G_{IIOD_0}^T(s) \\ G_{IIOR_0}^T(s) \\ G_{IIOU_0}^T(s) \\ G_{IIOY_0}^T(s) \end{bmatrix}^T, \qquad (6.19)$$

is the IIO plant (6.1), (6.2) *input to output (IO) full transfer function matrix,* the inverse Laplace transform of which is the plant IO *full fundamental matrix* $\Psi_{IIO}(t)$,

$$\Psi_{IIO}(t) = \mathcal{L}^{-1}\{F_{IIO}(s)\}, \qquad (6.20)$$

- $p_{IIO}(s)$,

$$p_{IIO}(s) = \det\left(E^{(\nu)}S_N^{(\nu)}(s)\right) \det\left(A^{(\alpha)}S_\rho^{(\alpha)}(s)\right), \qquad (6.21)$$

is the characteristic polynomial of the *IIO* plant (6.1), (6.2) and the denominator polynomial of all its transfer function matrices

- $G_{IIOD}(s)$,

$$G_{IIOD}(s) =$$

$$= \left[\begin{array}{c} \left(E^{(\nu)} S_N^{(\nu)}(s) \right)^{-1} R_y^{(\alpha-1)} S_\rho^{(\alpha)}(s) \left(A^{(\alpha)} S_\rho^{(\alpha)}(s) \right)^{-1} H^{(\mu)} S_d^{(\mu)}(s) \\ + \left(E^{(\nu)} S_N^{(\nu)}(s) \right)^{-1} V^{(\mu)} S_d^{(\mu)}(s) \end{array} \right] =$$

$$= p_{IIO}^{-1}(s) \bullet$$

$$\bullet \left[\begin{array}{c} adj \left(E^{(\nu)} S_N^{(\nu)}(s) \right) R_y^{(\alpha-1)} S_\rho^{(\alpha)}(s) adj \left(A^{(\alpha)} S_\rho^{(\alpha)}(s) \right) H^{(\mu)} S_d^{(\mu)}(s) \\ + \left[\det \left(A^{(\alpha)} S_\rho^{(\alpha)}(s) \right) \right] adj \left(E^{(\nu)} S_N^{(\nu)}(s) \right) V^{(\mu)} S_d^{(\mu)}(s) \end{array} \right],$$

$$\tag{6.22}$$

is the *IIO* plant (6.1), (6.2) *IO transfer function matrix relative to the disturbance* **D**,

- $G_{IIOU}(s)$,

$$G_{IIOU}(s) =$$

$$= \left[\begin{array}{c} \left(E^{(\nu)} S_N^{(\nu)}(s) \right)^{-1} R_y^{(\alpha-1)} S_\rho^{(\alpha)}(s) \left(A^{(\alpha)} S_\rho^{(\alpha)}(s) \right)^{-1} B^{(\mu)} S_r^{(\mu)}(s) + \\ + \left(E^{(\nu)} S_N^{(\nu)}(s) \right)^{-1} U^{(\mu)} S_r^{(\mu)}(s), \end{array} \right] =$$

$$p_{IIO}^{-1}(s) \bullet$$

$$\bullet \left[\begin{array}{c} adj \left(E^{(\nu)} S_N^{(\nu)}(s) \right) R_y^{(\alpha-1)} S_\rho^{(\alpha)}(s) adj \left(A^{(\alpha)} S_\rho^{(\alpha)}(s) \right) B^{(\mu)} S_r^{(\mu)}(s) \\ + \left[\det \left(A^{(\alpha)} S_\rho^{(\alpha)}(s) \right) \right] adj \left(E^{(\nu)} S_N^{(\nu)}(s) \right) U^{(\mu)} S_r^{(\mu)}(s), \end{array} \right]$$

$$\tag{6.23}$$

is the *IIO* plant (6.1), (6.2) *IO transfer function matrix relative to the control* **U**,

- $G_{IIOD_0}(s)$,

$$G_{IIOD_0}(s) = -p_{IIO}^{-1}(s) \bullet$$

$$\bullet \left[\begin{array}{c} adj \left(E^{(\nu)} S_N^{(\nu)}(s) \right) R_y^{(\alpha-1)} S_\rho^{(\alpha)}(s) adj \left(A^{(\alpha)} S_\rho^{(\alpha)}(s) \right) D^{(\mu)} Z_d^{(\mu-1)}(s) \\ + \left(\det A^{(\alpha)} S_\rho^{(\alpha)}(s) \right) adj \left(E^{(\nu)} S_N^{(\nu)}(s) \right) V^{(\mu)} Z_d^{(\mu-1)}(s), \end{array} \right]$$

$$\tag{6.24}$$

is the *IIO* plant (6.1), (6.2) *IO transfer function matrix relative to the initial extended disturbance* $\mathbf{D}_0^{\mu-1}$,

- $G_{IIOR_0}(s)$,

$$G_{IIOR_0}(s) = p_{IIO}^{-1}(s) \bullet$$

$$\bullet \left[\begin{array}{c} adj\left(E^{(\nu)}S_N^{(\nu)}(s)\right)R_y^{(\alpha-1)}S_\rho^{(\alpha)}(s)adj\left(A^{(\alpha)}S_\rho^{(\alpha)}(s)\right)A^{(\alpha)}Z_\rho^{(\alpha-1)}(s) \\ -\left[\det\left(A^{(\alpha)}S_\rho^{(\alpha)}(s)\right)\right]adj\left(E^{(\nu)}S_N^{(\nu)}(s)\right)R_y^{(\alpha-1)}Z_\rho^{(\alpha-1)}(s), \end{array} \right]$$

(6.25)

is the *IIO* plant (6.1), (6.2) *IO transfer function matrix relative to the initial internal state vector* $\mathbf{R}_0^{\alpha-1}$,

- $G_{IIOU_0}(s)$,

$$G_{IIOU_0}(s) = p_{IIO}^{-1}(s) \bullet$$

$$\bullet \left[\begin{array}{c} -adj\left(E^{(\nu)}S_N^{(\nu)}(s)\right)R_y^{(\alpha-1)}S_\rho^{(\alpha)}(s)adj\left(A^{(\alpha)}S_\rho^{(\alpha)}(s)\right)B^{(\mu)}Z_r^{(\mu-1)}(s) \\ -\left[\det\left(A^{(\alpha)}S_\rho^{(\alpha)}(s)\right)\right]adj\left(E^{(\nu)}S_N^{(\nu)}(s)\right)U^{(\mu)}Z_r^{(\mu-1)}(s) \end{array} \right],$$

(6.26)

is the *IO transfer function matrix relative to the initial extended control vector* $\mathbf{U}_0^{\mu-1}$ *of the IIO plant* (6.1), (6.2),

- $G_{IIOY_0}(s)$,

$$G_{IIOY_0}(s) = \left(E^{(\nu)}S_N^{(\nu)}(s)\right)^{-1}E^{(\nu)}Z_N^{(\nu-1)}(s)$$

(6.27)

is the *IIO* plant (6.1), (6.2) *IO transfer function matrix relative to the extended initial output state vector* $\mathbf{Y}_0^{\nu-1}$,

- $\mathbf{V}_{IIO}(s)$ and $\mathbf{C}_{IIO0}$,

$$\mathbf{V}_{IIO}(s) = \left[\begin{array}{c} \mathbf{I}_{IIO}(s) \\ \mathbf{C}_{IIO0} \end{array} \right], \ \mathbf{I}_{IIO}(s) = \left[\begin{array}{c} \mathbf{D}(s) \\ \mathbf{U}(s) \end{array} \right], \ \mathbf{C}_{IIO0} = \left[\begin{array}{c} \mathbf{D}_0^{\mu-1} \\ \mathbf{R}_0^{\alpha-1} \\ \mathbf{U}_0^{\mu-1} \\ \mathbf{Y}_0^{\nu-1} \end{array} \right],$$

(6.28)

are the Laplace transform of the action vector $\mathbf{V}_{IIO}(t)$ and the vector $\mathbf{C}_{IIO0}$ of all initial conditions, respectively.

Equations (6.9), (6.19), (6.21)-(6.23), (6.24)-(6.27) determine the Laplace transform $\mathbf{Y}(s)$ of the output vector $\mathbf{Y}(t)$,

$$\mathbf{Y}(s) = G_{IIOD}(s)\mathbf{D}(s) + G_{IIOU}(s)\mathbf{U}(s) + G_{IIOD_0}(s)\mathbf{D}_0^{\mu-1}+$$

$$+G_{IIOR_0}(s)\mathbf{R}_0^{\alpha-1} + G_{IIOU_0}(s)\mathbf{U}_0^{\mu-1} + G_{IIOY_0}(s)\mathbf{Y}_0^{\nu-1} =$$

$$= F_{IIO}(s)\mathbf{V}_{IIO}(s).$$

(6.29)

This can be set in a more compact form. Equations (6.28) and (6.29), together with

$$G_{IIO}(s) = \left[G_{IIOD}(s) \vdots G_{IIOU}(s)\right], \quad G_{IIOI_0}(s) = \left[G_{IIOD_0}(s) \vdots G_{IIOU_0}(s)\right],$$

give the compact form to the Laplace transform $\mathbf{Y}^{\mp}(s)$ of the system response $\mathbf{Y}(t; \mathbf{R}_{0-}^{\alpha-1}; \mathbf{Y}_0^{\nu-1}; \mathbf{I}^{\mu})$,

$$\mathbf{Y}^{\mp}(s) = G_{IIO}(s)\mathbf{I}(s) + G_{IIOI_0}(s)\mathbf{I}_0^{\mu-1} + G_{IIOR_0}(s)\,\mathbf{R}_{0-}^{\alpha-1} + G_{IIOY_0}(s)\,\mathbf{Y}_{0\mp}^{\nu-1}.$$
(6.30)

The inverse Laplace transform of this equation determines the IIO system (6.1), (6.2) response $\mathbf{Y}(t; \mathbf{R}_{0-}^{\alpha-1}; \mathbf{Y}_0^{\nu-1}; \mathbf{I}^{\mu})$,

$$\mathbf{Y}(t; \mathbf{R}_{0-}^{\alpha-1}; \mathbf{Y}_0^{\nu-1}; \mathbf{I}^{\mu}) = \mathcal{L}^{-1}\left\{\mathbf{Y}^{\mp}(s)\right\} = \int_{0-}^{t} \Gamma_{IIO}(\tau)\mathbf{I}(t-\tau)d\tau +$$

$$+ \Gamma_{IIOI_0}(t)\mathbf{I}_{0-}^{\mu-1} + \Gamma_{IIOR_0}(t)\mathbf{R}_{0-}^{\alpha-1} + \Gamma_{IIOY_0}(t)\mathbf{Y}_{0-}^{\nu-1}, \qquad (6.31)$$

$$\forall t \in \mathfrak{T}_0,$$

where

$$\Gamma_{IIO}(t) = \mathcal{L}^{-1}\left\{G_{IIO}(s)\right\} =$$

$$= \mathcal{L}^{-1}\left\{\Theta_{IIO}(s)\left[\begin{array}{c} E_\nu^{-1}R_y^{(\alpha)}S_\rho^{(\alpha)}(s)\left(A^{(\alpha)}S_\rho^{(\alpha)}(s)\right)^{-1}H^{(\mu)}S_M^{(\mu)}(s)+ \\ +E_\nu^{-1}Q^{(\mu)}S_M^{(\mu)}(s) \end{array}\right]\right\},$$
(6.32)

$$\Gamma_{IIOI_0}(t) = \mathcal{L}^{-1}\left\{G_{IIOI_0}(s)\right\} =$$

$$= \mathcal{L}^{-1}\left\{\Theta_{IIO}(s)\left[\begin{array}{c} -E_\nu^{-1}R_y^{(\alpha)}S_\rho^{(\alpha)}(s)\left(A^{(\alpha)}S_\rho^{(\alpha)}(s)\right)^{-1}H^{(\mu)}Z_M^{(\mu-1)}(s)- \\ -E_\nu^{-1}Q^{(\mu)}Z_M^{(\mu-1)}(s) \end{array}\right]\right\},$$
(6.33)

$$\Gamma_{IIOR_0}(t) = \mathcal{L}^{-1}\left\{G_{IIOR_0}(s)\right\} =$$

$$= \mathcal{L}^{-1}\left\{\Theta_{IIO}(s)\left[\begin{array}{c} E_\nu^{-1}R_y^{(\alpha)}S_\rho^{(\alpha)}(s)\left(A^{(\alpha)}S_\rho^{(\alpha)}(s)\right)^{-1}A^{(\alpha)}Z_\rho^{(\alpha-1)}(s)- \\ -E_\nu^{-1}R_y^{(\alpha)}Z_\rho^{(\alpha-1)}(s) \end{array}\right]\right\},$$
(6.34)

$$\Gamma_{IIOY_0}(t) = \mathcal{L}^{-1}\left\{G_{IIOY_0}(s)\right\} = \mathcal{L}^{-1}\left\{\Theta_{IIO}(s)E_\nu^{-1}E^{(\nu)}Z_N^{(\nu-1)}(s)\right\}. \quad (6.35)$$

Equations (6.32)-(6.35) define well the matrices $\Gamma_{IIO}(t)$, $\Gamma_{IIOI_0}(t)$, $\Gamma_{IIOR_0}(t)$ and $\Gamma_{IIOY_0}(t)$ in terms of the system transfer function matrices $G_{IIO}(s)$, $G_{IIOI_0}(s)$, $G_{IIOR_0}(s)$, and $G_{IIOY_0}(s)$, respectively.

Example 73 *Let the IIO system (6.1), (6.2) be defined by*

$$\begin{bmatrix} 1 & 2 \\ 3 & -4 \end{bmatrix}\begin{bmatrix} R_1^{(3)}(t) \\ R_2^{(3)}(t) \end{bmatrix} + \begin{bmatrix} 2 & 3 \\ 1 & 0 \end{bmatrix}\begin{bmatrix} R_1^{(1)}(t) \\ R_2^{(1)}(t) \end{bmatrix} + \begin{bmatrix} 4 & 0 \\ 1 & 1 \end{bmatrix}\begin{bmatrix} R_1(t) \\ R_2(t) \end{bmatrix} =$$

$$= \begin{bmatrix} 5 & 3 \\ 4 & 1 \end{bmatrix}\begin{bmatrix} I_1^{(2)}(t) \\ I_2^{(2)}(t) \end{bmatrix} + \begin{bmatrix} 1 & 0 \\ 0 & 3 \end{bmatrix}\begin{bmatrix} I_1^{(1)}(t) \\ I_2^{(1)}(t) \end{bmatrix} + \begin{bmatrix} 4 & 2 \\ 3 & 6 \end{bmatrix}\begin{bmatrix} I_1(t) \\ I_2(t) \end{bmatrix},$$

$$\qquad\qquad\qquad\qquad\qquad\qquad\qquad\qquad\qquad\qquad\qquad (6.36)$$

$$\begin{bmatrix} Y_1^{(2)}(t) \\ Y_2^{(2)}(t) \end{bmatrix} = \begin{bmatrix} 1 & 0 \\ 0 & 2 \end{bmatrix}\begin{bmatrix} R_1^{(2)}(t) \\ R_2^{(2)}(t) \end{bmatrix} + \begin{bmatrix} 1 & 2 \\ 2 & 1 \end{bmatrix}\begin{bmatrix} I_1^{(1)}(t) \\ I_2^{(1)}(t) \end{bmatrix} \quad (6.37)$$

In this case

$$N = 2, \ \alpha = 3, \ \rho = 2, \ R_{y0} = R_{y1} = R_{y3} = O_2, \ M = 2, \ \nu = 2,$$
$$E_0 = E_1 = O_2, \ E_2 = I_2, \ Q_0 = Q_2 = O_2, \ \ \mu = 2 < 3 = \alpha, \qquad (6.38)$$

$$R_y^{(\alpha)} = R_y^{(3)} = \begin{bmatrix} R_{y0} \ \vdots \ R_{y1} \ \vdots \ R_{y2} \ \vdots \ R_{y3} \end{bmatrix} =$$

$$= \begin{bmatrix} 0 & 0 & 0 & 0 & 1 & 0 & 0 & 0 \\ 0 & 0 & 0 & 0 & 0 & 2 & 0 & 0 \end{bmatrix}, \qquad (6.39)$$

$$R_y^{(3)}S_2^{(3)}(s) = \begin{bmatrix} 0 & 0 & 0 & 0 & 1 & 0 & 0 & 0 \\ 0 & 0 & 0 & 0 & 0 & 2 & 0 & 0 \end{bmatrix} \bullet$$

$$\bullet \begin{bmatrix} s^0 I_2 \ \vdots \ s^1 I_2 \ \vdots \ s^2 I_2 \ \vdots \ s^3 I_2 \end{bmatrix}^T \implies$$

$$R_y^{(3)}S_2^{(3)}(s) = \begin{bmatrix} s^2 & 0 \\ 0 & 2s^2 \end{bmatrix}, \qquad (6.40)$$

$$R_y^{(3)}Z_2^{(3-1)}(s) = \begin{bmatrix} 0 & 0 & 0 & 0 & 1 & 0 & 0 & 0 \\ 0 & 0 & 0 & 0 & 0 & 2 & 0 & 0 \end{bmatrix} \bullet$$

$$\bullet \begin{bmatrix} O_2 & O_2 & O_2 \\ s^{3-3}I_2 & O_2 & O_2 \\ s^{3-2}I_2 & s^{3-3}I_2 & O_2 \\ s^{3-1}I_2 & s^{3-2}I_2 & s^{3-3}I_2 \end{bmatrix} \implies$$

$$R_y^{(3)} Z_2^{(3-1)}(s) = \begin{bmatrix} s & 0 & 1 & 0 & 0 & 0 \\ 0 & 2s & 0 & 2 & 0 & 0 \end{bmatrix}. \tag{6.41}$$

Equation (6.36) is Equation (B.1) (Example 227, Subsection 2.1.2) so that Equation (B.14), (Example 227), holds for $\mathbf{Y}$ *replaced by* $\mathbf{R}$, $F_{IO}(s)$ *replaced by* $F_{IIOIS}(s)$, $\mathbf{V}_{IO}^{\mp}(s)$ *replaced by* $\mathbf{V}_{IIOIS}^{\mp}(s)$, *and it is denoted as (6.42):*

$$\mathbf{R}^{\mp}(s) = F_{IIOIS}(s)\, \mathbf{V}_{IIOIS}^{\mp}(s), \tag{6.42}$$

where in this context:

- $F_{IIOIS}(s)$, which is equal to $F_{IO}(s)$ (B.15) (Example 227),

$$\begin{bmatrix} G_{IIOIS}(s) & \vdots & G_{IIOISI_0}(s) & \vdots & G_{IIOISY_0}(s) \end{bmatrix}$$

$$F_{IIOIS}(s) = \begin{bmatrix} G_{IIOIS}(s) & \vdots & G_{IIOISI_0}(s) & \vdots & G_{IIOISY_0}(s) \end{bmatrix} =$$

$$= \begin{bmatrix} \frac{4s^3-1}{10s^6+19s^4+17s^3+3s^2+s-4} & \frac{2s^3+3s}{10s^6+19s^4+17s^3+3s^2+s-4} \\ \frac{3s^3+s+1}{10s^6+19s^4+17s^3+3s^2+s-4} & -\frac{s^3+2s+4}{10s^6+19s^4+17s^3+3s^2+s-4} \end{bmatrix} \bullet$$

$$\bullet \begin{bmatrix} \begin{bmatrix} 4+s+5s^2 & 2+3s^2 \\ 3+4s^2 & 6+3s+s^2 \end{bmatrix} \\ \begin{bmatrix} -1-5s & -3s & -5 & -3 \\ -4s & -3-s & -4 & -1 \end{bmatrix} \\ \begin{bmatrix} 2+s^2 & 3+2s^2 & s & 2s & 1 & 2 \\ 1+3s^2 & 3-4s^2 & 3s & -4s & 3 & -4 \end{bmatrix} \end{bmatrix}^T, \tag{6.43}$$

is the IS full transfer function matrix of the IIO system (6.36), (6.37),

- The system IS transfer function $G_{IIOIS}(s) = G_{IO}(s)$ (B.16) relative to the input vector $\mathbf{I}$:

$$G_{IIOIS}(s) =$$

$$\begin{bmatrix} \frac{-4+8s-5s^2+34s^3+4s^4+28s^5}{10s^6+19s^4+17s^3+3s^2+s-4} & \frac{-2+18s+6s^2+23s^3+6s^4+14s^5}{10s^6+19s^4+17s^3+3s^2+s-4} \\ \frac{-8-s+6s^2+6s^3+3s^4+11s^5}{10s^6+19s^4+17s^3+3s^2+s-4} & -\frac{26+26s+13s^2+17s^3+3s^4+10s^5}{10s^6+19s^4+17s^3+3s^2+s-4} \end{bmatrix}, \tag{6.44}$$

- The system IS transfer function $G_{IIOISI_0}(s) = G_{IOI_0}(s)$ (B.17) relative

to the extended initial input vector $\mathbf{I}_{0\mp}^{\mu-1}$:

$$G_{IIOISI_0}(s) =$$

$$= \begin{bmatrix} \dfrac{1+5s-12s^2-4s^3-28s^4}{10s^6+19s^4+17s^3+3s^2+s-4} & \dfrac{-1+10s+3s^2-3s^3-11s^4}{10s^6+19s^4+17s^3+3s^2+s-4} \\[2mm] \dfrac{-6s-3s^2-6s^3-14s^4}{10s^6+19s^4+17s^3+3s^2+s-4} & \dfrac{12+7s-s^2+3s^3-8s^4}{10s^6+19s^4+17s^3+3s^2+s-4} \\[2mm] \dfrac{5-12s\ -28s^3}{10s^6+19s^4+17s^3+3s^2+s-4} & \dfrac{11+3s-11s^3}{10s^6+19s^4+17s^3+3s^2+s-4} \\[2mm] \dfrac{3-3s+10s^3}{10s^6+19s^4+17s^3+3s^2+s-4} & \dfrac{1-s-8s^3}{10s^6+19s^4+17s^3+3s^2+s-4} \end{bmatrix}^T , \qquad (6.45)$$

- The system IS transfer function $G_{IIOISR_0}(s) = G_{IOY_0}(s)$ *(B.18) relative to the extended initial output vector* $R_{0\mp}^{\alpha-1}$:

$$G_{IIOISR_0} =$$

$$= \begin{bmatrix} \dfrac{-2+3s-s^2+19s^3+10s^5}{10s^6+19s^4+17s^3+3s^2+s-4} & \dfrac{-2-11s^2}{10s^6+19s^4+17s^3+3s^2+s-4} \\[2mm] \dfrac{-3+9s-2s^2+6s^3}{10s^6+19s^4+17s^3+3s^2+s-4} & \dfrac{-9-3s+18s^2+16s^3+20s^5}{10s^6+19s^4+17s^3+3s^2+s-4} \\[2mm] \dfrac{-s+9s^2+10s^4}{10s^6+19s^4+17s^3+3s^2+s-4} & \dfrac{-11s\ -5s^2}{10s^6+19s^4+17s^3+3s^2+s-4} \\[2mm] \dfrac{-2s-12s^2}{10s^6+19s^4+17s^3+3s^2+s-4} & \dfrac{-14s-6s^2+2s^4}{10s^6+19s^4+17s^3+3s^2+s-4} \\[2mm] \dfrac{-1+9s+10s^3}{10s^6+19s^4+17s^3+3s^2+s-4} & \dfrac{-11-5s}{10s^6+19s^4+17s^3+3s^2+s-4} \\[2mm] \dfrac{-2-12s}{10s^6+19s^4+17s^3+3s^2+s-4} & \dfrac{-14-6s+2s^3}{10s^6+19s^4+17s^3+3s^2+s-4} \end{bmatrix}^T , \qquad (6.46)$$

- The Laplace transform $\mathbf{V}_{IIOIS}(s) = \mathbf{V}_{IO}^{\mp}(s)$ *(B.19) of the system action vector* $\mathbf{V}_{IIOIS}(t)$ *reads*

$$\mathbf{V}_{IIOIS}(s) = \begin{bmatrix} \mathbf{I}^{\mp}(s) \\ \mathbf{I}_{0\mp}^1 \\ \mathbf{R}_{0\mp}^2 \end{bmatrix} = \begin{bmatrix} \mathbf{I}^{\mp}(s) \\ \mathbf{C}_{IIOIS0\mp} \end{bmatrix}, \qquad (6.47)$$

- And the vector $\mathbf{C}_{IIOIS0\mp} = \mathbf{C}_{IO0\mp}$ *(B.20) of all the initial conditions is found to be*

$$\mathbf{C}_{IIOIS0\mp} = \begin{bmatrix} \mathbf{I}_{0\mp}^1 \\ \mathbf{R}_{0\mp}^2 \end{bmatrix}. \qquad (6.48)$$

Equation (6.37) determines

$$E^{(\nu)} = E^{(2)} = \begin{bmatrix} E_0 \vdots E_1 \vdots E_2 \end{bmatrix} = \begin{bmatrix} 0 & 0 & 0 & 0 & 1 & 0 \\ 0 & 0 & 0 & 0 & 0 & 1 \end{bmatrix},$$

so that

$$E^{(2)}S_2^{(2)} = \begin{bmatrix} 0 & 0 & 0 & 0 & 1 & 0 \\ 0 & 0 & 0 & 0 & 0 & 1 \end{bmatrix} \begin{bmatrix} s^0 I_2 \\ s^1 I_2 \\ s^2 I_2 \end{bmatrix} = \begin{bmatrix} s^2 & 0 \\ 0 & s^2 \end{bmatrix}, \qquad (6.49)$$

$$\left(E^{(2)}S_2^{(2)}\right)^{-1} = \begin{bmatrix} \frac{1}{s^2} & 0 \\ 0 & \frac{1}{s^2} \end{bmatrix} = \frac{1}{s^2}I_2, \tag{6.50}$$

$$E^{(2)}Z_2^{(2-1)} = \begin{bmatrix} 0 & 0 & 0 & 0 & 1 & 0 \\ 0 & 0 & 0 & 0 & 0 & 1 \end{bmatrix} \begin{bmatrix} O_2 & O_2 \\ s^{2-2}I_2 & O_2 \\ s^{2-1}I_2 & s^{2-2}I_k \end{bmatrix} \Longrightarrow$$

$$E^{(2)}Z_2^{(2-1)} = \begin{bmatrix} s & 0 & 1 & 0 \\ 0 & s & 0 & 1 \end{bmatrix}. \tag{6.51}$$

From Equation (6.37) follow:

$$R^{(3)} = \begin{bmatrix} R_{y0} & \vdots & R_{y1} & \vdots & R_{y2} & \vdots & R_{y3} \end{bmatrix} =$$

$$= \begin{bmatrix} 0 & 0 & 0 & 0 & 1 & 0 & 0 & 0 \\ 0 & 0 & 0 & 0 & 0 & 2 & 0 & 0 \end{bmatrix}, \tag{6.52}$$

$$R^{(3)}S_2^{(3)} = \begin{bmatrix} 0 & 0 & 0 & 0 & 1 & 0 & 0 & 0 \\ 0 & 0 & 0 & 0 & 0 & 2 & 0 & 0 \end{bmatrix} \bullet$$

$$\bullet \begin{bmatrix} s^0I_2 & \vdots & s^1I_2 & \vdots & s^2I_2 & \vdots & s^3I_2 \end{bmatrix}^T \Longrightarrow$$

$$R^{(3)}S_2^{(3)} = \begin{bmatrix} s^2 & 0 \\ 0 & 2s^2 \end{bmatrix}, \tag{6.53}$$

$$R^{(3)}Z_2^{(3-1)} = \begin{bmatrix} 0 & 0 & 0 & 0 & 1 & 0 & 0 & 0 \\ 0 & 0 & 0 & 0 & 0 & 2 & 0 & 0 \end{bmatrix} \bullet$$

$$\bullet \begin{bmatrix} O_2 & O_2 & O_2 \\ s^{3-3}I_2 & O_2 & O_2 \\ s^{3-2}I_k & s^{3-3}I_k & O_2 \\ s^{3-1}I_2 & s^{3-2}I_2 & s^{3-3}I_2 \end{bmatrix} \Longrightarrow$$

$$R^{(3)}Z_2^{(3-1)} = \begin{bmatrix} s & 0 & 1 & 0 & 0 & 0 \\ 0 & 2s & 0 & 2 & 0 & 0 \end{bmatrix}, \tag{6.54}$$

$$Q^{(2)} = \begin{bmatrix} 0 & 0 & 1 & 2 & 0 & 0 \\ 0 & 0 & 2 & 1 & 0 & 0 \end{bmatrix},$$

$$Q^{(2)}S_2^{(2)} = \begin{bmatrix} 0 & 0 & 1 & 2 & 0 & 0 \\ 0 & 0 & 2 & 1 & 0 & 0 \end{bmatrix} \begin{bmatrix} s^0I_2 \\ s^1I_2 \\ s^2I_2 \end{bmatrix} = \begin{bmatrix} s & 2 \\ 2 & s \end{bmatrix} \tag{6.55}$$

In order to determine the system IO transfer function matrices we apply Equations (6.19), (6.23)-(6.27), and (6.28) to the system (6.36), (6.37):

$$F_{IIO}(s) = \left[G_{IIO}(s) \vdots G_{IIOI_0}(s) \vdots G_{IIOR_0}(s) \vdots G_{IIOY_0}(s) \right], \qquad (6.56)$$

where:

- $F_{IIO}(s)$ is the IO full transfer function matrix of the IIO system (6.36), (6.37), which, together with (6.49)-(6.55), determines:
- $G_{IIO}(s)$,

$$G_{IIO}(s) = \left(E^{(2)} S_2^{(2)}(s) \right)^{-1} \bullet$$
$$= \left[R^{(3)} S_2^{(3)}(s) G_{IIOIS}(s) + Q^{(2)} S_2^{(2)}(s) \right] \implies$$

$$G_{IIO}(s) =$$

$$= \left[\bullet \begin{bmatrix} \begin{bmatrix} \frac{1}{s^2} & 0 \\ 0 & \frac{1}{s^2} \end{bmatrix} \begin{bmatrix} s^2 & 0 \\ 0 & 2s^2 \end{bmatrix} \bullet \\ \begin{bmatrix} \frac{-4+8s-5s^2+34s^3+4s^4+28s^5}{10s^6+19s^4+17s^3+3s^2+s-4} & \frac{-2+18s+6s^2+23s^3+6s^4+14s^5}{10s^6+19s^4+17s^3+3s^2+s-4} \\ \frac{-8-s+6s^2+6s^3+3s^4+11s^5}{10s^6+19s^4+17s^3+3s^2+s-4} & -\frac{26+26s+13s^2+17s^3+3s^4+10s^5}{10s^6+19s^4+17s^3+3s^2+s-4} \end{bmatrix} \\ + \begin{bmatrix} \frac{1}{s^2} & 0 \\ 0 & \frac{1}{s^2} \end{bmatrix} \begin{bmatrix} s & 2 \\ 2 & s \end{bmatrix} \end{bmatrix} \right] + \implies$$

$$G_{IIO}(s) =$$

$$= \begin{bmatrix} \dfrac{\left(\begin{array}{c} -4s - 3s^2 + 11s^3 + \\ +12s^4 + 53s^5 \\ +4s^6 + 38s^7 \end{array}\right)}{10s^8+19s^6+17s^5+3s^4+s^3-4s^2} & \dfrac{\left(\begin{array}{c} -8 + 2s + 4s^2 + \\ +52s^3 44s^4 + 23s^5 + \\ +26s^6 + 14s^7 \end{array}\right)}{s^2(10s^6+19s^4+17s^3+3s^2+s-4)} \\ \dfrac{\left(\begin{array}{c} -8 + 2s - 10s^2 + \\ +32s^3 + 50s^4 + 12s^5 + \\ +26s^6 + +22s^7 \end{array}\right)}{10s^8+19s^6+17s^5+3s^4+s^3-4s^2} & \dfrac{\left(\begin{array}{c} -4s - 51s^2 - \\ -49s^3 - 9s^4 - 15s^5 - \\ -6s^6 - 10s^7 \end{array}\right)}{10s^8+19s^6+17s^5+3s^4+s^3-4s^2} \end{bmatrix}, \quad (6.57)$$

$G_{IIO}(s)$ is the IO transfer function matrix of the IIO system (6.36), (6.37) relative to the input vector $\mathbf{I}$,

- $G_{IIOI_0}(s)$,

$$G_{IIOI_0}(s) = R_y^{(3)} S_2^{(3)} G_{IIOISI_0}(s) =$$
$$= \begin{bmatrix} s & 0 \\ 0 & 2s \end{bmatrix} G_{IIOISI_0}(s) \implies$$

$$G_{IIOI_0}(s) = \begin{bmatrix} s & 0 \\ 0 & 2s \end{bmatrix} \bullet$$

$$\bullet \begin{bmatrix} \dfrac{1+5s-12s^2-4s^3-28s^4}{10s^6+19s^4+17s^3+3s^2+s-4} & \dfrac{-1+10s+3s^2-3s^3-3s^4}{10s^6+19s^4+17s^3+3s^2+s-4} \\ \dfrac{-6s-3s^2-6s^3-14s^4}{10s^6+19s^4+17s^3+3s^2+s-4} & \dfrac{12+7s-s^2+3s^3-8s^4}{10s^6+19s^4+17s^3+3s^2+s-4} \\ \dfrac{5-12s\ -28s^3}{10s^6+19s^4+17s^3+3s^2+s-4} & \dfrac{11+3s+4s^3}{10s^6+19s^4+17s^3+3s^2+s-4} \\ \dfrac{3-3s-14s^3}{10s^6+19s^4+17s^3+3s^2+s-4} & \dfrac{1-s-8s^3}{10s^6+19s^4+17s^3+3s^2+s-4} \end{bmatrix}^T \implies$$

$$G_{IIOI_0}(s) =$$

$$= \begin{bmatrix} \dfrac{s+5s^2-12s^3-4s^4-28s^5}{-4+s+3s^2+17s^3+19s^4+10s^6} & \dfrac{-2s+20s^2+6s^3-6s^4-6s^5}{-4+s+3s^2+17s^3+19s^4+10s^6} \\ \dfrac{-6s^2-3s^3-6s^4-14s^5}{-4+s+3s^2+17s^3+19s^4+10s^6} & \dfrac{24s+14s^2-2s^3+6s^4-16s^5}{-4+s+3s^2+17s^3+19s^4+10s^6} \\ \dfrac{5s^2-12s^3\ -28s^5}{-4+s+3s^2+17s^3+19s^4+10s^6} & \dfrac{22s+6s^2+8s^4}{-4+s+3s^2+17s^3+19s^4+10s^6} \\ \dfrac{3s-3s^2-14s^4}{-4+s+3s^2+17s^3+19s^4+10s^6} & \dfrac{2s-2s^2-16s^4}{-4+s+3s^2+17s^3+19s^4+10s^6} \end{bmatrix}^T , \qquad (6.58)$$

is the system IO transfer function matrix of the IIO system (6.36), (6.37) relative to the extended initial input vector $\mathbf{I}_{0\mp}^{\mu-1}$,

- $G_{IIOR_0}(s)$,

$$G_{IIOR_0}(s) = R_y^{(3)} S_2^{(3)} G_{IIOISR_0}(s) - R_y^{(3)} Z_2^{(3-1)}(s) =$$

$$= \left\{ \bullet \begin{bmatrix} \dfrac{-2+3s-s^2+19s^3+9s^5}{10s^6+19s^4+17s^3+3s^2+s-4} & \dfrac{-3+9s-2s^2+6s^3}{10s^6+19s^4+17s^3+3s^2+s-4} \\ \dfrac{-2-11s^2-2s^5}{10s^6+19s^4+17s^3+3s^2+s-4} & \dfrac{2-3s+18s^2+16s^3+18s^5}{10s^6+19s^4+17s^3+3s^2+s-4} \\ \dfrac{-s\ +9s^2+10s^4}{10s^6+19s^4+17s^3+3s^2+s-4} & \dfrac{-11s-5s^2}{10s^6+19s^4+17s^3+3s^2+s-4} \\ \dfrac{-2s-12s^2}{10s^6+19s^4+17s^3+3s^2+s-4} & \dfrac{18s+10s^2+10s^4}{10s^6+19s^4+17s^3+3s^2+s-4} \\ \dfrac{-1+9s+10s^3}{10s^6+19s^4+17s^3+3s^2+s-4} & \dfrac{-11-5s}{10s^6+19s^4+17s^3+3s^2+s-4} \\ \dfrac{-2-12s}{10s^6+19s^4+17s^3+3s^2+s-4} & \dfrac{18+10s+10s^3}{10s^6+19s^4+17s^3+3s^2+s-4} \end{bmatrix}^T - \begin{bmatrix} 1 & 0 & 0 & 0 & 0 & 0 \\ 0 & 2 & 0 & 0 & 0 & 0 \end{bmatrix} \right\}^T$$

$$\implies$$

$$G_{IIOR_0}(s) =$$

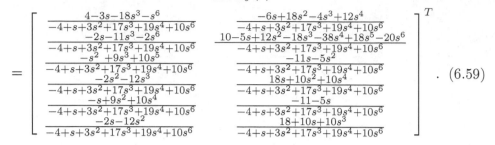

$$= \begin{bmatrix} \frac{4-3s-18s^3-s^6}{-4+s+3s^2+17s^3+19s^4+10s^6} & \frac{-6s+18s^2-4s^3+12s^4}{-4+s+3s^2+17s^3+19s^4+10s^6} \\ \frac{-2s-11s^3-2s^6}{-4+s+3s^2+17s^3+19s^4+10s^6} & \frac{10-5s+12s^2-18s^3-38s^4+18s^5-20s^6}{-4+s+3s^2+17s^3+19s^4+10s^6} \\ \frac{-s^2+9s^3+10s^5}{-4+s+3s^2+17s^3+19s^4+10s^6} & \frac{-11s-5s^2}{-4+s+3s^2+17s^3+19s^4+10s^6} \\ \frac{-2s^2-12s^3}{-4+s+3s^2+17s^3+19s^4+10s^6} & \frac{18s+10s^2+10s^4}{-4+s+3s^2+17s^3+19s^4+10s^6} \\ \frac{-s+9s^2+10s^4}{-4+s+3s^2+17s^3+19s^4+10s^6} & \frac{-11-5s}{-4+s+3s^2+17s^3+19s^4+10s^6} \\ \frac{-2s-12s^2}{-4+s+3s^2+17s^3+19s^4+10s^6} & \frac{18+10s+10s^3}{-4+s+3s^2+17s^3+19s^4+10s^6} \end{bmatrix}^T . \quad (6.59)$$

6.2 *IIO* plant desired regime

We adjust Definition 44, (Section 3.2), to the *IIO* plant (6.1), (6.2):

Definition 74 *A functional control-state pair* $\left[\mathbf{U}*(.), \mathbf{R}*^{\alpha-1}(.)\right]$ *is **nominal for the IIO plant (6.1), (6.2) relative to the functional vector pair*** $[\mathbf{D}(.), \mathbf{Y}_d(.)]$, *which is denoted by* $[\mathbf{U}_N(.), \mathbf{R}_N^{\alpha-1}(.)]$, *if and only if* $\left[\mathbf{I}(.), \mathbf{R}^{\alpha-1}(.)\right] = \left[\mathbf{I}*(.), \mathbf{R}*^{\alpha-1}(.)\right]$ *ensures that the corresponding real response* $\mathbf{Y}(.) = \mathbf{Y}*(.)$ *of the system obeys* $\mathbf{Y}*(t) = \mathbf{Y}_d(t)$ *all the time as soon as* $\mathbf{Y}_0^{\nu-1} = \mathbf{Y}_{d0}^{\nu-1}$,

$$\left[\mathbf{I}*(.), \mathbf{R}*^{\alpha-1}(.)\right] = \left[\mathbf{I}_N(.), \mathbf{R}_N^{\alpha-1}(.)\right] \Longleftrightarrow$$

$$\Longleftrightarrow \left\langle \mathbf{Y}_0^{\nu-1} = \mathbf{Y}_{d0}^{\nu-1} \Longrightarrow \mathbf{Y}*(t) = \mathbf{Y}_d(t), \ \forall t \in \mathfrak{T}_0 \right\rangle.$$

Let

$$\mathbf{w}_1(s) = \left\{ \begin{array}{c} -D^{(\mu)}S_d^{(\mu)}(s)\mathbf{D}(s) + D^{(\mu)}Z_d^{(\mu-1)}(s)\mathbf{D}_0^{\mu-1}+ \\ +B^{(\mu)}Z_r^{(\mu-1)}(s)\mathbf{U}_0^{*\mu-1} - A_P^{(\alpha)}Z_\rho^{(\alpha-1)}(s)\mathbf{R}_0^{*\alpha-1} \end{array} \right\}, \quad (6.60)$$

$$\mathbf{w}_2(s) = \left\{ \begin{array}{c} E^{(\nu)}S_N^{(\nu)}(s)\mathbf{Y}_d(s) - E^{(\nu)}Z_N^{(\nu-1)}(s)\mathbf{Y}_{d0}^{\nu-1}+ \\ +R_y^{(\alpha-1)}Z_\rho^{(\alpha-1)}(s)\mathbf{R}_0^{*\alpha-1} - V^{(\mu)}S_d^{(\mu)}(s)\mathbf{D}(s)+ \\ +V^{(\mu)}Z_d^{(\mu-1)}(s)\mathbf{D}_0^{\mu-1} + U^{(\mu)}Z_r^{(\nu-1)}(s)\mathbf{U}_0^{*\mu-1} \end{array} \right\}. \quad (6.61)$$

Definition 74 and the plant description (6.1), (6.2) imply the following:

Theorem 75 *In order for a functional vector pair* $\left[\mathbf{U}*(.), \mathbf{R}*^{\alpha-1}(.)\right]$ *to be nominal for the IIO plant (6.1), (6.2) relative to the functional vector pair* $[\mathbf{D}(.), \mathbf{Y}_d(.)]$,

$$\left[\mathbf{U}*(.), \mathbf{R}*^{\alpha-1}(.)\right] = \left[\mathbf{U}_N(.), \mathbf{R}_N^{\alpha-1}(.)\right],$$

it is necessary and sufficient that it obeys the following equations:

$$B^{(\mu)}\mathbf{U}^{*\mu}(t) - A^{(\alpha)}\mathbf{R}^{*\alpha}(t) = -D^{(\mu)}\mathbf{D}^{\mu}(t), \ \forall t \in \mathfrak{T}_0, \qquad (6.62)$$

$$U^{(\mu)}\mathbf{U}^{*\mu}(t) + R^{(\alpha)}\mathbf{R}^{*\alpha}(t) = E^{(\nu)}\mathbf{Y}_d^{\nu}(t) - V^{(\mu)}\mathbf{D}^{\mu}(t), \ \forall t \in \mathfrak{T}_0, \qquad (6.63)$$

or equivalently,

$$\begin{bmatrix} B^{(\mu)}S_r^{(\mu)}(s) & -A^{(\alpha)}S_\rho^{(\alpha)}(s) \\ U^{(\mu)}S_r^{(\mu)}(s) & R^{(\alpha)}S_\rho^{(\alpha)}(s) \end{bmatrix} \begin{bmatrix} \mathbf{U}^*(s) \\ \mathbf{R}^*(s) \end{bmatrix} = \begin{bmatrix} \mathbf{w}_1(s) \\ \mathbf{w}_2(s) \end{bmatrix}. \qquad (6.64)$$

What are the conditions for the existence of the solutions of the equations (6.62), (6.63), i.e., of (6.64)? There are $(r+\rho)$ unknown variables and $(N+\rho)$ equations. The unknown variables are the entries of $\mathbf{U}^*(s) \in \mathbb{C}^r$ and of $\mathbf{R}^*(s) \in \mathbb{C}^\rho$.

Claim 76 *In order to exist a nominal functional vector pair*

$$[\mathbf{U}_N(.), \mathbf{R}_N^{\alpha-1}(.)]$$

for the IIO plant (6.1), (6.2) relative to the functional pair $[\mathbf{D}(.), \mathbf{Y}_d(.)]$ it is necessary and sufficient that $N \leq r$.

The proof of this claim follows the proof of Claim 46 (Section 3.2).

Condition 77 *The desired output response of the IIO plant (6.1), (6.2) is realizable, i.e., $N \leq r$. The nominal control-state pair $[\mathbf{U}_N(.), \mathbf{R}_N^{\alpha-1}(.)]$ is known.*

The *time* domain description of the *IIO* plant in terms of the deviations reads:

$$A^{(\alpha)}\mathbf{r}^\alpha(t) = D^{(\mu)}\mathbf{d}^\mu(t) + B^{(\mu)}\mathbf{u}^\mu(t), \ \forall t \in \mathfrak{T}_0, \qquad (6.65)$$

$$E^{(\nu)}\mathbf{y}^\nu(t) = R_y^{(\alpha-1)}\mathbf{r}^{\alpha-1}(t) + V^{(\mu)}\mathbf{d}^\mu(t) + U^{(\mu)}\mathbf{u}^\mu(t), \ \forall t \in \mathfrak{T}_0. \qquad (6.66)$$

6.3   Exercises

Exercise 78 *1. Select a physical IIO plant.*
2. Determine its time domain IIO mathematical model.

Part II

OBSERVABILITY

Chapter 7

Mathematical preliminaries

7.1   Linear independence and matrices

This section gives a short outline on linear independence of (column and row, respectively) vectors ($\mathbf{k}_i \in \mathfrak{R}^\nu$ and $\mathbf{c}_i \in \mathfrak{C}^\nu$, $i = 1, 2, ..., \mu$, $\mathbf{r}_j \in \mathfrak{R}^{1 \times \mu}$ and $\mathbf{z}_j \in \mathbb{C}^{1 \times \mu}$, $j = 1, 2, ..., \nu$). The brief reminder of the linear (in)dependence and of the rank of the matrices is helpful to emphasize the subtle but crucial differences between the matrices and matrix functions.

Definition 79 *Linear independence and dependence of vectors*
A) Real valued μ vectors $\mathbf{k}_i \in \mathfrak{R}^\nu$ are linearly independent if and only if their linear combination (7.1),

$$\alpha_1 \mathbf{k}_1 + \alpha_2 \mathbf{k}_2 + ... + \alpha_\mu \mathbf{k}_\mu \tag{7.1}$$

vanishes, i.e.,

$$\alpha_1 \mathbf{k}_1 + \alpha_2 \mathbf{k}_2 + ... + \alpha_\mu \mathbf{k}_\mu = \mathbf{0}_\nu \tag{7.2}$$

only for zero values of all the real valued scalars $\alpha_1, \alpha_2, ..., \alpha_\mu$, i.e.,

$$\alpha_1 \mathbf{k}_1 + \alpha_2 \mathbf{k}_2 + ... + \alpha_\mu \mathbf{k}_\mu = \mathbf{0}_\nu \Longleftrightarrow \alpha_1 = \alpha_2 = ... = \alpha_\mu = 0. \tag{7.3}$$

Otherwise, i.e., if and only if (7.2) holds for at least one nonzero real number α_k, i.e.,

$$\exists\,(\alpha_k \neq 0) \in \mathfrak{R}, \ k \in \{1, 2, ..., \mu\} \Longrightarrow \alpha_1 \mathbf{k}_1 + \alpha_2 \mathbf{k}_2 + ... + \alpha_\mu \mathbf{k}_\mu = \mathbf{0}_\nu \tag{7.4}$$

real valued μ vectors $\mathbf{k}_i \in \mathfrak{R}^\nu$ are linearly dependent.

Real valued ν vectors $\mathbf{r}_j \in \mathfrak{R}^{1\times\mu}$ are linearly independent if and only if their linear combination (7.5),

$$b_1\mathbf{r}_1 + b_2\mathbf{r}_2 + ... + b_\nu\mathbf{r}_\nu \qquad (7.5)$$

vanishes, i.e.,

$$b_1\mathbf{r}_1 + b_2\mathbf{r}_2 + ... + b_\nu\mathbf{r}_\nu = \mathbf{0}_\mu^T \qquad (7.6)$$

only for zero values of all the real valued scalars $b_1, b_2, ..., b_\nu$, i.e.,

$$b_1\mathbf{r}_1 + b_2\mathbf{r}_2 + ... + b_\nu\mathbf{r}_\nu = \mathbf{0}_\mu^T \iff b_1 = b_2 = ... = b_\nu = 0. \qquad (7.7)$$

Otherwise, i.e., if and only if (7.5) holds for at least one nonzero real number b_m, i.e.,

$$\exists\,(b_m \neq 0) \in \mathfrak{R}, \ m \in \{1, 2, ..., \mu\} \implies b_1\mathbf{r}_1 + b_2\mathbf{r}_2 + ... + b_\nu\mathbf{r}_\nu = \mathbf{0}_\mu^T \qquad (7.8)$$

real valued ν vectors $\mathbf{r}_j \in \mathfrak{R}^{1\times\mu}$ are linearly dependent.

B) Complex valued μ vectors $\mathbf{c}_i \in \mathfrak{C}^\nu$ are linearly independent if and only if their linear combination (7.9),

$$\gamma_1\mathbf{c}_1 + \gamma_2\mathbf{c}_2 + ... + \gamma_\mu\mathbf{c}_\mu \qquad (7.9)$$

vanishes, i.e.,

$$\gamma_1\mathbf{c}_1 + \gamma_2\mathbf{c}_2 + ... + \gamma_\mu\mathbf{c}_\mu = \mathbf{0}_\nu \qquad (7.10)$$

only for zero values of all complex valued scalars $\gamma_1, \gamma_2, ..., \gamma_\mu$, i.e.,

$$\gamma_1\mathbf{c}_1 + \gamma_2\mathbf{c}_2 + ... + \gamma_\mu\mathbf{c}_\mu = \mathbf{0}_\nu \iff \gamma_1 = \gamma_2 = ... = \gamma_\mu = 0. \qquad (7.11)$$

Otherwise, i.e., if and only if (7.10) holds for at least one nonzero complex number γ_j, i.e.,

$$\exists\,(\gamma_j \neq 0) \in \mathfrak{C}, \ j \in \{1, 2, ..., \mu\} \implies$$
$$\gamma_1\mathbf{c}_1 + \gamma_2\mathbf{c}_2 + .. + \gamma_j\mathbf{c}_j + . + \gamma_\mu\mathbf{c}_\mu = \mathbf{0}_\nu, \qquad (7.12)$$

complex valued μ vectors $\mathbf{c}_i \in \mathfrak{C}^\nu$ are linearly dependent.

Complex valued ν vectors $\mathbf{z}_j \in \mathbb{C}^{1\times\mu}$ are linearly independent if and only if their linear combination (7.13),

$$\gamma_1\mathbf{z}_1 + \gamma_2\mathbf{z}_2 + ... + \gamma_\nu\mathbf{z}_\nu \qquad (7.13)$$

vanishes, i.e.,

$$\gamma_1\mathbf{z}_1 + \gamma_2\mathbf{z}_2 + ... + \gamma_\nu\mathbf{z}_\nu = \mathbf{0}_\mu^T \qquad (7.14)$$

only for zero values of all complex valued scalars $\gamma_1, \gamma_2, ..., \gamma_\nu$, i.e.,

$$\gamma_1 \mathbf{z}_1 + \gamma_2 \mathbf{z}_2 + ... + \gamma_\nu \mathbf{z}_\nu = \mathbf{0}_\mu^T \iff \gamma_1 = \gamma_2 = ... = \gamma_\nu = 0. \qquad (7.15)$$

Otherwise, i.e., if and only if (7.6) holds for at least one nonzero complex number γ_l, i.e.,

$$\exists (\gamma_l \neq 0) \in \mathfrak{R}, \ l \in \{1, 2, ..., \mu\} \implies \gamma_1 \mathbf{z}_1 + \gamma_2 \mathbf{z}_2 + ... + \gamma_\nu \mathbf{z}_\nu = \mathbf{0}_\mu^T \qquad (7.16)$$

complex valued ν vectors $\mathbf{z}_j \in \mathbb{C}^{1 \times \mu}$ are linearly dependent.

We can form a matrix $K \in \mathfrak{R}^{\nu \times \mu}$ of μ (column) vectors $\mathbf{k}_i \in \mathfrak{R}^\nu$ and the matrix $C \in \mathbb{C}^{\nu \times \mu}$ of μ (column) vectors $\mathbf{c}_i \in \mathbb{C}^\nu$:

$$K = \begin{bmatrix} \mathbf{k}_1 & \mathbf{k}_2 & ... & \mathbf{k}_\mu \end{bmatrix}, \ C = \begin{bmatrix} \mathbf{c}_1 & \mathbf{c}_2 & ... & \mathbf{c}_\mu \end{bmatrix}. \qquad (7.17)$$

Let

$$\mathbf{k}_i = \begin{bmatrix} k_{1i} \\ : \\ k_{\nu i} \end{bmatrix} \ and \ \mathbf{c}_i = \begin{bmatrix} c_{1i} \\ : \\ c_{\nu i} \end{bmatrix}, \ i = 1, 2, ..., \mu, \qquad (7.18)$$

then, e.g., for $\nu < \mu$:

$$K = \begin{bmatrix} k_{11} & k_{12} & ... & k_{1i} & ... & k_{1\mu} \\ : & : & : ... : & : & : ... : & : \\ k_{\nu 1} & k_{\nu 2} & ... & k_{\nu i} & ... & k_{\nu \mu} \end{bmatrix} \in \mathfrak{R}^{\nu \times \mu}, \qquad (7.19)$$

$$C = \begin{bmatrix} c_{11} & c_{12} & ... & c_{1i} & ... & c_{1\mu} \\ : & : & : ... : & : & : ... : & : \\ c_{\nu 1} & c_{\nu 2} & ... & c_{\nu i} & ... & c_{\nu \mu} \end{bmatrix} \in \mathfrak{C}^{\nu \times \mu}. \qquad (7.20)$$

Equations (7.17) through (7.20) establish the link between the vectors and the matrices. The vectors $\mathbf{k}_i$ and $\mathbf{c}_i$ represent the $i-th$ column of the matrices K and C, respectively, for every $i = 1, 2, .., \mu$. The number ν is the number of the rows of the matrices K and C, and the number μ is the number of the columns of the matrices K and C.

7.2   Matrix range, null space, and rank

Let a matrix $M \in \{\mathfrak{R}^{\nu \times \mu}, \ \mathbb{C}^{\nu \times \mu}\}$. The following sets are determined by the properties of the matrix M:

-        The set denoted by R(M) called **the range of the matrix** M is the set of all vectors $\mathbf{y} \in \{\mathfrak{R}^{\nu}, \mathbb{C}^{\nu}\}$ for which there exists a vector $\mathbf{x} \in \{\mathfrak{R}^{\mu}, \mathbb{C}^{\mu}\}$ such that $\mathbf{y} = M\mathbf{x}$,

$$\mathcal{R}(M) = \{\mathbf{y} : \exists \mathbf{x} \in \{\mathfrak{R}^{\mu}, \mathbb{C}^{\mu}\} \implies \mathbf{y} = M\mathbf{x} \in \{\mathfrak{R}^{\nu}, \mathbb{C}^{\nu}\}\} \subseteq .\mathfrak{R}^{\nu}.$$

The dimension denoted by dimR(M) of R(M) is the largest number of linearly independent vectors in R(M) such that every vector in R(M) can be represented as their linear combination.

-        The set denoted by $\mathcal{N}(M)$ called **the null space of the matrix** M is the set of all vectors $\mathbf{x} \in \{\mathfrak{R}^{\mu}, \mathbb{C}^{\mu}\}$ such that $M\mathbf{x} = \mathbf{0}_{\nu}$,

$$\mathcal{N}(M) = \{\mathbf{x} \in \{\mathfrak{R}^{\mu}, \mathbb{C}^{\mu}\} : M\mathbf{x} = \mathbf{0}_{\nu}\} \subseteq \mathfrak{R}^{\mu}.$$

The dimension of $\mathcal{N}(M)$ denoted by $dim\mathcal{N}(M)$ and called **the nullity of** M is the largest number of linearly independent vectors in $\mathcal{N}(M)$ such that every vector in $\mathcal{N}(M)$ can be represented as their linear combination.

dimR(M) and $dim\mathcal{N}(M)$ obey

$$dim\mathcal{R}(M) + dim\mathcal{N}(M) = \mu.$$

Definition 80 *Rank of a matrix*
    The following definitions of the rank of a $\nu \times \mu$ matrix $M \in \{\mathfrak{R}^{\nu \times \mu}, \mathbb{C}^{\nu \times \mu}\}$ are equivalent:

-        *The rank of a $\nu \times \mu$ matrix $M \in \{\mathfrak{R}^{\nu \times \mu}, \mathbb{C}^{\nu \times \mu}\}$ is the number ρ such that its minor M_{ρ} of the order ρ is nonzero and every its minor M_k of the order k bigger than ρ is zero,*

$$rank \ M = \rho \iff \left\{ \begin{array}{c} M_k \neq 0, \ \forall k = 1, 2, ..., \rho, \\ M_k \neq 0, \ \forall k = \rho + 1, \rho + 2, ..., \min(\nu, \mu). \end{array} \right\} \quad (7.21)$$

-        *The rank of a $\nu \times \mu$ matrix $M \in \{\mathfrak{R}^{\nu \times \mu}, \mathbb{C}^{\nu \times \mu}\}$ is the dimension ρ of the largest nonsingular square submatrix of the matrix M.*

-        *The rank of a $\nu \times \mu$ matrix $M \in \{\mathfrak{R}^{\nu \times \mu}, \mathbb{C}^{\nu \times \mu}\}$ is the maximal number ρ of the linearly independent rows of the matrix M. This is also called **the row rank** of the matrix M.*

-        *The rank of a $\nu \times \mu$ matrix $M \in \{\mathfrak{R}^{\nu \times \mu}, \mathbb{C}^{\nu \times \mu}\}$ is the maximal number ρ of the linearly independent columns of the matrix M. This is also called **the column rank** of the matrix M.*

Claim 81 *Maximal numbers of linearly independent columns, of linearly independent rows and rank of a $\nu \times \mu$ matrix M*

The maximal number $n_{ir\,max}$ of linearly independent rows, the maximal number $n_{ic\,max}$ of linearly independent columns, and the rank ρ of $\nu \times \mu$ matrix $M \in \{\mathfrak{R}^{\nu \times \mu}, \mathbb{C}^{\nu \times \mu}\}$ are all equal and are not bigger than $min(\nu, \mu)$:

$$n_{ic\,max} = n_{ir\,max} = rankM = \rho \leq min\,(\nu, \mu)\,. \tag{7.22}$$

By applying the elementary transformations to any $\nu \times \mu$ matrix $M \in \{\mathfrak{R}^{\nu \times \mu}, \mathbb{C}^{\nu \times \mu}\}$ it can be transformed into its equivalent ($\check{}$) matrix of the following structure (e.g., [2, pp. 107, 108]):

$$M \check{} \begin{bmatrix} 1 & 0 & 0 & 0 \\ 0 & 1 & 0 & 0 \\ 0 & 0 & 1 & 0 \\ 0 & 0 & 0 & 0 \end{bmatrix} \Longrightarrow$$

$$\Longrightarrow n_{ic\,max} = n_{ir\,max} = rankM = 3. \tag{7.23}$$

This explains and illustrates Claim 81.

Example 82 *Let*

$$M = \begin{bmatrix} 2 & 1 \\ -2 & -1 \\ 4 & 3 \end{bmatrix} \in \mathfrak{R}^{3 \times 2} \Longrightarrow \nu = 3,\ \mu = 2.$$

The minor $M_{3,2}^{2,2}$ composed of the entries of the second row and the third row of the matrix M is nonsingular, which implies rank $M = 2$ because there is not a third order minor of the matrix M:

$$M_{3,2}^{2,2} = \begin{vmatrix} -2 & -1 \\ 4 & 3 \end{vmatrix} = (-2)\,3 - (-1)\,4 = -2 \neq 0 \Longrightarrow rankM = 2,$$

In order to determine the number of linearly independent columns by definition we test the existence of at least one nonzero number between α_1 and α_2 such that the linear combination of the columns of the matrix M vanishes:

$$\alpha_1 \begin{bmatrix} 2 \\ -2 \\ 4 \end{bmatrix} + \alpha_2 \begin{bmatrix} 1 \\ -1 \\ 3 \end{bmatrix} = \begin{bmatrix} 2 & 1 \\ -2 & -1 \\ 4 & 3 \end{bmatrix} \begin{bmatrix} \alpha_1 \\ \alpha_2 \end{bmatrix} = \begin{bmatrix} 0 \\ 0 \\ 0 \end{bmatrix} \Longleftrightarrow$$

$$\Longleftrightarrow \begin{matrix} 2\alpha_1 + \alpha_2 = 0 \\ -2\alpha_1 - \alpha_2 = 0 \\ 4\alpha_1 + 3\alpha_2 = 0 \end{matrix} \Longleftrightarrow \begin{matrix} \alpha_2 = -2\alpha_1 \\ \alpha_2 = -2\alpha_1 \\ \alpha_2 = -\frac{4}{3}\alpha_1 \end{matrix} \Longleftrightarrow \alpha_1 = \alpha_2 = 0.$$

The linear combination of the columns of the matrix M vanishes only if the numbers α_1 and α_2 are equal to zero. The number of linearly independent rows is 2, and the number of linearly independent columns is 2.

We apply the sequence of elementary transformations to the matrix M in order to verify its rank:

$$
M = \begin{bmatrix} 2 & 1 \\ -2 & -1 \\ 4 & 3 \end{bmatrix} \sim \begin{bmatrix} 2 & 1 \\ -2 & -1 \\ -2 & 0 \end{bmatrix} \sim \begin{bmatrix} 2 & 2 \\ -2 & -2 \\ -2 & 0 \end{bmatrix} \sim \begin{bmatrix} 2 & 2 \\ 0 & 0 \\ -2 & 0 \end{bmatrix} \sim
$$

$$
\sim \begin{bmatrix} 2 & 2 \\ -2 & 0 \\ 0 & 0 \end{bmatrix} \sim \begin{bmatrix} 1 & 1 \\ 1 & 0 \\ 0 & 0 \end{bmatrix} \sim \begin{bmatrix} 0 & 1 \\ 1 & 0 \\ 0 & 0 \end{bmatrix} \sim \begin{bmatrix} 1 & 0 \\ 0 & 1 \\ 0 & 0 \end{bmatrix} \implies rank M = 2.
$$

The number of linearly independent columns is 2, and the number of linearly independent rows is also 2. They are equal to the rank of M.

Lemma 83 *Linear dependence of columns/rows of the matrix K*

1) In order for columns $\mathbf{k}_i \in \mathfrak{R}^\nu$, $i = 1, 2, ..., \mu$, of the matrix K (7.19) to be linearly dependent it is necessary and sufficient that there exists an $\mu \times 1$ nonzero vector $\mathbf{a} = [\alpha_1 \ \alpha_2 ... \alpha_\mu]^T$, $(\mathbf{a} \neq \mathbf{0}_\mu) \in \mathfrak{R}^\mu$, such that the product $K\mathbf{a}$ is zero vector:

Linear dependence of the columns $\mathbf{k}_i (.)$ of $K (.)$ (7.19)
$$\iff (\exists (\mathbf{a} \neq \mathbf{0}_\mu) \in \mathfrak{R}^\mu \implies K\mathbf{a} = \mathbf{0}_\nu). \qquad (7.24)$$

2) In order for rows $\mathbf{r}_i$, $\mathbf{r}_i \in \mathfrak{R}^{1 \times \mu}$, $i = 1, 2, ..., \nu$, of the matrix K (7.19) to be linearly dependent it is necessary and sufficient that there exists an $1 \times \nu$ nonzero vector $\mathbf{g} = [\gamma_1 \ \gamma_2 ... \gamma_\nu]$, $(\mathbf{g} \neq \mathbf{0}_\nu^T) \in \mathfrak{R}^{1 \times \nu}$, such that the product $\mathbf{g}K$ is zero vector:

Linear dependence of the rows $\mathbf{r}_i (.)$ of K (7.19)
$$\iff (\exists (\mathbf{g} \neq \mathbf{0}_\nu^T) \in \mathfrak{R}^{1 \times \nu} \implies \mathbf{g}K = \mathbf{0}_\mu^T). \qquad (7.25)$$

3) The statements under 1) and 2) analogously hold for the matrix $C \in \mathbb{C}^{\nu \times \mu}$.

Proof. 1) The necessity and sufficiency follow directly from A-1) of Definition 79 when (7.4) is written in the vector-matrix form:

$$K\mathbf{a} = \mathbf{0}_\nu \ for \ \mathbf{a} = \begin{bmatrix} \alpha_1 & ... & \alpha_\mu \end{bmatrix}^T,$$

and from Definition 80 of the matrix rank in terms of the linear dependence of the columns of the matrix K (7.19).

2) The necessity and sufficiency follow directly from A-2) of Definition 79 when (7.8) is written in the vector-matrix form:

$$\mathbf{b}K = \mathbf{0}_\mu^T \ for \ \mathbf{b} = \left[\begin{array}{ccc} b_1 & ... & b_\nu \end{array} \right],$$

and from Definition 80 of the matrix rank in terms of the linear dependence of the rows of the matrix K (7.19).

3) The proof of 3) is literally analogous to the proofs of 1) and 2). ■

Lemma 84 *Linear independence of column vectors of a matrix*

For the columns $\mathbf{k}_i(.)$, $i = 1, 2, ..., \mu$, of the matrix K (7.19) to be linearly independent it is necessary and sufficient that for every $\mu \times 1$ nonzero vector **a,**

$$\mathbf{a} = [\alpha_1 \ \alpha_2 ... \alpha_\mu]^T, \left(\mathbf{a} \neq \mathbf{0}_\mu\right) \in \mathfrak{R}^\mu,$$

the product $K\mathbf{a}$ is nonzero vector:

$$Linear \ independence \ of \ the \ columns \ \mathbf{k}_i(.) \ of \ K \ (7.19)$$
$$\Longleftrightarrow \forall \left(\mathbf{a} \neq \mathbf{0}_\mu\right) \in \mathfrak{R}^\mu \Longrightarrow K\mathbf{a} \neq \mathbf{0}_\mu. \tag{7.26}$$

*The preceding statement analogously holds for the matrix C **(7.20)**, i.e., for its columns $\mathbf{c}_i \in \mathbb{C}^{\nu \times 1}$, $i = 1, 2, ..., \mu$.*

Proof. The statement of this lemma follows directly from (7.3) written in the vector-matrix form. ■

Lemma 85 *Linear independence of row vectors of a matrix*

For the rows $\mathbf{r}_i$, $i = 1, 2, ..., \nu$, of the matrix K (7.19) to be linearly independent it is necessary and sufficient that for every $1 \times \nu$ nonzero vector $\mathbf{g} = [\gamma_1 \ \gamma_2 ... \gamma_\nu]$, $\left(\mathbf{g} \neq \mathbf{0}_\nu^T\right) \in \mathfrak{R}^{1 \times \nu}$, the product $\mathbf{g}K$ is nonzero vector:

$$Linear \ independence \ of \ the \ rows \ \mathbf{r}_i(.) \ of \ K \ (7.19)$$
$$\Longleftrightarrow \forall \left(\mathbf{g} \neq \mathbf{0}_\nu^T\right) \in \mathfrak{R}^{1 \times \nu} \Longrightarrow \mathbf{g}K \neq \mathbf{0}_\nu^T. \tag{7.27}$$

The preceding statement analogously holds for the matrix C (7.20), i.e., for its rows $\mathbf{z}_i \in \mathbb{C}^{1 \times \mu}$, $i = 1, 2, ..., \nu$.

Proof. The statement of this lemma follows directly from (7.7) written in the vector-matrix form. ■

Theorem 86 *Linear independence of the rows of a matrix product*

Let $M \in \mathbb{C}^{p \times p}$ be nonsingular, $det M \neq 0$, and $N \in \mathbb{C}^{p \times q}$, where p and q are natural numbers.

Let k be a natural number not bigger than $\min(p, q)$, $0 < k \leq \min(p, q)$.

a) In order for k rows of the matrix product MN to be linearly independent it is necessary and sufficient that k rows of the matrix N are linearly independent.

b) In order for all p rows of the matrix product MN to be linearly independent it is necessary and sufficient that all p rows of the matrix N are linearly independent, equivalently that $rank N = p \leq q$.

Proof. Let $M(t) \in \mathbb{C}^{p \times p}$ be nonsingular at every $t \in \mathfrak{T}_0$, $det M(t) \neq 0$, $\forall t \in \mathfrak{T}_0$, which implies the linear independence of its rows (and of its columns) and $rank M(t) = p$ at every $t \in \mathfrak{T}_0$. Let $N \in \mathbb{C}^{p \times q}$, where p and q are natural numbers.

Let any $\tau \in \mathfrak{T}_0$ be chosen and fixed. The condition $det M(t) \neq 0, \forall t \in \mathfrak{T}_0$, implies $det M(\tau) \neq 0$.

Necessity.

a) Let k be a natural number not bigger than p, $0 < k \leq p$. Let $rank M(t) N = k$ at every $t \in \mathfrak{T}_0$. Let at most $k - 1$ rows of the matrix N be assumed linearly independent. Then $rank N \leq k - 1$. Since $M(t)$ is nonsingular at every $t \in \mathfrak{T}_0$, its $rank M(\tau) = p \neq 0$. The assumption that $rank N \leq k - 1$ and the well-known rule on the rank of the matrix product of a nonsingular matrix $M(\tau)$ multiplying the rectangular matrix N, which reads $rank M(\tau) N = rank N$, yield $rank M(\tau) N = rank N \leq k - 1$. This contradicts $rank M(\tau) N = k$. The contradiction is the result of the assumption on the linear independence of at most $k - 1$ rows of the matrix N. The assumption fails, which implies that k rows of the matrix N are linearly independent, i.e.,, $rank N = k$.

b) The necessity of the statement under b) results from the necessity of a) for $k = p$.

Sufficiency.

a) Let $rank N = k$. The well-known rule on the rank of the matrix product in which the premultiplying matrix M is nonsingular ensures $rank M(\tau) N = rank N = k$. This implies the linear independence of the k rows of the matrix product $M(\tau) N$.

b) The sufficiency of the statement under b) results from the sufficiency of a) for $k = p$, which completes the proof by recalling an arbitrary choice of $\tau \in \mathfrak{T}_0$. ∎

Example 87 *Let*

$$M = \begin{bmatrix} 1 & 3 \\ 2 & 2 \end{bmatrix}, \ N = \begin{bmatrix} 0 \\ 1 \end{bmatrix}, \ rank M = 2, \ rank N = 1,$$

and

$$MN = \begin{bmatrix} 3 \\ 2 \end{bmatrix} \Longrightarrow rank MN = 1 = rank N.$$

This agrees with Theorem 86.

Example 88 *Let*

$$M = \begin{bmatrix} 1 & 0 & 3 \\ 6 & 7 & 11 \\ 0 & 9 & 0 \end{bmatrix}, \ N = \begin{bmatrix} 2 & 4 & 7 & 1 \\ 3 & 3 & 6 & 8 \\ 4 & 8 & 14 & 2 \end{bmatrix}, \ rank M = 3, \ rank N = 2,$$

and

$$MN = \begin{bmatrix} 1 & 0 & 3 \\ 6 & 7 & 11 \\ 0 & 9 & 0 \end{bmatrix} \begin{bmatrix} 2 & 4 & 7 & 1 \\ 3 & 3 & 6 & 8 \\ 4 & 8 & 14 & 2 \end{bmatrix} = \begin{bmatrix} 14 & 28 & 49 & 7 \\ 77 & 133 & 238 & 84 \\ 27 & 27 & 54 & 72 \end{bmatrix} \Longrightarrow$$

$$rank MN = 2.$$

This verifies Theorem 86.

Note 89 *Theorem 86 fails in the framework of matrix functions (Note 103 in the sequel).*

Comment 90 *If a matrix $M \in \{\mathfrak{R}^{\nu \times \mu}, \mathbb{C}^{\nu \times \mu}\}$:*
- *Has linearly independent columns then $\nu \geq \mu$,*
- *Has linearly independent rows then $\nu \leq \mu$.*
- *Has all linearly independent columns then $rank M = \mu \leq \nu$,*
- *Has all linearly independent rows then $rank M = \nu \leq \mu$.*

Definition 91 *Pseudo inverse matrix of a given matrix*
  If a matrix $M \in \{\mathfrak{R}^{\nu \times \mu}, \mathbb{C}^{\nu \times \mu}\}$ then:
- *A matrix $L \in \{\mathfrak{R}^{\nu \times \mu}, \mathbb{C}^{\nu \times \mu}\}$ is its **left pseudo inverse matrix** denoted by $M_L^\dagger$, $L = M_L^\dagger$, if and only if $LM = I_\mu$,*
- *A matrix $R \in \{\mathfrak{R}^{\nu \times \mu}, \mathbb{C}^{\nu \times \mu}\}$ is its **right pseudo inverse matrix** denoted by $M_R^\dagger$, $R = M_R^\dagger$, if and only if $MR = I_\nu$.*

Comment 90 and Definition 91 imply:

Lemma 92 *Pseudo inverse matrix and matrix rank*

If a matrix $M \in \{\mathfrak{R}^{\nu \times \mu}, \mathbb{C}^{\nu \times \mu}\}$:

- Has the full column rank, $rankM = \mu \leq \nu$, then it has the left pseudo inverse matrix $M_L^{\dagger}$:

$$M_L^{\dagger} = \left(M^T M\right)^{-1} M^T \in \mathfrak{R}^{\mu \times \nu}, \tag{7.28}$$

- Has the full row rank, $rankM = \nu \leq \mu$, then it has the right pseudo inverse matrix $M_R^{\dagger}$:

$$M_R^{\dagger} = M^T \left(M M^T\right)^{-1} \in \mathfrak{R}^{\mu \times \nu}, \tag{7.29}$$

- Is right (left) invertible then its rows (columns) are linearly independent, respectively,
- Has both the left and the right pseudo inverse then $\nu = \mu$, the matrix M is square and nonsingular matrix, and its pseudo inverse matrices are equal to its inverse matrix M^{-1}:

$$M \in \left\{\mathfrak{R}^{\nu \times \mu}, \mathbb{C}^{\nu \times \mu}\right\} : \; \exists M_L^{\dagger}, \; \exists M_R^{\dagger} \Longrightarrow \nu = \mu, \; M_L^{\dagger} = M_R^{\dagger} = M^{-1}.$$

7.3   Linear independence and scalar functions

The definitions of the linear (in)dependence of constant vectors and of constant matrices are not directly applicable to the linear (in)dependence of scalar functions.

Definition 93 *Linear dependence and independence of scalar functions*

A) Real valued scalar functions $\kappa_i (.) : \mathfrak{T} \longrightarrow \mathfrak{R}$, $i = 1, 2, ..., \mu$, are:

A-1) **Linearly dependent on** $\mathfrak{T}_0$ if and only if their exist real numbers α_i, $i = 1, 2, ..., \mu$, at least one of which is nonzero,

$$|\alpha_1| + |\alpha_2| + ... + |\alpha_\mu| \neq 0, \tag{7.30}$$

such that the linear combination (7.31),

$$\alpha_1 \kappa_1 (t) + \alpha_2 \kappa_2 (t) + ... + \alpha_\mu \kappa_\mu (t) \tag{7.31}$$

of the scalar functions $\kappa_i (.)$ vanishes at every $t \in \mathfrak{T}_0$,

$$\alpha_1 \kappa_1 (t) + \alpha_2 \kappa_2 (t) + ... + \alpha_\mu \kappa_\mu (t) = 0, \; \forall t \in \mathfrak{T}_0. \tag{7.32}$$

A-2) **Linearly independent on** $\mathfrak{T}_0$ *if and only if their linear combination (7.33) vanishes at every* $t \in \mathfrak{T}_0$ *if and only if all numbers* α_i *are equal to zero,*

$$\forall t \in \mathfrak{T}_0 : \alpha_1 \kappa_1(t) + \alpha_2 \kappa_2(t) + ... + \alpha_\mu \kappa_\mu(t) = 0 \Longleftrightarrow$$
$$\Longleftrightarrow |\alpha_1| + |\alpha_2| + ... + |\alpha_\mu| = 0. \tag{7.33}$$

B) Complex valued scalar functions $\xi_i(.) : \mathbb{C} \longrightarrow \mathbb{C}$, $i = 1, 2, ..., \mu$, *are:*

B-1) **Linearly dependent on** $\mathbb{C}$ *if and only if there exist complex numbers* β_i, $i = 1, 2, ..., \mu$, *at least one of which is nonzero,*

$$|\beta_1| + |\beta_2| + ... + |\beta_\mu| \neq 0, \tag{7.34}$$

such that the linear combination (7.35) of the functions $\xi_i(.)$,

$$\beta_1 \xi_1(s) + \beta_2 \xi_2(s) + ... + \beta_\mu \xi_\mu(s) \tag{7.35}$$

vanishes for every $s \in \mathbb{C}$,

$$\beta_1 \xi_1(s) + \beta_2 \xi_2(s) + ... + \beta_\mu \xi_\mu(s) = 0, \ \forall s \in \mathbb{C}. \tag{7.36}$$

B-2) **Linearly independent on** $\mathbb{C}$ *if and only if their linear combination (7.35) vanishes at every* $s \in \mathbb{C}$ *if and only if all complex numbers* β_i *are equal to zero,*

$$\forall s \in \mathfrak{C} \Longrightarrow \beta_1 \xi_1(s) + \beta_2 \xi_2(s) + ... + \beta_\mu \xi_\mu(s) = 0 \Longleftrightarrow$$
$$\Longleftrightarrow |\beta_1| + |\beta_2| + ... + |\beta_\mu| = 0. \tag{7.37}$$

Example 94 *Let* $\kappa_1(t) = 1$, $\kappa_2(t) = 2t$ *and* $\kappa_3(t) = e^t$. *Their linear combination*

$$\alpha_1 \kappa_1(t) + \alpha_2 \kappa_2(t) + \alpha_3 \kappa_3(t) =$$
$$= \alpha_1 + 2\alpha_2 t + \alpha_3 e^t = 0, \ \forall t \in \mathfrak{T}_0 \Longleftrightarrow \alpha_1 = \alpha_2 = \alpha_3 = 0.$$

The functions $\kappa_1(.)$, $\kappa_2(.)$ *and* $\kappa_3(.)$ *are linearly independent on* $\mathfrak{T}_0$.

Example 95 *Let* $\kappa_1(t) = t$, $\kappa_2(t) = 2t$ *and* $\kappa_3(t) = -5t$. *Their linear combination*

$$\alpha_1 t + \alpha_2 2t - \alpha_3 5t = (\alpha_1 + 2\alpha_2 - 5\alpha_3)t = 0, \ \forall t \in \mathfrak{T}_0,$$

vanishes at every $t \in \mathfrak{T}_0$ *for*

$$\alpha_1 = -2\alpha_2 + 5\alpha_3, \ \forall (\alpha_2 \neq 0, \ \alpha_3 \neq 0) \in \mathfrak{R}.$$

The functions $\kappa_1(.)$, $\kappa_2(.)$ *and* $\kappa_3(.)$ *are linearly dependent on* $\mathfrak{T}_0$.

Definition 96 *Wronskian of (n-1) times differentiable scalar functions*

Let n scalar functions $\kappa_i(.) : \mathfrak{R} \longrightarrow \mathfrak{R}$ be *(n-1) times differentiable on a connected subset $\mathfrak{T}^*$ of $\mathfrak{T}$ with the nonempty interior, $In\mathfrak{T}^* \neq \phi$, $\kappa_i(t) \in \mathfrak{C}^{n-1}(\mathfrak{T}^*)$, for every $i = 1, 2, ..., n$. Their Wronskian $W(t; \kappa_1, \kappa_2, .., \kappa_n)$ at $t \in \mathfrak{T}^*$ is the determinant composed so that entries of its first row are the values of the functions $\kappa_i(.)$, $i = 1, 2, ..., n$, at $t \in \mathfrak{T}^*$ and every next row is the elementwise derivative of the preceding row at $t \in \mathfrak{T}^*$ by finishing with the (n-1) differentiation:*

$$W(t; \kappa_1, \kappa_2, .., \kappa_n) = \begin{vmatrix} \kappa_1(t) & \kappa_2(t) & ... & \kappa_n(t) \\ \kappa_1^{(1)}(t) & \kappa_2^{(1)}(t) & ... & \kappa_n^{(1)}(t) \\ ... & ... & ... & ... \\ \kappa_1^{(n-1)}(t) & \kappa_2^{(n-1)}(t) & ... & \kappa_n^{(n-1)}(t) \end{vmatrix}. \quad (7.38)$$

The following theorem is very useful to test linear independence of the functions $\kappa_i(.)$, $i = 1, 2, ..., n$:

Theorem 97 *Wronskian of (n-1) times differentiable n scalar functions and their (in)dependence*

Let n scalar functions $\kappa_i(.) : \mathfrak{R} \longrightarrow \mathfrak{R}$ be *(n-1) times differentiable on a connected subset $\mathfrak{T}^*$ of $\mathfrak{T}$ with the nonempty interior, $In\mathfrak{T}^* \neq \phi$, $\kappa_i(t) \in \mathfrak{C}^{n-1}(\mathfrak{T}^*)$, for every $i = 1, 2, ..., n$.*

1) For the functions $\kappa_i(t) \in \mathfrak{C}^{n-1}(\mathfrak{T}^*)$, $i = 1, 2, ..., n$, to be linearly independent on the set $\mathfrak{T}^*$ it is sufficient that there is $\tau \in \mathfrak{T}^*$ such their Wronskian (7.38) is different from zero at $t = \tau$:

$$\exists \tau \in \mathfrak{T}^* \Longrightarrow W(\tau; \kappa_1, \kappa_2, .., \kappa_n) \neq 0. \quad (7.39)$$

2) If the functions $\kappa_i(t) \in \mathfrak{C}^{n-1}(\mathfrak{T}^*)$, $i = 1, 2, ..., n$, are linearly dependent on the set $\mathfrak{T}^*$ then their Wronskian is equal to zero at every $t \in \mathfrak{T}^*$.

Proof. Let n scalar functions $\kappa_i(.) : \mathfrak{R} \longrightarrow \mathfrak{R}$ be $(n-1)$ times differentiable on a connected subset $\mathfrak{T}^*$ of $\mathfrak{T}$ with the nonempty interior, $In\mathfrak{T}^* \neq \phi$, $\kappa_i(t) \in \mathfrak{C}^{n-1}(\mathfrak{T}^*)$, for every $i = 1, 2, ..., n$.

1) Let the condition be wrong, i.e., let there be $\tau \in \mathfrak{T}^*$ such that (7.39), (Section 7.3), holds and the functions $\kappa_i(.)$, $i = 1, 2, ..., n$, are linearly dependent. This means that there is a non zero vector $\mathbf{v}$,

$$(\mathbf{v} \neq \mathbf{0}_n) \in \mathfrak{R}^n, \quad \mathbf{v} = \begin{bmatrix} v_1 \vdots v_2 \vdots .. \vdots v_n \end{bmatrix}, \quad (7.40)$$

such that

$$\kappa_1(t) v_1 + \kappa_2(t) v_1 + .. + \kappa_n(t) v_1 = 0, \ \forall t \in \mathfrak{T}^*.$$

After differentiating this equation $(n-1)$ times the following system of the equations results:

$$\kappa_1(t) v_1 + \kappa_2(t) v_1 + ... + \kappa_n(t) v_1 = 0, \ \forall t \in \mathfrak{T}^*$$

$$\kappa_1^{(1)}(t) v_1 + \kappa_2^{(1)}(t) v_1 + ... + \kappa_n^{(1)}(t) v_1 = 0, \ \forall t \in \mathfrak{T}^*$$

$$...$$

$$\kappa_1^{(n-1)}(t) v_1 + \kappa_2^{(n-1)}(t) v_1 + ... + \kappa_n^{(n-1)}(t) v_1 = 0, \ \forall t \in \mathfrak{T}^*,$$

or in the matrix-vector form:

$$V(t) \mathbf{v} = \mathbf{0}_n, \ \forall t \in \mathfrak{T}^*, \tag{7.41}$$

where

$$V(t) = \begin{bmatrix} \kappa_1(t) & \kappa_2(t) & ... & \kappa_n(t) \\ \kappa_1^{(1)}(t) & \kappa_2^{(1)}(t) & ... & \kappa_n^{(1)}(t) \\ ... & ... & ... & ... \\ \kappa_1^{(n-1)}(t) & \kappa_2^{(n-1)}(t) & ... & \kappa_n^{(n-1)}(t) \end{bmatrix}, \ \forall t \in \mathfrak{T}^*.$$

Since the solution vector $\mathbf{v}$ of the homogenous linear algebraic equation (7.41) is non zero vector then the matrix $V(t)$ is singular on $\mathfrak{T}^*$:

$$det V(t) = 0, \ \forall t \in \mathfrak{T}^*,$$

which contradicts the condition (7.39) because

$$det V(t) = W(t; \kappa_1, \kappa_2, .., \kappa_n), \ \forall t \in \mathfrak{T}^*, \tag{7.42}$$

and proves the statement under 1).

2) If the functions $\kappa_i(t) \in \mathfrak{C}^{n-1}(\mathfrak{T}^*)$, $i = 1, 2, ..., n$, are linearly dependent on the set $\mathfrak{T}^*$ then there is a non zero vector $\mathbf{v}$, (7.40), such that it obeys (7.41) that implies $det V(t) = 0, \ \forall t \in \mathfrak{T}^*$. This and (7.42) prove the statement under 2). ∎

7.4 Linear independence and matrix functions

We introduce the matrix function $R(.) : \mathfrak{T} \longrightarrow \mathfrak{R}^{\nu \times \mu}$ of μ column vector functions $\kappa_i(.) : \mathfrak{T} \longrightarrow \mathfrak{R}^{\nu}$, $i = 1, 2, ..., \mu$, and the matrix function $C(.) :$

$\mathbb{C} \longrightarrow \mathbb{C}^{\nu \times \mu}$ of μ column vector functions $\xi_i\left(.\right) : \mathbb{C} \longrightarrow \mathbb{C}^\nu$, $i = 1, 2, ..., \mu$:

$$R\left(t\right) = \left[\begin{array}{cccc} \kappa_1\left(t\right) & \kappa_2\left(t\right) & ... & \kappa_\mu\left(t\right) \end{array}\right],$$
$$C\left(s\right) = \left[\begin{array}{cccc} \xi_1\left(s\right) & \xi_2\left(s\right) & ... & \xi_\mu\left(s\right) \end{array}\right]. \tag{7.43}$$

Let

$$\kappa_i\left(t\right) = \left[\begin{array}{c} \kappa_{1i}\left(t\right) \\ : \\ \kappa_{\nu i}\left(t\right) \end{array}\right] \text{ and } \xi_i\left(s\right) = \left[\begin{array}{c} \xi_{1i}\left(s\right) \\ : \\ \xi_{\nu i}\left(s\right) \end{array}\right], \ i = 1, 2, ..., \mu, \tag{7.44}$$

The vector functions $\kappa_i\left(.\right) : \mathfrak{T} \longrightarrow \mathfrak{R}^\nu$ and $\xi_i\left(.\right) : \mathbb{C} \longrightarrow \mathbb{C}^\nu$ represent the $i - th$ column of the matrix functions $R\left(.\right) : \mathfrak{T} \longrightarrow \mathfrak{R}^{\nu \times \mu}$ and $C\left(.\right) : \mathbb{C} \longrightarrow \mathbb{C}^{\nu \times \mu}$, respectively, for every $i = 1, 2, .., \mu$. The number ν is the number of the rows of the matrix functions $R\left(.\right)$ and $C\left(.\right)$, and the number μ is the number of their columns.

The matrix functions $R\left(.\right)$ and $C\left(.\right)$ can be represented also in terms of their row vector functions $\rho_k\left(.\right) : \mathfrak{T} \longrightarrow \mathfrak{R}^{1 \times \mu}$ and $\zeta_k\left(.\right) : \mathbb{C} \longrightarrow \mathbb{C}^{1 \times \mu}$, $k = 1, 2, ..., \nu$, respectively, e.g., for $\nu < \mu$:

$$R\left(t\right) = \left[\begin{array}{ccccccc} \rho_{11}\left(t\right) & ... & \rho_{1i}\left(t\right) & ... & \rho_{1\mu}\left(t\right) \\ : & : ... : & : & : ... : & : \\ \rho_{\nu 1}\left(t\right) & ... & \rho_{\nu i}\left(t\right) & ... & \rho_{\nu \mu}\left(t\right) \end{array}\right] = \left[\begin{array}{c} \rho_1\left(t\right) \\ : \\ \rho_\nu\left(t\right) \end{array}\right],$$
$$\rho_k\left(t\right) = \left[\begin{array}{ccccc} \rho_{k1}\left(t\right) & ... & \rho_{ki}\left(t\right) & ... & \rho_{k\mu}\left(t\right) \end{array}\right] \in \mathfrak{R}^{1 \times \mu}, \ k = 1, 2, ..., \nu, \tag{7.45}$$

$$C\left(s\right) = \left[\begin{array}{ccccc} \zeta_{11}\left(s\right) & ... & \zeta_{1i}\left(s\right) & ... & \zeta_{1\mu}\left(s\right) \\ : & : ... : & : & : ... : & : \\ \zeta_{\nu 1}\left(s\right) & ... & \zeta_{\nu i}\left(s\right) & ... & \zeta_{\nu \mu}\left(s\right) \end{array}\right] = \left[\begin{array}{c} \zeta_1\left(s\right) \\ \zeta_2\left(s\right) \\ : \\ \zeta_\nu\left(s\right) \end{array}\right],$$
$$\zeta_k\left(s\right) = \left[\begin{array}{ccccc} \zeta_{k1}\left(s\right) & ... & \zeta_{ki}\left(s\right) & ... & \zeta_{k\mu}\left(s\right) \end{array}\right] \in \mathbb{C}^{1 \times \mu}, \ k = 1, 2, ..., \nu. \tag{7.46}$$

Definition 79 does not specify the linear independence of (column) vector functions $\kappa_i\left(.\right) : \mathfrak{T} \longrightarrow \mathfrak{R}^\nu$ or $\xi_i\left(.\right) : \mathbb{C} \longrightarrow \mathbb{C}^\nu$, $i = 1, 2, ..., \mu$.

Definition 98 *Linear dependence and independence of vector functions and of rows/columns of a matrix function*

A) Real valued row vector functions $\rho_i\left(.\right) : \mathfrak{T} \longrightarrow \mathfrak{R}^{1 \times \mu}$, $i = 1, 2, ..., \nu$, which are rows of a matrix function $R\left(.\right) : \mathfrak{T} \longrightarrow \mathfrak{R}^{\nu \times \mu}$, are:

*A-1) **Linearly dependent on** $\mathfrak{T}_0$ if and only if their exist real numbers γ_i, $i = 1, 2, ..., \nu$, at least one of which is nonzero,*

$$\left|\gamma_1\right| + \left|\gamma_2\right| + ... + \left|\gamma_\nu\right| \neq 0, \tag{7.47}$$

such that the linear combination (7.48),

$$\gamma_1 \rho_1\left(t\right) + \gamma_2 \rho_2\left(t\right) + ... + \gamma_\nu \rho_\nu\left(t\right) \tag{7.48}$$

of the vector functions $\rho_i\left(.\right)$ vanishes at every $t \in \mathfrak{T}_0$,

$$\gamma_1 \rho_1\left(t\right) + \gamma_2 \rho_2\left(t\right) + ... + \gamma_\nu \rho_\nu\left(t\right) = \mathbf{0}_\mu, \ \forall t \in \mathfrak{T}_0. \tag{7.49}$$

*A-2) **Linearly independent on** $\mathfrak{T}_0$ if and only if their linear combination (7.48) vanishes at every $t \in \mathfrak{T}_0$ if and only if all numbers γ_i are equal to zero,*

$$\forall t \in \mathfrak{T}_0: \ \gamma_1 \rho_1\left(t\right) + \gamma_2 \rho_2\left(t\right) + ... + \gamma_\nu \rho_\nu\left(t\right) = \mathbf{0}_\mu \Longleftrightarrow$$
$$\Longleftrightarrow |\gamma_1| + |\gamma_2| + ... + |\gamma_\nu| = 0. \tag{7.50}$$

B) Complex valued row vector functions $\zeta_i\left(.\right): \mathbb{C} \longrightarrow \mathbb{C}^\nu$, $i = 1, 2, ..., \nu$, which are rows of a matrix function $C\left(.\right): \mathbb{C} \longrightarrow \mathbb{C}^{\nu \times \mu}$, are:

*B-1) **Linearly dependent on** $\mathbb{C}$ if and only if there exist complex numbers θ_i, $i = 1, 2, ..., \nu$, at least one of which is nonzero,*

$$|\theta_1| + |\theta_2| + ... + |\theta_\nu| \neq 0, \tag{7.51}$$

such that the linear combination (7.52),

$$\theta_1 \zeta_1\left(s\right) + \theta_2 \zeta_2\left(s\right) + ... + \theta_\nu \zeta_\nu\left(s\right) \tag{7.52}$$

of the vector functions $\zeta_i\left(.\right)$ vanishes at every $s \in \mathbb{C}$,

$$\theta_1 \zeta_1\left(s\right) + \theta_2 \zeta_2\left(s\right) + ... + \theta_\nu \zeta_\nu\left(s\right) = \mathbf{0}_\nu, \ \forall s \in \mathbb{C}. \tag{7.53}$$

*B-2) **Linearly independent on** $\mathbb{C}$ if and only if their linear combination (7.52) vanishes at every $s \in \mathbb{C}$ if and only if all complex numbers θ_i are equal to zero,*

$$\forall s \in \mathbb{C}: \ \theta_1 \zeta_1\left(s\right) + \theta_2 \zeta_2\left(s\right) + ... + \theta_\nu \zeta_\nu\left(s\right) = \mathbf{0}_\nu \Longleftrightarrow$$
$$\Longleftrightarrow |\theta_1| + |\theta_2| + ... + |\theta_\nu| = 0. \tag{7.54}$$

C) Real valued column vector functions $\kappa_i\left(.\right): \mathfrak{T} \longrightarrow \mathfrak{R}^\nu$, $i = 1, 2, ..., \mu$, which are columns of a matrix function $R\left(.\right): \mathfrak{T} \longrightarrow \mathfrak{R}^{\nu \times \mu}$, are:

*C-1) **Linearly dependent on** $\mathfrak{T}_0$ if and only if their exist real numbers α_i, $i = 1, 2, ..., \mu$, at least one of which is nonzero,*

$$|\alpha_1| + |\alpha_2| + ... + |\alpha_\mu| \neq 0, \tag{7.55}$$

such that the linear combination (7.56),

$$\alpha_1 \kappa_1 (t) + \alpha_2 \kappa_2 (t) + ... + \alpha_\mu \kappa_\mu (t) \tag{7.56}$$

of the vector functions $\kappa_i (.)$ *vanishes at every* $t \in \mathfrak{T}_0$,

$$\alpha_1 \kappa_1 (t) + \alpha_2 \kappa_2 (t) + ... + \alpha_\mu \kappa_\mu (t) = \mathbf{0}_\nu, \ \forall t \in \mathfrak{T}_0. \tag{7.57}$$

C-2) **Linearly independent on** $\mathfrak{T}_0$ *if and only if their linear combination (7.56) vanishes at every* $t \in \mathfrak{T}_0$ *if and only if all numbers* α_i *are equal to zero,*

$$\forall t \in \mathfrak{T}_0 : \ \alpha_1 \kappa_1 (t) + \alpha_2 \kappa_2 (t) + ... + \alpha_\mu \kappa_\mu (t) = \mathbf{0}_\nu \Longleftrightarrow$$
$$\Longleftrightarrow |\alpha_1| + |\alpha_2| + ... + |\alpha_\mu| = 0. \tag{7.58}$$

D) Complex valued column vector functions $\xi_i (.) : \mathbb{C} \longrightarrow \mathbb{C}^\nu$, $i = 1, 2, ..., \mu$, *which are columns of a matrix function* $C (.) : \mathbb{C} \longrightarrow \mathbb{C}^{\nu \times \mu}$, *are:*

D-1) **Linearly dependent on** $\mathbb{C}$ *if and only if there exist complex numbers* β_i, $i = 1, 2, ..., \mu$, *at least one of which is nonzero,*

$$|\beta_1| + |\beta_2| + ... + |\beta_\mu| \neq 0, \tag{7.59}$$

such that the linear combination (7.60),

$$\beta_1 \xi_1 (s) + \beta_2 \xi_2 (s) + ... + \beta_\mu \xi_\mu (s) \tag{7.60}$$

vanishes at every $s \in \mathbb{C}$,

$$\beta_1 \xi_1 (s) + \beta_2 \xi_2 (s) + ... + \beta_\mu \xi_\mu (s) = \mathbf{0}_\nu, \ \forall s \in \mathbb{C}. \tag{7.61}$$

D-2) **Linearly independent on** $\mathbb{C}$ *if and only if their linear combination (7.60) vanishes at every* $s \in \mathbb{C}$ *if and only if all complex numbers* β_i *are equal to zero,*

$$\forall s \in \mathbb{C}: \beta_1 \xi_1 (s) + \beta_2 \xi_2 (s) + ... + \beta_\mu \xi_\mu (s) = \mathbf{0}_\nu \Longleftrightarrow$$
$$\Longleftrightarrow |\beta_1| + |\beta_2| + ... + |\beta_\mu| = 0. \tag{7.62}$$

Example 99 illustrates Definition 98.

Example 99 *Let*

$$\kappa_1 (t) = \begin{bmatrix} 6e^t \\ -6e^t \\ 6e^t - 2e^{-2t} \end{bmatrix}, \ \kappa_2 (t) = \begin{bmatrix} e^t \\ -e^t \\ 7e^t + 2e^{-2t} \end{bmatrix}.$$

Their linear combination reads:

$$\alpha_1 \kappa_1(t) + \alpha_2 \kappa_2(t) = \alpha_1 \begin{bmatrix} 6e^t \\ -6e^t \\ 6e^t - 2e^{-2t} \end{bmatrix} + \alpha_2 \begin{bmatrix} e^t \\ -e^t \\ 7e^t + 2e^{-2t} \end{bmatrix}.$$

We test the existence of at least one real number α_i such that the linear combination $\alpha_1 \kappa_1(t) + \alpha_2 \kappa_2(t)$ is equal to the zero vector $\mathbf{0}_3$ at every moment $t \in \mathfrak{T}$ for $|\alpha_1| + |\alpha_2| \neq 0$:

$$\exists \alpha_1, \alpha_2 \in \mathfrak{R}, \quad |\alpha_1| + |\alpha_2| \neq 0 \Longrightarrow$$

$$\alpha_1 \begin{bmatrix} 6e^t \\ -6e^t \\ 6e^t - 2e^{-2t} \end{bmatrix} + \alpha_2 \begin{bmatrix} e^t \\ -e^t \\ 7e^t + 2e^{-2t} \end{bmatrix} = \mathbf{0}_3, \ \forall t \in \mathfrak{T}?$$

If they exist then

$$\begin{bmatrix} 6\alpha_1 e^t + \alpha_2 e^t \\ -6\alpha_1 e^t - \alpha_2 e^t \\ 6\alpha_1 e^t - 2\alpha_1 e^{-2t} + 7\alpha_2 e^t + 2\alpha_2 e^{-2t} \end{bmatrix} = \begin{bmatrix} 0 \\ 0 \\ 0 \end{bmatrix}, \forall t \in \mathfrak{T}$$

holds. However, the first two equations demand

$$\alpha_2 = -6\alpha_1 \ and \ \alpha_2 = -\frac{6e^t - 2e^{-2t}}{7e^t + 2e^{-2t}} = \frac{2 - 6e^{3t}}{2 + 7e^{3t}} = \alpha_2(t).$$

The result is that α_1 and α_2 should be either equal to zero or should be time dependent functions in order for the linear combination of the vector functions $\kappa_1(.)$ and $\kappa_2(.)$ to be equal to the zero vector $\mathbf{0}_3$ at every moment $t \in \mathfrak{T}$. There do not exist real numbers α_1 and α_2 obeying $|\alpha_1| + |\alpha_2| \neq 0$ so that the linear combination of the vector functions $\kappa_2(.)$ and $\kappa_2(.)$ is equal to the zero vector $\mathbf{0}_3$ at every moment $t \in \mathfrak{T}$. In view of A) of Definition 98, (Section 7.4), the vector functions $\kappa_1(.)$ and $\kappa_2(.)$ are linearly independent on $\mathfrak{T}$.

Lemma 100 *Linear dependence of rows/columns of the matrix function $R(.)$ on $\mathfrak{T}_0$ or the matrix function $C(.)$ on $\mathbb{C}$*

1) In order for rows $\rho_i(.)$, $\rho_i(.) : \mathfrak{T} \longrightarrow \mathfrak{T}^{1 \times \mu}$, $i = 1, 2, ..., \nu$, of the matrix function $R(.)$ (7.45) to be linearly dependent on $\mathfrak{T}_0$ it is necessary and sufficient that there exists an $1 \times \nu$ nonzero vector $\mathbf{g} = [g_1 \ g_2 ... g_\nu]$, $(\mathbf{g} \neq \mathbf{0}_\nu^T) \in \mathfrak{R}^{1 \times \nu}$, such that for every $t \in \mathfrak{T}_0$ the product $\mathbf{g}R(t)$ is zero vector:

Linear dependence of the rows $\rho_i(.)$ of $R(.)$ (7.45)

$$on \ \mathfrak{T}_0 \Longleftrightarrow (\exists (\mathbf{g} \neq \mathbf{0}_\nu^T) \in \mathfrak{R}^{1 \times \nu} \Longrightarrow \mathbf{g}R(t) = \mathbf{0}_\mu^T, \ \forall t \in \mathfrak{T}_0). \qquad (7.63)$$

2) *In order for columns* $\kappa_i(.)$, $\kappa_i(.) : \mathfrak{T} \longrightarrow \mathfrak{T}^\nu$, $i = 1, 2, ..., \mu$, *of the matrix function* $R(.)$ *(7.43) to be linearly dependent on* $\mathfrak{T}_0$ *it is necessary and sufficient that there exists an* $\mu \times 1$ *nonzero vector* $\mathbf{a} = [\alpha_1 \ \alpha_2...\alpha_\mu]^T$, $(\mathbf{a} \neq \mathbf{0}_\mu) \in \mathfrak{R}^{\mu \times 1}$, *such that for every* $t \in \mathfrak{T}_0$ *the product* $R(t)\mathbf{a}$ *is zero vector:*

$$\text{Linear dependence of the columns } \kappa_i(.) \text{ of } R(.) \quad (7.45)$$
$$\text{on } \mathfrak{T}_0 \Longleftrightarrow (\exists (\mathbf{a} \neq \mathbf{0}_\mu) \in \mathfrak{R}^{\mu \times 1} \Longrightarrow R(t)\mathbf{a} = \mathbf{0}_\nu, \forall t \in \mathfrak{T}_0). \quad (7.64)$$

3) *If* $C(.) : \mathbb{C} \longrightarrow \mathbb{C}^{\nu \times \mu}$ *then the statements under 1) and 2) analogously hold.*

Proof. 1) The necessity and sufficiency follow directly from A-1) of Definition 98, i.e.,, from Equation 7.49 of the linear dependence of the rows of the matrix function $R(.)$ (7.45) when it is written in the vector-matrix form $\mathbf{g}R(t) = \mathbf{0}_\mu^T$, $\forall t \in \mathfrak{T}_0$.

2) The necessity and sufficiency follow directly from C-1) of Definition 98, i.e.,, from Equation 7.57 of the linear dependence of the columns of the matrix function $R(.)$ (7.45) written in the matrix-vector form $R(t)\mathbf{a} = \mathbf{0}_\nu$, $\forall t \in \mathfrak{T}_0$.

3) The proof of 3) is literally analogous to the proofs of 1) and 2). ∎

Lemma 101 *Linear independence of row vectors of a matrix function on* $\mathfrak{T}_0$ *or on* $\mathbb{C}$

In order for the rows $\rho_i(.)$, $i = 1, 2, ..., \nu$, *of the matrix function* $R(.)$ *(7.45) to be linearly independent on* $\mathfrak{T}_0$ *it is necessary and sufficient that for every* $1 \times \nu$ *nonzero vector* $\mathbf{g} = [\gamma_1 \ \gamma_2...\gamma_\nu]$, $(\mathbf{g} \neq \mathbf{0}_\nu^T) \in \mathfrak{R}^{1 \times \nu}$, *there exists* $\sigma \in \mathfrak{T}_0$ *for which* $\mathbf{g}R(\sigma)$ *is nonzero vector:*

$$\text{Linear independence of the rows } \rho_i(.) \text{ of } R(.) \quad (7.45)$$
$$\text{on } \mathfrak{T}_0 \Longrightarrow \forall (\mathbf{g} \neq \mathbf{0}_\nu^T) \in \mathfrak{R}^{1 \times \nu}, \exists \sigma \in \mathfrak{T}_0 \Longrightarrow \mathbf{g}R(\sigma) \neq \mathbf{0}_\nu^T. \quad (7.65)$$

If $C(.) : \mathbb{C} \longrightarrow \mathbb{C}^{\nu \times \mu}$, *i.e.,, its rows* $\zeta_i(.) : \mathbb{C} \longrightarrow \mathbb{C}^{1 \times \mu}$, $i = 1, 2, ..., \nu$, *then the statement analogously holds.*

Proof. The necessity and sufficiency follow directly from A-2) of Definition 98, i.e.,, from the condition 7.50 of the linear dependence of the rows of the matrix function $R(.)$ (7.45) when it is written in the vector-matrix form $\mathbf{g}R(t) = \mathbf{0}_\mu^T$, $\forall t \in \mathfrak{T}_0$. ∎

Appendix D.1 provides the more detailed proof of Lemma 101.

Lemma 102 *Linear independence of column vectors of a matrix function on $\mathfrak{T}_0$ or on $\mathbb{C}$*

In order for the columns $\kappa_i(.)$, $i = 1, 2, ..., \mu$, of the matrix function $R(.)$ (7.45) to be linearly independent on $\mathfrak{T}_0$ it is necessary and sufficient that for every $\mu \times 1$ nonzero vector

$$\mathbf{a} = [\alpha_1 \ \alpha_2...\alpha_\mu]^T, (\mathbf{a} \neq \mathbf{0}_\mu) \in \mathfrak{R}^{\mu \times 1},$$

there exists $\sigma \in \mathfrak{T}_0$ for which $R(\sigma)\mathbf{a}$ is nonzero vector:

$$Linear\ independence\ of\ the\ columns\ \kappa_i(.)\ of\ R(.)\ (7.45)$$
$$on\ \mathfrak{T}_0 \Longrightarrow \forall (\mathbf{a} \neq \mathbf{0}_\mu) \in \mathfrak{R}^{\mu \times 1},\ \exists \sigma \in \mathfrak{T}_0 \Longrightarrow R(\sigma)\mathbf{a} \neq \mathbf{0}_\mu. \qquad (7.66)$$

If $C(.) : \mathbb{C} \longrightarrow \mathbb{C}^{\nu \times \mu}$, i.e.,, its columns $\xi_i(.) : \mathbb{C} \longrightarrow \mathbb{C}^{\nu \times 1}$, $i = 1, 2, ..., \mu$, then the preceding statement analogously holds.

The proof of this theorem is the full analogy of the proof of Theorem 101.

Note 103 *Theorem 86 and the nonsingular matrix function*

Theorem 86 is not valid on $\mathfrak{T}_0$ despite the nonsingular matrix is a matrix function $M(.) : \mathfrak{T} \longrightarrow \mathfrak{R}^{p \times p}$ or $M(.) : \mathbb{C} \longrightarrow \mathbb{C}^{p \times p}$.

In this regard we present the following:

Example 104 *Let*

$$M(t) = \begin{bmatrix} 1 & e^{-t} \\ 2t & 2 \end{bmatrix}, \ N = \begin{bmatrix} 0 \\ 1 \end{bmatrix}, \ rankN = 1,\ ,\ M(t)N = \begin{bmatrix} e^{-t} \\ 2 \end{bmatrix} \Longrightarrow$$

$$a_1 e^{-t} + a_2 2 = 0 \Longleftrightarrow a_1 = -a_2 2 e^t \neq const. \Longleftrightarrow a_1 = a_2 = 0.$$

The rows of $M(t)N$ are linearly independent on $\mathfrak{T}_0$. Notice that

$$rankM(t)N \equiv 1 < 2\ ,\ \forall t \in \mathfrak{T}_0,$$

in spite of the linear independence of the rows of $M(t)N$ on $\mathfrak{T}_0$.

At any fixed moment $\tau \in \mathfrak{T}_0$,

$$a_1 \neq 0\ and\ a_2 = -2^{-1} a_1 e^{-\tau} \Longrightarrow$$
$$a_1 e^{-\tau} + a_2 2 = a_1 e^{-\tau} + \left(-2^{-1} a_1 e^{-\tau}\right) 2 = a_1 e^{-\tau} - a_1 e^{-\tau} = 0,$$

the rows of $M(t)N$ are linearly dependent. However, the value of the coefficient $a_2 = a_2(\tau)$ is not a number, but it depends on $\tau \in \mathfrak{T}_0$, which implies the linear independence of the rows of $M(t)N$ on $\mathfrak{T}_0$. Notice also that

$$det M(t) = \begin{vmatrix} 1 & e^{-t} \\ 2t & 2 \end{vmatrix} = 2\left(1 - te^{-t}\right) \neq 0, \ \forall t \in \mathfrak{T}.$$

The following theorem explains the difference between the rank and the independence of the rows of a time varying matrices product:

Theorem 105 *Linear independence of the rows and the rank of a time varying matrix product*

Let $M(t) \in \mathbb{C}^{p \times p}$ be nonsingular at every $t \in \mathfrak{T}_0$, $det M(t) \neq 0$, $\forall t \in \mathfrak{T}_0$, and $N \in \mathbb{C}^{p \times q}$, where p and q, $q \geq p$, are natural numbers.

Let k be a natural number not bigger than p, $0 < k \leq p$.

a) In order for the rank of the matrix product $M(t)N$ to be k at every $t \in \mathfrak{T}_0$ it is necessary and sufficient that k rows of the matrix N are linearly independent, equivalently, $rank N = k$.

b) In order for the rank of the matrix product $M(t)N$ to be full, i.e., p, at every $t \in \mathfrak{T}_0$ it is necessary and sufficient that all p rows of the matrix N are linearly independent, equivalently $rank N = p$.

Proof. Let $M(t) \in \mathbb{C}^{p \times p}$ be nonsingular at every $t \in \mathfrak{T}_0$, $det M(t) \neq 0$, $\forall t \in \mathfrak{T}_0$, which implies the linear independence of its rows (and of its columns) and $rank M(t) = p$ at every $t \in \mathfrak{T}_0$. Let $N \in \mathbb{C}^{p \times q}$, where p and q are natural numbers.

Let any $\tau \in \mathfrak{T}_0$ be chosen and fixed. The condition $det M(t) \neq 0$, $\forall t \in \mathfrak{T}_0$, implies $det M(\tau) \neq 0$.

Necessity.

a) Let k be a natural number not bigger than p, $0 < k \leq p$. Let $rank M(t)N = k$ at every $t \in \mathfrak{T}_0$. Let at most $k - 1$ rows of the matrix N be assumed linearly independent. Then $rank N \leq k - 1$. Since $M(t)$ is nonsingular at every $t \in \mathfrak{T}_0$, its $rank M(\tau) = p \neq 0$. The assumption that $rank N \leq k - 1$ and the well-known rule on the rank of the matrix product of a nonsingular matrix $M(\tau)$ multiplying the rectangular matrix N, which reads $rank M(\tau)N = rank N$, yield $rank M(\tau)N = rank N \leq k - 1$. This contradicts $rank M(\tau)N = k$. The contradiction is the result of the assumption on the linear independence of at most $k - 1$ rows of the matrix N. The assumption fails, which implies that k rows of the matrix N are linearly independent, i.e., $rank N = k$.

b) The necessity of the statement under b) results from the necessity of a) for $k = p$.

Sufficiency.

a) Let $rank N = k$. The well-known rule on the rank of the matrix product in which the premultiplying matrix M is nonsingular ensures $rank M(\tau) N = rank N = k$. This implies the linear independence of the k rows of the matrix product $M(\tau) N$.

b) The sufficiency of the statement under b) results from the sufficiency of a) for $k = p$, which completes the proof by recalling an arbitrary choice of $\tau \in \mathfrak{T}_0$. ∎

Let us consider an illustrative example:

Example 106 *Let*

$$M(t) = \begin{bmatrix} 1 & e^{-t} \\ 2t & 2 \end{bmatrix}, \quad N = \begin{bmatrix} 4 & 2 & 3 \\ 8 & 4 & 6 \end{bmatrix} \Longrightarrow rank N = 1 < 2,$$

$$det M(t) = \begin{vmatrix} 1 & e^{-t} \\ 2t & 2 \end{vmatrix} = 2\left(1 - te^{-t}\right) \neq 0, \ \forall t \in \mathfrak{T},$$

$$M(t)N = \begin{bmatrix} 1 & e^{-t} \\ 2t & 2 \end{bmatrix} \begin{bmatrix} 4 & 2 & 3 \\ 8 & 4 & 6 \end{bmatrix} = \begin{bmatrix} 4 + 8e^{-t} & 2 + 4e^{-t} & 3 + 6e^{-t} \\ 8t + 16 & 4t + 8 & 6t + 12 \end{bmatrix} \Longrightarrow$$

1. *The rows of N are linearly dependent, i.e.,, its rank is defective.*
2. *The rows of $M(t)N$ are linearly independent on $\mathfrak{T}$ but*
3. *$rank M(t)N = 1 < 2, \ \forall t \in \mathfrak{T}$.*

What can we say about the linear (in)dependence of rows of a matrix product in which a matrix function $P(.): \mathfrak{T} \longrightarrow \mathfrak{R}^{\nu \times \mu}$ is postmultiplied by a matrix function $Q(.): \mathfrak{T} \longrightarrow \mathfrak{R}^{\mu \times \xi}$?

Lemma 107 *Linear (in)dependence of the matrix rows and a matrix product*

Let the matrix functions

$$P(.): \mathfrak{T} \longrightarrow \mathfrak{R}^{\nu \times \mu} \text{ and } Q(.): \mathfrak{T} \longrightarrow \mathfrak{R}^{\mu \times \xi}$$

form the matrix function product $P(.)Q(.): \mathfrak{T} \longrightarrow \mathfrak{R}^{\nu \times \xi}$.

1) If the rows of the matrix function product $P(.)Q(.): \mathfrak{T} \longrightarrow \mathfrak{R}^{\nu \times \xi}$ are linearly independent on a nonempty connected subset $\mathfrak{T}^*$ of $\mathfrak{T}$, $\mathfrak{T}^* \subseteq \mathfrak{T}$, with the nonempty interior, $In\mathfrak{T}^* \neq \phi$, then the rows of the matrix function $P(.)$ are also linearly independent on the set $\mathfrak{T}^*$.

2) If the rows of the matrix function $P(.)$ are linearly dependent on the set $\mathfrak{T}^*$ then the rows of the matrix function product $P(.)Q(.) : \mathfrak{T} \longrightarrow \mathfrak{R}^{\nu \times \xi}$ are also linearly dependent on the set $\mathfrak{T}^*$.

3) Let $M \in \mathfrak{R}^{\mu \times \mu}$ be nonsingular constant matrix, $\det M \neq 0$. If the rows of the matrix function $P(.)$ are linearly independent on the set $\mathfrak{T}^*$ then the rows of the matrix function product $P(.)M$ are also linearly independent on the set $\mathfrak{T}^*$.

Proof. Let the matrix functions

$$P(.) : \mathfrak{T} \longrightarrow \mathfrak{R}^{\nu \times \mu} \text{ and } Q(.) : \mathfrak{T} \longrightarrow \mathfrak{R}^{\mu \times \xi}$$

form the matrix function product $P(.)Q(.) : \mathfrak{T} \longrightarrow \mathfrak{R}^{\nu \times \xi}$. Let the set $\mathfrak{T}^*$ be a nonempty connected subset of $\mathfrak{T}$, $\mathfrak{T}^* \subseteq \mathfrak{T}$, with the nonempty interior, $In\mathfrak{T}^* \neq \phi$.

1) Let the rows of the matrix function product $P(.)Q(.) : \mathfrak{T} \longrightarrow \mathfrak{R}^{\nu \times \xi}$ be linearly independent on the set $\mathfrak{T}^*$. Let be assumed that the rows of the matrix function $P(.)$ are linearly dependent on the set $\mathfrak{T}^*$. Then there is a vector $\left(\mathbf{a} \neq \mathbf{0}_\nu^T\right) \in \mathfrak{R}^{1 \times \nu}$ such that $\mathbf{a}P(t) = \mathbf{0}_\mu^T$ for every $t \in \mathfrak{T}^*$. The multiplication of the matrix function product $P(t)Q(t)$ by the vector $\mathbf{a}$ results in the following:

$$\mathbf{a}P(t)Q(t) = [\mathbf{a}P(t)]Q(t) = \mathbf{0}_\mu^T Q(t) = \mathbf{0}_\xi^T, \ \forall t \in \mathfrak{T}^*. \qquad (7.67)$$

This, by the definition (Definition 98), i.e.,, due to Lemma 100, means that the rows of the matrix function product $P(t)Q(t)$ are linearly dependent on $\mathfrak{T}^*$, which contradicts to their linear independence. The contradiction is the consequence of the assumption on the linear dependence of the rows of $P(t)$ on $\mathfrak{T}^*$. The failure of the assumption proves that the rows of $P(t)$ are linearly independent on $\mathfrak{T}^*$.

2) The proof of the statement under 1), i.e.,, Equations (7.67) prove the statement under 2).

3) Let $M \in \mathfrak{R}^{\mu \times \mu}$ be nonsingular constant matrix, $\det C \neq 0$, and the rows of the matrix function $P(.)$ be linearly independent on the set $\mathfrak{T}^*$. Let be assumed that the rows of the matrix function product $P(.)M$ are linearly dependent on the set $\mathfrak{T}^*$. There exist $1 \times \nu$ constant non zero vector $\mathbf{a}$, $\left(\mathbf{a} \neq \mathbf{0}_\nu^T\right) \in \mathfrak{R}^{1 \times \nu}$, such that

$$\mathbf{a}P(t)M = \mathbf{0}_\mu^T, \ \forall t \in \mathfrak{T}^*.$$

We multiply this equation on the right by M^{-1}. The result reads:

$$\mathbf{a}P(t) = \mathbf{0}_\mu^T, \ \forall t \in \mathfrak{T}^*.$$

This means that the rows of the matrix $P(t)$ are linearly dependent on the set $\mathfrak{T}^*$, which contradicts their linear independence on the set $\mathfrak{T}^*$. The assumed constant non zero vector $\mathbf{a}$, $(\mathbf{a} \neq \mathbf{0}_\nu^T) \in \mathfrak{R}^{1 \times \nu}$, does not exist. This proves that the rows of the matrix function product $P(.)M$ are also linearly independent on the set $\mathfrak{T}^*$. ∎

Lemma 108 *Linear (in)dependence of the matrix columns and a matrix product*

Let the matrix functions $P(.) : \mathfrak{T} \longrightarrow \mathfrak{R}^{\nu \times \mu}$ and $Q(.) : \mathfrak{T} \longrightarrow \mathfrak{R}^{\mu \times \xi}$ form the matrix function product $P(.)Q(.) : \mathfrak{T} \longrightarrow \mathfrak{R}^{\nu \times \xi}$.

1) If the columns of the matrix function product $P(.)Q(.) : \mathfrak{T} \longrightarrow \mathfrak{R}^{\nu \times \xi}$ are linearly independent on a nonempty connected subset $\mathfrak{T}^*$ of $\mathfrak{T}$, $\mathfrak{T}^* \subseteq \mathfrak{T}$, with the nonempty interior, $In\mathfrak{T}^* \neq \phi$, then the columns of the matrix function $Q(.)$ are also linearly independent on the set $\mathfrak{T}^*$.

2) If the columns of the matrix function $Q(.)$ are linearly dependent on the set $\mathfrak{T}^*$ then the columns of the matrix function product $P(.)Q(.) : \mathfrak{T} \longrightarrow \mathfrak{R}^{\nu \times \xi}$ are also linearly dependent on the set $\mathfrak{T}^*$.

3) Let $D \in \mathfrak{R}^{\nu \times \nu}$ be nonsingular constant matrix, $detD \neq 0$. If the columns of the matrix function $P(.)$ are linearly independent on the set $\mathfrak{T}^*$ then the columns of the matrix function product $DP(.)$ are also linearly independent on the set $\mathfrak{T}^*$.

The proof is essential repetition of the proof of Theorem 107.

Note 109 *Lemma 100 through Lemma 108 hold for complex matrix functions with exactly one change that is the replacement of the argument t by s.*

Definition 110 *Rank of a matrix function at a particular value of its argument*

The rank of a $\nu \times \mu$ matrix function $M(.)$,

$$M(.) \in \left\{ \mathfrak{T} \longrightarrow \mathfrak{R}^{\nu \times \mu}, \mathbb{C} \longrightarrow \mathbb{C}^{\nu \times \mu} \right\},$$

at $\sigma \in \{t, s\}$ is the integer ρ such that the matrix minor $M_k(\sigma)$ is nonzero and every matrix minor $M_k(\sigma)$ of the order k bigger than ρ is zero,

$$rank\ M(\sigma) = \rho,\ \sigma \in \{t, s\} \Longleftrightarrow$$

$$\Longleftrightarrow \left\{ \begin{array}{c} M_k(\sigma) \neq 0,\ \forall k = 1, 2, ..., \rho, \\ M_k(\sigma) = 0,\ \forall k = \rho + 1, \rho + 2, ...,\ \min(\nu, \mu). \end{array} \right\} \quad (7.68)$$

This definition specifies that the rank of a matrix function depends on its independent variable $\sigma \in \{t, s\}$. The consequence is the invalidity of Claim 81 for the matrix functions on any nonsingleton nonempty set $\mathfrak{J} \in \{\mathfrak{T}, \mathbb{C}\}$ because Definition 110 determines the rank of a matrix function at any fixed value of its argument, but not on a nonsingleton nonempty set $\mathfrak{J} \in \{\mathfrak{T}, \mathbb{C}\}$.

Example 111 *Let*

$$R(t) = \frac{1}{3} \begin{bmatrix} 6e^t & e^t \\ -6e^t & -e^t \\ 6e^t - 2e^{-2t} & 7e^t + 2e^{-2t} \end{bmatrix}.$$

$R(t)$ has the second order minor $R_2(t)$,

$$R_2(t) = \frac{1}{3} \begin{vmatrix} -6e^t & -e^t \\ 6e^t - 2e^{-2t} & 7e^t + 2e^{-2t} \end{vmatrix} = -16e^{2t} \neq 0, \ \forall t \in \mathfrak{T}.$$

which is nonsingular. The rank of $R(t)$ is 2 for every $t \in \mathfrak{T}$. Its rank on $\mathfrak{T}$ is equal to 2.

The equivalent form of $R(t)$ follows after multiple application of the elementary transformations,

$$\frac{1}{3} \begin{bmatrix} 6e^t & e^t \\ -6e^t & -e^t \\ 6e^t - 2e^{-2t} & 7e^t + 2e^{-2t} \end{bmatrix} \sim \frac{1}{3} \begin{bmatrix} 0 & 0 \\ 0 & -e^t \\ -6e^{-2t} & 7e^t + 2e^{-2t} \end{bmatrix} \sim$$

$$\sim \frac{1}{3} \begin{bmatrix} 0 & 0 \\ 0 & -e^t \\ -6e^{-2t} & 2e^{-2t} \end{bmatrix} \sim \frac{1}{3} \begin{bmatrix} -e^t & 0 \\ -6e^{-2t} & 2e^{-2t} \\ 0 & 0 \end{bmatrix} \sim \frac{1}{3} \begin{bmatrix} -e^t & 0 \\ 0 & 2e^{-2t} \\ 0 & 0 \end{bmatrix}.$$

and it reads:

$$R(t) \sim \frac{1}{3} \begin{bmatrix} -e^t & 0 \\ 0 & 2e^{-2t} \\ 0 & 0 \end{bmatrix}, \ \forall t \in \mathfrak{T}.$$

This verifies that $R(t)$ has two linearly independent columns on $\mathfrak{T}$ and shows that its rank is 2 on $\mathfrak{T}$. In this example the rank of the matrix function on $\mathfrak{T}$ and the number of its linearly independent columns and rows on $\mathfrak{T}$ are equal (to 2).

Example 112 *Let*

$$sign t = \frac{t}{|t|} \iff t \neq 0, \ sign t = 0 \iff t = 0,$$

and

$$R\left(t\right) = \begin{bmatrix} tsignt & 0 & 0 \\ 0 & sign^2t & 0 \\ 0 & 0 & t \end{bmatrix} \Longrightarrow$$

$$t < 0 \Longrightarrow$$

$$R\left(t\right) = \begin{bmatrix} -t & 0 & 0 \\ 0 & 1 & 0 \\ 0 & 0 & t \end{bmatrix} \Longrightarrow$$

3 *linearly independent rows and columns*, $rankF\left(t\right) = 3,$

$$t = 0 \Longrightarrow$$

$$R\left(0\right) = \begin{bmatrix} 0 & 0 & 0 \\ 0 & 0 & 0 \\ 0 & 0 & 0 \end{bmatrix} \Longrightarrow$$

0 *linearly independent rows and columns*, $rankF\left(0\right) = 0$

$$t > 0 \Longrightarrow$$

$$R\left(t\right) = \begin{bmatrix} t & 0 & 0 \\ 0 & 1 & 0 \\ 0 & 0 & t \end{bmatrix} \Longrightarrow$$

3 *linearly independent rows and columns*, $rankF\left(t\right) = 3.$

This analysis shows that $R\left(t\right)$ has 3 linearly independent rows and columns on $\mathfrak{T}$ in view of Definition 98, (Section 7.4). Its rank varies in terms of $t \in \mathfrak{T}$ so that it is either 0 (at $t = 0$) or 3 (for $t \neq 0$). Its rank on $\mathfrak{T}$ is time-varying, but the number of linearly independent columns and rows on $\mathfrak{T}$ is constant (3) in view of Definition 110, (Section 7.4).

Example 113 *Let*

$$R\left(t\right) = \begin{bmatrix} -2 & e^t & e^{-t} \end{bmatrix} \Longrightarrow t = 0 \Longrightarrow R\left(0\right) = \begin{bmatrix} -2 & 1 & 1 \end{bmatrix}.$$
$$R\left(0\right) \ has\ one\ linearly\ independent\ row,$$
$$one\ linearly\ independent\ column,\ and\ rankR\left(0\right) = 1$$
$$-2a_1 + e^t a_2 + e^{-t} a_3 = 0,\ \forall t \in \mathfrak{T} \Longleftrightarrow a_1 = a_2 = a_3 = 0 \Longrightarrow$$
$$R\left(t\right) \ has\ three\ linearly\ independent\ columns\ on\ \mathfrak{T},$$
$$one\ independent\ row\ on\ \mathfrak{T},\ and\ rankR\left(t\right) = 1,\ \forall t \in \mathfrak{T}.$$

Let a matrix function $G\left(.\right):\mathfrak{T}\longrightarrow\mathfrak{R}^{\nu\times\nu}$ be the definite integral of the matrix function $R\left(.\right):\mathfrak{T}\longrightarrow\mathfrak{R}^{\nu\times\mu}$ (7.45) and of its transpose:

$$G\left(t\right)=\int_{0}^{t}R\left(\tau\right)R^{T}\left(\tau\right)d\tau. \tag{7.69}$$

The matrix $G\left(t\right)$ (7.69) is *the Gram matrix (or, grammian) of the matrix* $R\left(.\right)$ (7.45). It enables us to test the linear independence of the rows $\rho_{i}\left(.\right):\mathfrak{T}\longrightarrow\mathfrak{T}^{1\times\mu}$ of the matrix function $R\left(.\right)$ (7.45).

Theorem 114 *Gram matrix criterion for the linear independence of the rows of the inegrable matrix function $R\left(.\right)$ (7.45)*
For the rows $\rho_{i}\left(.\right):\mathfrak{T}\longrightarrow\mathfrak{T}^{1\times\mu}$, $i=1,\,2,\,\ldots,\,\nu$, of the integrable matrix function $R\left(.\right)$ (7.45), $R\left(t\right)\in\mathfrak{C}\left(\mathfrak{T}_{0}\right)$, to be linearly independent on $\mathfrak{T}_{0}$ it is necessary and sufficient that their Gram matrix $G\left(t\right)\in\mathfrak{R}^{\nu\times\nu}$ (7.69) is nonsingular on $\mathfrak{T}_{0}$.

For the proof see Appendix D.2.

7.5   Polynomial matrices. Matrix polynomials

A matrix function

$$P\left(.\right):\mathbb{C}\longrightarrow\mathbb{C}^{N\times r},\;P\left(s\right)=\left[p_{j,k}\left(s\right)\right]\in\mathbb{C}^{N\times r}, \tag{7.70}$$

is *polynomial matrix if and only if every its entry $p_{j,k}\left(s\right)$ is a polynomial in the complex variable $s\in\mathbb{C}$,*

$$p_{j,k}\left(s\right)=\sum_{i=0}^{i=\mu}p_{j,k}^{i}s^{i},\;p_{j,k}^{i}\in\mathfrak{R},\;\forall j=1,2,..,N,\;\forall k=1,2,..,r. \tag{7.71}$$

The polynomial matrix $P\left(s\right)$ can be set in the form of *the matrix polynomial*

$$P\left(s\right)=\sum_{i=0}^{i=\mu}P_{i}s^{i},\;P_{i}\in\mathfrak{R}^{N\times r},\;\forall i=0,1,...,\mu, \tag{7.72}$$

or in the compact form by applying $P^{\left(\mu\right)}$,

$$P^{\left(\mu\right)}=\left[P_{0}\vdots P_{1}\vdots...\vdots P_{\mu}\right]\in\mathfrak{R}^{N\times\left(\mu+1\right)r}, \tag{7.73}$$

by using $S_r^{(\mu)}(s)$,

$$S_r^{(\mu)}(s) = \left[I_r \;\vdots\; sI_r \;\vdots\; s^2 I_r \;\vdots\; \cdots \;\vdots\; s^\mu I_r \right]^T \in \mathbb{C}^{(\mu+1)r \times r} \implies \qquad (7.74)$$

$$rank S_r^{(\mu)}(s) \equiv r, \qquad (7.75)$$

and the identity matrix $I_r \in \mathfrak{R}^{r \times r}$, so that

$$P(s) = \sum_{i=0}^{i=\mu} P_i s^i = P^{(\mu)} S_r^{(\mu)}(s). \qquad (7.76)$$

The matrix $P^{(\mu)}$ (7.73) is *the generating matrix* of both the matrix polynomial (7.72) and the polynomial matrix $P(s)$ (7.70).

A square polynomial matrix is nonsingular if and only if it has the full rank.

By referring to [34, Definition B.6] and to [84] we will use the following definition:

Definition 115 *Rank of a polynomial matrix*
*   **The normal rank** (for short: **rank**) of the polynomial matrix $P(s)$ is the number of its linearly independent rows (columns).*

*   The rank of a polynomial matrix $P(s)$ (7.70) on $\mathbb{C}$ is the rank of the matrix almost everywhere in $s \in \mathbb{C}$.*

*   The polynomial matrix $P(s)$ has full column /row/ rank on $\mathbb{C}$ if it has full column /row/ rank everywhere in the complex plane $\mathbb{C}$ except at a finite number of points $s \in \mathbb{C}$, respectively.*

*   The full column and the full row rank of the polynomial matrix $P(s)$ on $\mathbb{C}$ are mutually equal and equal to its full rank on $\mathbb{C}$:*

$$full\ column\ rank\ P(s)\ on\ \mathbb{C} = full\ row\ rank\ P(s)\ on\ \mathbb{C} =$$
$$= full\ rank\ P(s)\ on\ \mathbb{C} = \rho = \min(N, r).$$

*   The rank of a polynomial matrix $P(s)$ obeys*

$$rank P(s) = \max\left[rank P(s) : s \in \mathbb{C} \right] \leq \min(N, r).$$

A square polynomial matrix is nonsingular if and only if it has full rank. The expression "on $\mathbb{C}$" will be sometimes omitted in the sequel.

The roots or zeros of a polynomial matrix $P(s)$ are those points $s \in \mathbb{C}$ where $P(s)$ loses rank.

If $P(s)$ is square then its roots are the roots of its determinant $det\ P(s)$, including multiplicity. A square polynomial matrix $P(s)$ is *unimodular* if its determinant $det\ P(s)$ is a nonzero constant. The inverse of a unimodular polynomial matrix $P(s)$ is again a polynomial matrix.

Matrix pencils are matrix polynomials of degree 1, such as $P(s) = P_0 + P_1 s$. Matrix pencils are often represented as polynomial matrices of the special form, e.g., *the characteristic matrix $sI - A$ of the matrix A.*

Elementary operations hold for polynomial matrices.

Let s^0 be any complex number s, $s^0 \in \mathbb{C}$, for which $rankP(s^0) < \min(N, r)$.

Comment 116 *Throughout this book $rankP(s)$ signifies "$rankP(s)$ on $\mathbb{C}$" meaning "for almost every $s \in \mathbb{C}$",*

$$rankP(s) = \{rankP(s), \forall(s \neq s^0) \in \mathbb{C}\} = \{rankP(s)\ on\ \mathbb{C}\}. \quad (7.77)$$

Theorem 117 *Rank of a polynomial matrix*

Let $N \leq r$.

1) In order for the polynomial matrix $P(s)$ (7.70), $P(s) \in \mathbb{C}^{N \times r}$, to have the full rank $\rho = \min(N, r) = N$ it is necessary and sufficient that there is $s^* \in \mathbb{C}$ such that $rankP(s^*) = N$:

$$full\ rankP(s) = N\ on\ \mathbb{C} \Longleftrightarrow \exists s^* \in \mathbb{C},\ rankP(s^*) = N, \quad (7.78)$$

2) If the polynomial matrix $P(s)$ (7.70), $P(s) \in \mathbb{C}^{N \times r}$, has the full rank $\rho = \min(N, r) = N$ then its generating matrix $P^{(\mu)} \in \mathfrak{R}^{N \times (\mu+1)r}$ has also the full rank $\rho = N$,

$$full\ rankP(s) = N \Longrightarrow rankP^{(\mu)} = full\ rankP^{(\mu)} = N. \quad (7.79)$$

Proof. Let $N \leq r$.

1) *Necessity.* Let the polynomial matrix $P(s) \in \mathbb{C}^{N \times r}$ have the full rank $\rho = N$.

There is a $\rho \times \rho$ polynomial submatrix $P_{\rho,\rho}(s)$ of $P(s)$ that is nonsingular on $\mathbb{C}$ almost for all $s \in \mathbb{C}$, i.e.,, for any $s \in \mathbb{C}$ except for finite number of values of $s^0 \in \mathbb{C}$ for which $detP_{\rho,\rho}(s^0) = 0$. There exist infinitely many $(s^* \neq s^0) \in \mathbb{C}$ for which $rankP(s^*) = \rho = N$. This proves the necessity of the condition under 1).

Sufficiency. Let there exist $s^* \in \mathbb{C}$ such that $rankP(s^*) = \rho = N$. The polynomial matrix $P(s)$ has the full rank for $s = s^*$. There is a $\rho \times$

ρ polynomial submatrix $P_{\rho,\rho}(s^*)$ of $P(s^*)$ that is nonsingular. It can be singular, i.e.,, $P(s)$ can have a defective rank, only for a finite number of s-values $s^0 \in \mathbb{C}$. For all other complex numbers s it has the same rank as for s^*, i.e.,, it has the full rank $\rho = N = min(N, r)$ on $\mathbb{C}$. This proves the sufficiency of the condition under 1).

2) Equation (7.76) implies the following:

$$rank P(s) = \rho = N = \min(N, r) = \min(N, (\mu + 1) r) = rank P^{(\mu)} \Longrightarrow$$
$$rank P^{(\mu)} = N.$$

This completes the proof. ∎

The condition 2) of this theorem has the following consequence:

Corollary 118 *1) For the polynomial matrix $P(s)$ (7.70) to have the full rank N it is necessary (but not sufficient) that its generating matrix $P^{(\mu)}$ has the full rank N.*

2) The polynomial matrix $P(s)$ (7.70) cannot have the full rank N if its generating matrix $P^{(\mu)}$ does not have the full rank N:

$$rank P^{(\mu)} < N \Longrightarrow rank P(s) < N \text{ on } \mathbb{C}. \tag{7.80}$$

It is easier to test $rank P^{(\mu)}$ than $rank P(s)$. If $P^{(\mu)}$ has a defective rank, i.e.,, the rank less than N, then there is not a need to test whether $P(s)$ has the full rank.

Example 119 *Let a 2×3 polynomial matrix $P(s)$ be*

$$P(s) = \begin{bmatrix} 2s^2 + 3s & 4s - 1 & s^3 \\ 0 & 4s^3 + s^2 + 3 & 6s - 6 \end{bmatrix} \in \mathbb{C}^{2 \times 3}.$$

It has the full rank 2, $rank P(s) = 2 = $ full $rank P(s)$. It induces the following matrix polynomial

$$P(s) = \begin{bmatrix} 2s^2 + 3s & 4s - 1 & s^3 \\ 0 & 4s^3 + s^2 + 3 & 6s - 6 \end{bmatrix}$$

$$= \begin{bmatrix} 0 & -1 & 0 \\ 0 & 3 & -6 \end{bmatrix} + \begin{bmatrix} 3 & 4 & 0 \\ 0 & 0 & 6 \end{bmatrix} s + \begin{bmatrix} 2 & 0 & 0 \\ 0 & 1 & 0 \end{bmatrix} s^2 + \begin{bmatrix} 0 & 0 & 1 \\ 0 & 4 & 0 \end{bmatrix} s^3 =$$

$$= \begin{bmatrix} 0 & -1 & 0 & 3 & 4 & 0 & 2 & 0 & 0 & 0 & 0 & 1 \\ 0 & 3 & -6 & 0 & 0 & 6 & 0 & 1 & 0 & 0 & 4 & 0 \end{bmatrix} S_3^{(12)}(s) = P^{(12)} S_3^{(12)}(s).$$

The generic matrix $P^{(12)}$ of the matrix polynomial is found to read

$$P^{(12)} = \begin{bmatrix} 0 & -1 & 0 & 3 & 4 & 0 & 2 & 0 & 0 & 0 & 0 & 1 \\ 0 & 3 & -6 & 0 & 0 & 6 & 0 & 1 & 0 & 0 & 4 & 0 \end{bmatrix} \Longrightarrow$$
$$rank P^{(12)} = 2 = full\ rank P^{(12)}.$$

This illustrates the statement under 1) of Corollary 118 that the full rank of the polynomial matrix $P(s)$ implies the full rank of the generic matrix $P^{(12)}$ of the matrix polynomial $P(s)$. Equivalently, the full rank of $P^{(12)}$ is necessary for $P(s)$ to have the full rank.

Example 120 Let another 2×3 polynomial matrix $P(s)$ be

$$P(s) = \begin{bmatrix} 2s^2 + 3s & 4s + 6 & 0 \\ 0.5s & 1 & 0 \end{bmatrix} \Longrightarrow$$
$$rank P(s) = 1 < 2 = full\ rank\ of\ P(s).$$

The matrix polynomial form of the polynomial matrix $P(s)$ reads:

$$P(s) = \begin{bmatrix} 0 & 6 & 0 \\ 0 & 1 & 0 \end{bmatrix} + \begin{bmatrix} 3 & 4 & 0 \\ 0.5 & 0 & 0 \end{bmatrix} s + \begin{bmatrix} 2 & 0 & 0 \\ 0 & 0 & 0 \end{bmatrix} s^2 =$$
$$= \begin{bmatrix} 0 & 6 & 0 & 3 & 4 & 0 & 2 & 0 & 0 \\ 0 & 1 & 0 & 0.5 & 0 & 0 & 0 & 0 & 0 \end{bmatrix} S_3^{(9)}(s).$$

Its generic matrix $P^{(9)}$ follows:

$$P^{(9)} = \begin{bmatrix} 0 & 6 & 0 & 3 & 4 & 0 & 2 & 0 & 0 \\ 0 & 1 & 0 & 0.5 & 0 & 0 & 0 & 0 & 0 \end{bmatrix} \Longrightarrow$$
$$rank P^{(9)} = 2 = full\ rank P^{(9)}.$$

This illustrates that the full rank of $P^{(9)}$ does not imply the full rank of $P(s)$.

Problem 121 *Unsolved matrix polynomial problem*
      Under what conditions the full rank N of the generic matrix $P^{(\mu)}$ of the matrix polynomial $P(s) = P^{(\mu)} S_r^{(\mu)}(s)$, Equation (7.76), implies the full rank N of the matrix polynomial $P(s)$, Equations (7.70) and (7.71), i.e., of the polynomial matrix $P(s) = P^{(\mu)} S_r^{(\mu)}(s)$, Equation (7.76)?

7.6 Rational matrices

A matrix function

$$R\left(.\right):\mathbb{C}\longrightarrow\mathbb{C}^{m\times n},\,R\left(s\right)=\left[f_{j,k}\left(s\right)\right],\tag{7.81}$$

is *rational matrix (function)* if and only if every its entry $r_{j,k}\left(s\right)$ is a rational function,

$$r_{j,k}\left(s\right)=\frac{\displaystyle\sum_{i=0}^{i=\mu_{j,k}}a_{j,k}^{i}s^{i}}{\displaystyle\sum_{i=0}^{i=\eta_{j,k}}b_{j,k}^{i}s^{i}},\quad a_{j,k}^{i}\in\mathfrak{R},\;\forall j=1,2,..,m,\;\forall k=1,2,..,n$$

$$b_{j,k}^{i}\in\mathfrak{R},\;\forall j=1,2,..,m,\;\forall k=1,2,..,n.$$

Definition 122 *Rank of a rational matrix*

The rank of a rational matrix $R\left(s\right)$ (7.81) on $\mathbb{C}$ is the rank of the matrix almost everywhere in $s\in\mathbb{C}$.

The rational matrix $R\left(s\right)$ has the full column/row/ rank on $\mathbb{C}$ if it has the full column/row/rank everywhere in the complex plane $\mathbb{C}$ except at a finite number of points $s\in\mathbb{C}$, respectively.

The full column and the full row rank of the rational matrix $R\left(s\right)$ on $\mathbb{C}$ are mutually equal and equal to its full rank on $\mathbb{C}$:

full column rank $R\left(s\right)$ on $\mathbb{C}=$ full row rank $R\left(s\right)$ on $\mathbb{C}=$
$=$ full rank $R\left(s\right)$ on $\mathbb{C}=\min\left(m,n\right).$

The rank of the rational matrix $R\left(s\right)$ obeys

$$rankR\left(s\right)=\max\left[rankR\left(s\right):s\in\mathbb{C}\right]\leq\min\left(m,n\right).$$

A square rational matrix is nonsingular if and only if it has the full rank.

The expression "on $\mathbb{C}$" will be sometimes omitted in the sequel. Let

$$M\left(s\right)\;\in\;\mathbb{C}^{m\times m},\;detM\left(s\right)\neq0\;on\;\mathbb{C},\tag{7.82}$$
$$N\left(s\right)\;\in\;\mathbb{C}^{n\times n},\;detN\left(s\right)\neq0\;on\;\mathbb{C}.\tag{7.83}$$

be rational matrices and

$$R\left(s\right)=M\left(s\right)P\left(s\right)N\left(s\right),\tag{7.84}$$

where $P\left(s\right)$ is the matrix polynomial, Equation (7.70).

Definition 98 and Definition 122 imply directly the following:

Theorem 123 *Linear independence of rows (columns) of a rational matrix and its rank*

If a rational matrix $R(s)$ has the full row (column) rank then it has the maximal number of linearly independent rows (columns) equal to its full row (column) rank, respectively, but the inverse statement does not hold.

This theorem is important for the relationship between trackability and output controllability.

The following well-known result is also important for trackability conditions.

Theorem 124 *Rank of the matrix product (7.82)-(7.84)*

The rank of the rational matrix $R(s)$ (7.82)-(7.84) is equal to the rank of the polynomial matrix $P(s)$ (7.70),

$$rankR(s) = rankM(s)P(s)N(s) = rankP(s) \leq \min(m,n). \qquad (7.85)$$

Proof. Rank of every nonsingular matrix does not influence the rank of the product of that matrix and another matrix, i.e.,, the rank of the matrix product is equal to the rank of that another matrix. In the case of the rational matrix $R(s)$ (7.84) that rule yields the following:

$$rankR(s) = rankM(s)P(s)N(s) = rankP(s)$$

due to the nonsingularity of the square matrices $M(s)$ (7.82) and $N(s)$ (7.83). ∎

The resolvent matrix $(sI - A)^{-1}$ of a square matrix $A \in \mathfrak{R}^{n \times n}$ is rational matrix,

$$(sI - A)^{-1} = \frac{adj(sI - A)}{det(sI - A)}, \ \ det[adj(sI - A)] = [det(sI - A)]^{n-1}, \quad (7.86)$$

$$det\left[(sI - A)^{-1}\right] = [det(sI - A)]^{-1} \qquad (7.87)$$

where *the adjoint matrix* $adj(sI - A)$ of the matrix A is matrix polynomial, and the determinant $det(sI - A)$ is the characteristic polynomial of the matrix A. Equations (7.86), (7.87) imply the following:

$$rank(sI - A)^{-1} = rank \ adj(sI - A). \qquad (7.88)$$

Let $C \in \mathfrak{R}^{N \times n}$, $A \in \mathfrak{R}^{n \times n}$, and $B \in \mathfrak{R}^{n \times r}$ and $U \in \mathfrak{R}^{N \times r}$. They define the rational matrix $R(s)$ by

$$R(s) = \left[C(sI - A)^{-1}B + U\right] \in \mathbb{C}^{N \times r}. \qquad (7.89)$$

Let

$$p(s) = \det(sI - A),\qquad(7.90)$$

and

$$L(s) = Cadj(sI - A)B + p(s)U,\ \ L(s) \in \mathbb{C}^{N\times r}.\qquad(7.91)$$

$L(s)$ is polynomial matrix,

$$L(s) =$$

$$= \begin{cases} \displaystyle\sum_{i=0}^{i=n} L_i s^i = L^{(n)} S_n^{(n)}(s),\ L_i \in \mathfrak{R}^{N\times r}, \\[2mm] \forall i = 0,1,...,n,\ \iff\ U \neq O_{N,r} \\[2mm] \displaystyle\sum_{i=0}^{i=n-1} L_i s^i = L^{(n-1)} S_n^{(n-1)}(s),\ L_i \in \mathfrak{R}^{N\times r}, \\[2mm] \forall i = 0,1,...,n-1, \iff\ U = O_{N,r}, \end{cases}\qquad(7.92)$$

Equations (7.89)-(7.92) simplify the definition of $R(s)$:

$$R(s) = \frac{L(s)}{p(s)} = p^{-1}(s)L(s).\qquad(7.93)$$

The following theorem results directly form Theorem 117:

Theorem 125 *Rank of a rational matrix*
   The rank of the rational matrix $R(s)$ (7.89) is the rank of its numerator polynomial matrix $L(s)$ (7.91),

$$rankF(s) = rank\left[C(sI-A)^{-1}B + U\right] =$$
$$= rank\left[p^{-1}(s)L(s)\right] = rankL(s) \leq \min(N,r).\qquad(7.94)$$

Proof. Let Equation (7.89) hold. It can be set into the form (7.93) determined by Equations (7.89)-(7.92):

$$R(s) = \frac{L(s)}{p(s)},\ \ L(s) = Cadj(sI - A)B + p(s)U.$$

The scalar polynomial $p(s)$ does not influence the rank of $p^{-1}(s)L(s)$. This fact and the preceding equations, i.e.,, Equations (7.93), furnish

$$rankF(s) = rankL(s),$$

which proves (7.94). ■

Notice that Theorem 125 is a special case of Theorem 124, which is characterized with

$$M\left(s\right) = p^{-1}\left(s\right)I_m, \ detM\left(s\right) = p^{-m}\left(s\right) \neq 0, \ and \ N\left(s\right) = I_n,$$
$$rankP\left(s\right) = rankL\left(s\right).$$

Example 126 *Let*

$$R\left(s\right) = C\left(sI_2 - A\right)^{-1}B + U = p^{-1}(s)L(s) \in C^{1 \times 3}$$

for

$$A = \begin{bmatrix} 0 & 1 \\ -1 & -2 \end{bmatrix}, \ C = \begin{bmatrix} 1 \vdots 2 \end{bmatrix}, \ B = \begin{bmatrix} 1 & 2 & 0 \\ 2 & 1 & 2 \end{bmatrix}, \ U = \begin{bmatrix} 1 & 0 & 3 \end{bmatrix}.$$

The matrix A is stable matrix with two eigenvalues, both equal to −1. The matrix A does not have an eigenvalue at the origin.
    The above data yield:

$$sI_2 - A = \begin{bmatrix} s & -1 \\ 1 & s+2 \end{bmatrix} \Longrightarrow \left(sI_2 - A\right)^{-1} = \frac{\begin{bmatrix} s+2 & 1 \\ -1 & s \end{bmatrix}}{s^2 + 2s + 1} \Longrightarrow$$

$$p(s) = s^2 + 2s + 1 = (s+1)^2, \ adj\left(sI_2 - A\right)^{-1} = \begin{bmatrix} s+2 & 1 \\ -1 & s \end{bmatrix},$$

so that

$$R\left(s\right) = C\left(sI_2 - A\right)^{-1}B + U =$$

$$= \begin{bmatrix} 1 \vdots 2 \end{bmatrix} \frac{\begin{bmatrix} s+2 & 1 \\ -1 & s \end{bmatrix}}{s^2 + 2s + 1} \begin{bmatrix} 1 & 2 & 0 \\ 2 & 1 & 2 \end{bmatrix} + \begin{bmatrix} 1 & 0 & 3 \end{bmatrix} =$$

$$= \frac{\begin{bmatrix} s^2 + 11s + 1 \vdots 4s + 1 \vdots 3s^2 + 10s + 5 \end{bmatrix}}{s^2 + 2s + 1} = p^{-1}(s)L(s) \Longrightarrow$$

$$L(s) = \begin{bmatrix} 1 + 11s + s^2 & 1 + 4s & 5 + 10s + 3s^2 \end{bmatrix} \in \mathbb{C}^{1 \times 3} \Longrightarrow$$
$$L(s) = \underbrace{\begin{bmatrix} 1 & 1 & 5 \end{bmatrix}}_{L_0} + \underbrace{\begin{bmatrix} 11 & 4 & 10 \end{bmatrix}}_{L_1} s + \underbrace{\begin{bmatrix} 1 & 0 & 3 \end{bmatrix}}_{L_2} s^2 =$$
$$= L^{(2)}S_3^{(2)}(s) \Longrightarrow$$
$$rankR(s) = rankL(s) = 1,$$

$$S_3^{(2)}(s) = \begin{bmatrix} 1 & 0 & 0 & s & 0 & 0 & s^2 & 0 & 0 \\ 0 & 1 & 0 & 0 & s & 0 & 0 & s^2 & 0 \\ 0 & 0 & 1 & 0 & 0 & s & 0 & 0 & s^2 \end{bmatrix}^T \in \mathbb{C}^{9\times 3}, \ rankS_3^{(2)}(s) = 3,$$

$$L^{(2)} = \begin{bmatrix} L_0 \ \vdots \ L_1 \ \vdots \ L_2 \end{bmatrix} \in \mathfrak{R}^{1\times 9},$$

$$L^{(2)} = \begin{bmatrix} 1 & 1 & 5 & 11 & 4 & 10 & 1 & 0 & 3 \end{bmatrix} \Longrightarrow$$

$$rankL^{(2)} = 1 = rankL(s) = rankR(s).$$

This illustrates Theorem 117.

Notice that in this example, $s^* = 0$ is not an eigenvalue of the matrix A, and we test $rankR(0)$:

$$rankR(0) = rank\begin{bmatrix} -CA^{-1}B + U \end{bmatrix} =$$

$$= rankL(0) = rankL_0 = rank\begin{bmatrix} 1 & 1 & 5 \end{bmatrix} = 1 = full \ rankF(s),$$

$$and \ rankL^{(2)} = full \ rankL^{(2)} = 1.$$

Example 127 *Let*

$$R(s) = C(sI_3 - A)^{-1}B + U = p^{-1}(s)L(s) \in C^{2\times 12}$$

for

$$A = \begin{bmatrix} 0 & 1 & 0 \\ 0 & 0 & 1 \\ -3 & -7 & -5 \end{bmatrix}, \ C = \begin{bmatrix} 1 & 0 & 2 \\ 3 & 1 & 2 \end{bmatrix},$$

$$B = \begin{bmatrix} 1 & -1 & 2 \\ 3 & 2 & -2 \\ 1 & 3 & 1 \end{bmatrix}, \ U = \begin{bmatrix} 1 & 2 & 3 \\ 3 & 2 & 1 \end{bmatrix}.$$

The matrix A is stable matrix, i.e.,, Hurwitz matrix,

$$\lambda_1(A) = \lambda_2(A) = -1 < 0, \ \lambda_3(A) = -3 < 0.$$

The matrices A, C, B and U imply:

$$sI_3 - A = \begin{bmatrix} s & -1 & 0 \\ 0 & s & -1 \\ 3 & 7 & s+5 \end{bmatrix},$$

$$(sI_3 - A)^{-1} = \frac{\begin{bmatrix} s^2 + 5s + 7 & s + 5 & 1 \\ -3 & s^2 + 5s & s \\ -3s & -7s - 3 & s^2 \end{bmatrix}}{s^3 + 5s^2 + 7s + 3} \Longrightarrow$$

$$p(s) = s^3 + 5s^2 + 7s + 3 = (s+1)^2(s+3),$$

$$R(s) = C(sI_3 - A)^{-1} B + U =$$

$$= \frac{\begin{bmatrix} 1 & 0 & 2 \\ 3 & 1 & 2 \end{bmatrix} \begin{bmatrix} s^2 + 5s + 7 & s + 5 & 1 \\ -3 & s^2 + 5s & s \\ -3s & -7s - 3 & s^2 \end{bmatrix}}{s^3 + 5s^2 + 7s + 3} + \begin{bmatrix} 1 & 2 & 3 \\ 3 & 2 & 1 \end{bmatrix},$$

so that

$$R(s) = C(sI_3 - A)^{-1} B + U =$$

$$= \begin{bmatrix} \frac{s^3+6s^2+6s+10}{s^3+5s^2+7s+3} & \frac{2s^3+10s^2+s+5}{s^3+5s^2+7s+3} & \frac{3s^3+17s^2+21s+10}{s^3+5s^2+7s+3} \\ \frac{3s^3+18s^2+30s+27}{s^3+5s^2+7s+3} & \frac{2s^3+11s^2+8s+15}{s^3+5s^2+7s+3} & \frac{s^3+7s^2+8s+6}{s^3+5s^2+7s+3} \end{bmatrix} =$$

$$= \frac{\begin{bmatrix} \begin{pmatrix} s^3 + 6s^2 + \\ +6s + 10 \end{pmatrix} & \begin{pmatrix} 2s^3 + 10s^2 + \\ +s + 5 \end{pmatrix} & \begin{pmatrix} 3s^3 + 17s^2 + \\ +21s + 10 \end{pmatrix} \\ \begin{pmatrix} 3s^3 + 18s^2 + \\ +30s + 27 \end{pmatrix} & \begin{pmatrix} 2s^3 + 11s^2 + \\ +8s + 15 \end{pmatrix} & \begin{pmatrix} s^3 + 7s^2 + \\ +8s + 6 \end{pmatrix} \end{bmatrix}}{s^3 + 5s^2 + 7s + 3}$$

has the full rank 2:

$$rankR(s) = 2.$$

This leads to the following results:

$$L(s) =$$

$$= \begin{bmatrix} \begin{pmatrix} s^3 + 6s^2 + \\ +6s + 10 \end{pmatrix} & \begin{pmatrix} 2s^3 + 10s^2 + \\ +s + 5 \end{pmatrix} & \begin{pmatrix} 3s^3 + 17s^2 + \\ +21s + 10 \end{pmatrix} \\ \begin{pmatrix} 3s^3 + 18s^2 + \\ +30s + 27 \end{pmatrix} & \begin{pmatrix} 2s^3 + 11s^2 + \\ +8s + 15 \end{pmatrix} & \begin{pmatrix} s^3 + 7s^2 + \\ +8s + 6 \end{pmatrix} \end{bmatrix},$$

$$L(s) \in \mathbb{C}^{2 \times 3}.$$

Let us verify the nonsingularity of the minor $M_{2,2}(s)$ of $L(s)$:

$$M_{2,2}(s) = \begin{vmatrix} s^3 + 6s^2 + 6s + 10 & 2s^3 + 10s^2 + s + 5 \\ 3s^3 + 18s^2 + 30s + 27 & 2s^3 + 11s^2 + 8s + 15 \end{vmatrix} =$$

$$= -4s^6 - 43s^5 - 157s^4 - 238s^3 - 142s^2 - 7s + 15 \neq 0,$$

$$\forall s \notin \{0.25, \ -1.0, \ -3.0, \ -5.0\} \Longrightarrow$$

$$rank L(s) = 2 = \rho = \min(2,3) \Longrightarrow rank R(s) = 2.$$

This illustrates Theorem 117.

We accept $s = 0$ because it is not eigenvalue of the matrix A in order to check whether $s = 0$ can be adopted for s^* of the statement under 1) of Theorem 117. Hence,

$$L(0) = \begin{bmatrix} 10 & 5 & 10 \\ 27 & 15 & 6 \end{bmatrix} \Longrightarrow rank L(0) = 2 = \min(2,3) = \rho.$$

$L(0)$ has the full rank $\rho = 2$, which permits accept $s^* = 0$. Verification,

$$rank R(0) = rank\left(U - GA^{-1}B\right) = p^{-1}(0) rank L(0) =$$

$$= rank \frac{\begin{bmatrix} 10 & 5 & 10 \\ 27 & 15 & 6 \end{bmatrix}}{3} = 2 = rank R(s).$$

This example illustratively verifies Theorem 117 and Theorem 125.

Chapter 8

Observability and stability

8.1 Observability and system regime

Kalman introduced the concept of the system observability [59], [61]. He introduced it in the framework of the *time*-invariant linear systems. It has become one of the fundamental dynamical systems and control concepts [2], [13], [15], [18], [48]-[52], [67], [79], [82], [90].

System observability reflects the system property to permit the determination of its initial state $\mathbf{S}_0$, i.e., of its state $\mathbf{S}(t)$ at the initial moment $t = t_0 = 0$, from the system response $\mathbf{Y}(t; \mathbf{S}_0)$.

The relationship between the system response and the initial state is independent of the input actions on the system over the *time* interval $\mathfrak{T}_0$. This is due to the fact that the initial state is independent of the input action that starts at the initial moment. This enables us to consider the system in the free regime, i.e., for the zero input vector. This is fully meaningful if we use the system mathematical model in terms of the deviations. The zero input deviation means that the total input vector is nominal if we use the system mathematical model in terms of the total coordinates. This assumes that we know also the total initial nominal state vector $\mathbf{S}_{N0}$. Once we determine the initial state deviation vector $\mathbf{s}_0 = \mathbf{S}_0 - \mathbf{S}_{N0}$ from the system response in the free regime, we easily determine the total initial vector from $\mathbf{S}_0 = \mathbf{S}_{N0} + \mathbf{s}_0$ and vice versa.

Let $\mathbf{S} \in \mathfrak{R}^n$, $A \in \mathfrak{R}^{n \times n}$, $P^{(\mu)} \in \mathfrak{R}^{n \times (\mu+1)M}$, $\mathbf{Y} \in \mathfrak{R}^N$, $C \in \mathfrak{R}^{N \times n}$, and $\mathbf{I}^\mu \in \mathfrak{R}^{(\mu+1)M}$ be an input vector of the system described in terms of the total coordinates in the *ISO* form (3.1), (3.2) by (8.1) and (8.2):

$$\frac{d\mathbf{S}(t)}{dt} = A\mathbf{S}(t) + P^{(\mu)}\mathbf{I}^{\mu}(t), \tag{8.1}$$

$$\mathbf{Y}(t) = C\mathbf{S}(t) + Q\mathbf{I}(t), \tag{8.2}$$

The fundamental matrix function $\Phi\left(.,t_0\right) : \mathfrak{T} \longrightarrow \mathfrak{R}^{n \times n}$ of the system (8.1), (8.2) is determined in (3.3) by

$$\Phi\left(t\right) = e^{At} \in \mathfrak{R}^{n \times n}, \forall t \in \mathfrak{T}, \tag{8.3}$$

and has the following property:

$$det\Phi\left(t\right) \neq 0, \ \forall t \in \mathfrak{T}, \ \Phi\left(0\right) = e^{A0} = I. \tag{8.4}$$

The system motion results from (8.1) after its integration:

$$\mathbf{S}(t; \mathbf{S}_0; \mathbf{I}^{\mu}) = e^{At}\mathbf{S}_0 + \int_0^t e^{A(t-\tau)} P^{(\mu)}\mathbf{I}^{\mu}(\tau)d\tau, \ \mathbf{S}_0 = \mathbf{S}(0). \tag{8.5}$$

This and (8.2) determine the system response:

$$\mathbf{Y}(t; \mathbf{S}_0; \mathbf{I}^{\mu}) = Ce^{At}\mathbf{S}_0 + C\int_0^t e^{A(t-\tau)} P^{(\mu)}\mathbf{I}^{\mu}(\tau)d\tau + Q\mathbf{I}(t). \tag{8.6}$$

Let us recall Conditions 38, 47, 60, 67, and 77. They are accepted and summarized as:

Claim 128 *The nominal data are known*
   For any given system desired total output $\mathbf{Y}_d(t)$ *the nominal state vector* $\mathbf{S}_N\left(t\right)$ *and the nominal input vector* $\mathbf{I}_N(t)$ *are known at every* $t \in \mathfrak{T}_0$.

This assumption is valid in all what follows in the book.

8.2   Observability definition in general

We present the observability definitions in general valid for all types of the systems studied herein.

Definition 129 *Observability of the linear time-invariant continuous-time dynamical system under the action of the known input vector*

An initial state $\mathbf{S}_0 \in \mathfrak{R}^n$ at $t_0 = 0$ of a linear time-invariant continuous-time dynamical system is observable if and only if the system output response $\mathcal{Y}(.; \mathbf{S}_0; \mathbf{I}^\mu)$ on $\mathfrak{T}_0$ to the known extended input vector $\mathbf{I}^\mu(t)$ on $\mathfrak{T}_0$ uniquely determines $\mathbf{S}_0$.

The system is observable if and only if every its state is observable.

Since the definition of the system observability does not care about the input action then it permits us to accept that $\mathbf{I}(t)$ is known at every $t \in \mathfrak{T}_0$, which we do. Then the left hand sides of the following equalities obtained from (8.6):

$$\widetilde{\mathbf{Y}}(t; \mathbf{S}_0; \mathbf{I}^\mu) = \mathbf{Y}(t; \mathbf{S}_0; \mathbf{I}^\mu) - C \int_0^t e^{A(t-\tau)} P^{(\mu)} \mathbf{I}^\mu(\tau) d\tau - Q\mathbf{I}(t) = Ce^{At}\mathbf{S}_0,$$
$$(8.7)$$

is also known as soon as the response $\mathbf{Y}(t; \mathbf{S}_0; \mathbf{I}^\mu)$ is known for a given $\mathbf{I}(t)$, i.e., $\widetilde{\mathbf{Y}}(t; \mathbf{S}_0; \mathbf{I}^\mu)$ is known. Its relationship with $\mathbf{S}_0$ is defined exclusively by Ce^{At} independently of $\mathbf{I}(t)$. The sense of the left hand side of Equation (8.7) is that the system complete response is decreased by that part of the response resulted from the input action. Their difference expresses the part of the response resulting from the initial state influence on the response. It is shown by the right hand side of Equation (8.7). If the input $\mathbf{I}(t)$ is known at every $t \in \mathfrak{T}_0$ then all what holds for $\mathbf{Y}(t; \mathbf{S}_0; \mathbf{0}_{(\mu+1)M})$ holds also for $\widetilde{\mathbf{Y}}(t; \mathbf{S}_0; \mathbf{I}^\mu)$.

We accepted the validity of Claim 128 that the nominal vector functions $\mathbf{S}_N(t)$ and $\mathbf{I}_N(t)$ are determined for a given nominal $\mathbf{Y}_d(t)$. We can select $\mathbf{I}(t) \equiv \mathbf{I}_N(t)$. Then the knowledge of the deviations $\mathbf{s}$, $\mathbf{i}$ and $\mathbf{y}$,

$$\mathbf{s} = \mathbf{S} - \mathbf{S}_N, \quad \mathbf{i} = \mathbf{I} - \mathbf{I}_N, \quad \mathbf{y} = \mathbf{Y} - \mathbf{Y}_d, \qquad (8.8)$$

determines the total values $\mathbf{S}$, $\mathbf{I}$ and $\mathbf{Y}$.

The system (8.1), (8.2) in the nominal regime takes the following form:

$$\frac{d\mathbf{S}_N(t)}{dt} = A\mathbf{S}_N(t) + P^{(\mu)}\mathbf{I}_N^\mu(t), \qquad (8.9)$$

$$\mathbf{Y}_d(t) = C\mathbf{S}_N(t) + Q\mathbf{I}_N(t), \qquad (8.10)$$

In terms of the deviations $\mathbf{s}$, $\mathbf{v}$ and $\mathbf{y}$ Equations (8.1), (8.2), (8.5), and (8.6) become, respectively, for $\mathbf{I}(t) \equiv \mathbf{I}_N(t)$ that implies $\mathbf{i}(t) \equiv \mathbf{0}_M$:

$$\frac{d\mathbf{s}(t)}{dt} = A\mathbf{s}(t), \qquad (8.11)$$

$$\mathbf{y}(t) = C\mathbf{s}(t), \qquad (8.12)$$

$$\mathbf{s}(t; \mathbf{s}_0; \mathbf{0}_{(\mu+1)M}) = \Phi(t)\mathbf{s}_0 = e^{At}\mathbf{s}_0, \ \mathbf{s}_0 = \mathbf{s}(0), \qquad (8.13)$$

$$\mathbf{y}(t; \mathbf{s}_0; \mathbf{0}_{(\mu+1)M}) = C\Phi(t)\mathbf{s}_0 = Ce^{At}\mathbf{s}_0. \qquad (8.14)$$

The comparison of (8.11) and (8.12) with (8.1) and (8.2), or (8.13), (8.14) with (8.5), (8.6), respectively, verifies that they have the same matrices, they are of the same order, structure and form so that they have the same qualitative dynamical properties in general, and the same observability property in particular. Only the matrices A and C determine the relationship between the system response and the initial state.

This analysis justifies to consider the system in terms of the deviations in the free regime if we analyze the system observability.

With this in mind we will mainly study the system observability by using the system model in the free regime and expressed in terms of the deviations.

Definition 130 *Observability of the linear time-invariant continuous-time dynamical system*

An initial state $\mathbf{s}_0 \in \mathfrak{R}^n$ at $t_0 = 0$ of a linear time-invariant continuous-time dynamical system is observable if and only if the system output response $\mathcal{Y}(.; \mathbf{s}_0; \mathbf{0}_{(\mu+1)M})$ in the free regime on $\mathfrak{T}_0$ uniquely determines the initial state vector $\mathbf{s}_0$.

The system is observable if and only if every its state is observable.

8.3   Observability criterion in general

The general observability criterion reads:

Theorem 131 *Observability criterion in general*

For the linear dynamical system (8.1), (8.2) to be observable it is necessary and sufficient that any of the following equivalent conditions holds:

1. The observability Gram matrix $G_{OB}(t)$,

$$G_{OB}(t) = \int_0^t \Phi^T(t) C^T C \Phi(t) dt, \qquad (8.15)$$

is nonsingular for any $t \in In\mathfrak{T}_0$.

2. All columns of $C\Phi(t)$ are linearly independent on $[0,t]$ for any $t \in In\mathfrak{T}_0$.

3. All columns of $C(sI_N - A)^{-1}$ are linearly independent on $\mathbb{C}$.

4. The $nN \times n$ observability matrix O_{OBS},

$$O_{OBS} = \begin{bmatrix} C \\ CA \\ CA^2 \\ \cdots \\ CA^{n-1} \end{bmatrix} \in \mathfrak{R}^{nN \times n}, \tag{8.16}$$

has the full rank n,

$$rank O_{OBS} = n. \tag{8.17}$$

5. For every complex number $s \in \mathbb{C}$, equivalently for every eigenvalue $s_i(A)$ of the matrix A, $\forall i = 1, 2, ..., n$, the $(n + N) \times n$ complex matrix $O_O(s)$,

$$O_O(s) = \begin{bmatrix} sI - A \\ C \end{bmatrix} \in \mathfrak{R}^{(n+N) \times n}, \tag{8.18}$$

has the full rank n,

$$rank O_O(s) = n, \ \forall s \in \mathbb{C}. \tag{8.19}$$

The proof of this theorem is given in Appendix D.3.

Note 132 *The form of Theorem 131 has been well known. The novelty of this theorem is its validity proved in the sequel for all five classes of the systems studied herein instead of only for the ISO systems and for the IO systems formally transformed into ISO systems by loosing the physical sense of the state variables if $\mu > 0$ as it has been known so far. For more details see in the sequel Comment 144, Section 9.1.*

Comment 133 *Let the columns of the matrix C be linearly dependent. Let at a moment $\tau \in \mathfrak{T}_0$ all columns of $C\Phi(\tau)$ be linearly independent. The linearity and the time invariance of the linear time-invariant continuous-time system imply the continuity of the system motion and of the system response in the free regime. This implies the existence of $\varepsilon > 0$ such that all columns of $C\Phi(t)$ are linearly independent at any moment $t \in]\tau - \varepsilon, \tau + \varepsilon[$. However, this does not guarantee their independence at every $t \in \mathfrak{T}_0$. Since all columns of the matrix C are not linearly independent then all columns of $C\Phi(t)$ at the initial instant, i.e., for $t = t_0 = 0$ they are not linearly independent because $\Phi(0) = I$ and $C\Phi(0) = CI = C$. In spite of that, all columns of $C\Phi(\tau)$ are linearly independent on $\mathfrak{T}_0$.*

Comment 134 *System response and its derivatives*

Let the system (8.1), (8.2) be in the free regime:

$$\frac{d\mathbf{s}(t)}{dt} = A\mathbf{s}(t), \tag{8.20}$$

$$\mathbf{y}(t) = C\mathbf{s}(t). \tag{8.21}$$

Its motion $\mathbf{s}(t; \mathbf{s}_0; \mathbf{0}_M)$ *and its output response* $\mathbf{y}(t; \mathbf{s}_0; \mathbf{0}_M)$ *have the following well known forms in view of (8.3):*

$$\mathbf{s}\left(t; \mathbf{s}_0; \mathbf{0}_{(\mu+1)M}\right) = e^{At}\mathbf{s}_0 = \Phi\left(t, t_0\right)\mathbf{s}_0, \tag{8.22}$$

$$\mathbf{y}\left(t; \mathbf{s}_0; \mathbf{0}_{(\mu+1)M}\right) = C\mathbf{s}(t) = Ce^{At}\mathbf{s}_0 = C\Phi\left(t, t_0\right)\mathbf{s}_0. \tag{8.23}$$

The system response (D.8) determines directly and uniquely its derivatives:

$$\mathbf{y}^{(i)}\left(t; \mathbf{s}_0; \mathbf{0}_{(\mu+1)M}\right) = C\Phi^{(i)}\left(t, t_0\right)\mathbf{s}_0 = CA^i e^{At}\mathbf{s}_0,$$

$$\forall i = 0, 1, ..., n-1, ..., \forall t \in [t_0, t_1], \tag{8.24}$$

It is now clear that for any initial state vector $\mathbf{s}_0 \in \mathfrak{R}^n$ *the knowledge of the matrices* A *and* C *determines completely, in the free regime, the system motion* $\mathbf{s}\left(t; \mathbf{s}_0; \mathbf{0}_{(\mu+1)M}\right)$, *the system response* $\mathbf{y}\left(t; \mathbf{s}_0; \mathbf{0}_{(\mu+1)M}\right)$, *all their derivatives* $\mathbf{s}^{(k)}\left(t; \mathbf{s}_0; \mathbf{0}_{(\mu+1)M}\right)$ *and* $\mathbf{y}^{(k)}\left(t; \mathbf{s}_0; \mathbf{0}_{(\mu+1)M}\right)$ *at every moment* $t \in \mathfrak{T}_0$.

This analysis permits the following:

Claim 135 *Simple proof of the condition 4. of Theorem 131,*

Proof. Let the system response

$$\mathbf{y}(t; \mathbf{s}_0; \mathbf{0}_M) = Ce^{At}\mathbf{s}_0 = C\Phi(t)\mathbf{s}_0, \ \forall(t, \mathbf{s}_0) \in \mathfrak{T}_0 \times \mathfrak{R}^n,$$

in (D.8) be known at every $t \in \mathfrak{T}_0$.

Necessity. Let the system (8.1), (8.2), (Sections 8.1), be observable. Definition 130, (Section 8.2), holds. The vector form of Equations (8.24) at the initial moment $t = t_0 = 0$ reads for every $\mathbf{s}_0 \in \mathfrak{R}^n$:

$$\begin{bmatrix} \mathbf{y}\left(0; \mathbf{s}_0; \mathbf{0}_M\right) \\ \mathbf{y}^{(1)}\left(0; \mathbf{s}_0; \mathbf{0}_M\right) \\ ... \\ \mathbf{y}^{(n-1)}\left(0; \mathbf{s}_0; \mathbf{0}_M\right) \end{bmatrix} = \mathbf{y}_0^{n-1} = \underbrace{\begin{bmatrix} C \\ CA \\ ... \\ CA^{n-1} \end{bmatrix}}_{O_{OBS}} \mathbf{s}_0 = O_{OBS}\mathbf{s}_0. \tag{8.25}$$

The system observability guarantees both the knowledge of the left-hand side of this vector equation and the existence of its unique solution $\mathbf{s}_0$, hence the system observability guarantees the full rank n of the matrix $O_{OBS} \in \mathfrak{R}^{nN \times n}$ that proves the necessity of the condition 4 of Theorem 131.

Sufficiency. Let the condition 4 of Theorem 131 hold. The full rank n of the matrix O_{OBS} guarantees that the matrix $O_{OBS}^T O_{OBS} \in \mathfrak{R}^{n \times n}$ has also the full rank n:

$$rank O_{OBS}^T O_{OBS} = n, \quad i.e., \quad det O_{OBS}^T O_{OBS} \neq 0.$$

This justifies to introduce and to use the left inverse

$$O_{OBS}^+ = \left(O_{OBS}^T O_{OBS} \right)^{-1} O_{OBS}^T \in \mathfrak{R}^{n \times nN} \qquad (8.26)$$

of the matrix O_{OBS}. It has the following property:

$$O_{OBS}^+ O_{OBS} = I. \qquad (8.27)$$

After the multiplication of Equation (8.25) on the left by O_{OBS}^+ (8.26) the equation transforms into:

$$O_{OBS}^+ O_{OBS} \mathbf{s}_0 = O_{OBS}^+ \mathbf{y}_0^{n-1}, \quad \forall \mathbf{s}_0 \in \mathfrak{R}^n,$$

the final form of which reads

$$\mathbf{s}_0 = O_{OBS}^+ \mathbf{y}_0^{n-1}, \quad \forall \mathbf{s}_0 \in \mathfrak{R}^n,$$

due to (8.27). The preceding resulting equation proves the observability of every initial state vector $\mathbf{s}_0 \in \mathfrak{R}^n$. The system is observable in view of (Definition 130). ∎

Theorem 131 resolves completely the problem of the necessary and sufficient conditions for the observability of all systems that can be set in the form (8.1), (8.2) as shown in the sequel.

8.4 Observability and stability

8.4.1 System stability

The Lyapunov stability theory is established and developed directly only for the *ISO* systems (3.1), (3.2). The book [40, Part III, Chapter 13: Lyapunov stability] broadness the Lyapunov stability concept, properties, method, methodology and theorems directly to the *IO* systems (2.1) and to the *IIO* systems (6.1), (6.2). They incorporate the direct application of the Lyapunov stability theory also to the $HISO$ systems (5.1), (5.2) and to the $EISO$ systems (4.1), (4.2). This is due to the facts that in the free regime the state equation of the $HISO$ system (5.1), (5.2) coincides with the state equation of the IO system (2.1), and that in the free regime the state equation of the $EISO$ system (4.1), (4.2) coincides with the state equation of the ISO system (3.1), (3.2).

The Lyapunov stability properties concern the system behavior in the nominal regime in the total coordinates, i.e., in the free regime in terms of the deviations of all variables. The mathematical models of all these systems can be transformed into the system (8.1), (8.2), i.e., into

$$\frac{d\mathbf{S}(t)}{dt} = A\mathbf{S}(t) + P^{(\mu)}\mathbf{I}_N^\mu(t), \tag{8.28}$$

$$\mathbf{Y}_d(t) = C\mathbf{S}(t) + Q\mathbf{I}_N(t), \tag{8.29}$$

in terms of the total coordinates, and into the system (8.11), (8.12)

$$\frac{d\mathbf{s}(t)}{dt} = A\mathbf{s}(t), \tag{8.30}$$

$$\mathbf{y}(t) = C\mathbf{s}(t), \tag{8.31}$$

in terms of the deviations of all variables in the free regime. The solution of the system (8.32), (8.33),

$$\frac{d\mathbf{S}_N(t)}{dt} = A\mathbf{S}_N(t) + P^{(\mu)}\mathbf{I}_N^\mu(t), \tag{8.32}$$

$$\mathbf{Y}_d(t) = C\mathbf{S}_N(t) + Q\mathbf{I}_N(t), \tag{8.33}$$

for the nominal both the initial state $\mathbf{S}_N(0) = \mathbf{S}_{N0}$ and the input vector $\mathbf{I}_N(t)$ is the system nominal motion. The system forms (8.28), (8.28) and (8.30), (8.31) describe the system under the action of the nominal input $\mathbf{I}_N(t)$ and under the influence of arbitrary initial conditions, (8.33). They are appropriate for the analysis of both stability properties of the system nominal motion and stability properties of the system zero equilibrium state $\mathbf{s}_e = \mathbf{0}n$. The nominal motion $\mathbf{S}_N(t)$ of the system (8.1), (8.2) and the zero equilibrium state $\mathbf{s}_e = \mathbf{0}n$ of the system (8.30), (8.31) have the same stability properties.

Definition 136 *A square matrix $A \in \Re^{n \times n}$ is **stable** (or **stability**, or **Hurwitz**) **matrix** if and only if the real parts of all its eigenvalues are negative.*

The fundamental result of the Lyapunov stability theory for the *time-invariant* continuous-*time* systems (8.28), (8.29) and (8.30), (8.31) is the well known Lyapunov matrix theorem [40, Part III, Chapter 13: Lyapunov stability]:

Theorem 137 *Lyapunov matrix theorem*
For the zero equilibrium state $\mathbf{s}_e = \mathbf{0}n$ of the system (8.30), (8.31) to be (globally) asymptotically stable, or equivalently, for the square matrix $A \in \mathfrak{R}^{n \times n}$ to be stable, it is both necessary and sufficient that that for any positive definite symmetric matrix G, $G = G^T \in \mathfrak{R}^{n \times n}$, the matrix solution H of the Lyapunov matrix equation:

$$A^T H + H A = -G \tag{8.34}$$

is also positive definite symmetric matrix and the unique solution to (8.34).

8.4.2 System stability and observability

The condition for the positive definiteness of the matrix G in the Lyapunov matrix equation (8.34) can be sometimes restrictive. Kalman relaxed for the observable systems (8.28), (8.29) and (8.30), (8.31).

The pair (A,C) is observable pair if and only if the system (8.28), (8.29), equivalently, if and only if the system (8.30), (8.31), is observable.

Theorem 138 *Kalman relaxation of the Lyapunov matrix theorem*
For the zero equilibrium state $\mathbf{s}_e = \mathbf{0}_n$ of the observable system (8.30), (8.31) to be (globally) asymptotically stable, or equivalently, for the square matrix $A \in \mathfrak{R}^{n \times n}$ to be stable in the case (A,C) is observable pair, it is both necessary and sufficient that that the matrix solution H of the relaxed Lyapunov matrix equation (8.35),

$$A^T H + H A = -C^T C, \tag{8.35}$$

is positive definite symmetric matrix and the unique solution to (8.35).

The proof of this theorem is in Appendix D.4.

Comment 139 *While in Theorem 137 the matrix G is positive definite symmetric matrix, in Theorem 138 the matrix $C^T C$ is only positive semidefinite symmetric. Theorem 138 relaxes slightly the condition of Theorem 137. The relaxation is possible due to the system observability.*

Chapter 9

Various systems observability

9.1   *IO* systems observability

In order to facilitate reading and following the text and formulae we will use the compact form (2.15):

$$A^{(\nu)}\mathbf{Y}^\nu(t) = H^{(\mu)}\mathbf{I}^\mu(t), \ \forall t \in \mathfrak{T}_0, \ \nu \geq 1, \ 0 \leq \mu \leq \nu, \qquad (9.1)$$

of the *IO* system (2.15) in terms of the total coordinates and in its compact form, or

$$A^{(\nu)}\mathbf{y}^\nu(t) = H^{(\mu)}\mathbf{i}^\mu(t), \ \forall t \in \mathfrak{T}_0, \ \nu \geq 1, \ 0 \leq \mu \leq \nu, \qquad (9.2)$$

in terms of the deviations (2.54), (9.3),

$$\mathbf{i} = \mathbf{I} - \mathbf{I}_N, \qquad (9.3)$$

$$\mathbf{y} = \mathbf{Y} - \mathbf{Y}_d,$$

of all variables and also in the compact form. For the observability study we can consider the system in the free regime (Section 8.1), in which (9.2) becomes

$$A^{(\nu)}\mathbf{y}^\nu(t) = \mathbf{0}_N, \ \forall t \in \mathfrak{T}_0, \ \nu \geq 1, \ 0 \leq \mu \leq \nu, \qquad (9.4)$$

In view of Note 28, Equation (2.16) (Section 2.1), the state vector deviation $\mathbf{s} \in \mathfrak{R}^n$ of the system (9.2), hence of its form (9.4) in the free regime, has the following form:

$$\mathbf{s} = \mathbf{y}^{\nu-1} = \left[\mathbf{y}^T \vdots \mathbf{y}^{(1)^T} \vdots \dots \vdots \mathbf{y}^{(\nu-1)^T}\right]^T \in \mathfrak{R}^n, \ n = \nu N. \qquad (9.5)$$

This permits us to apply directly Definition 129 to the observability of the system (9.1), equivalently, to apply directly Definition 130 to the system (9.2).

The observability criterion for the IO system (9.1), equivalently for the system (9.2), has the simplest possible form that is the same for both because they have the same properties:

Theorem 140 *Observability criterion for the IO system (2.15), equivalently, for (9.2), (9.4)*

The IO system (2.15), equivalently (9.2), in total coordinates and its form (9.2), i.e., (9.4), in terms of deviations of all variables, is invariably observable; i.e., it is observable independently of all its matrices A_k, $k = 0, 1, ..., \nu$, and B_k, $k = 0, 1, ..., \mu$, i.e., independently of its matrices $A^{(\nu)}$ and $B^{(\mu)}$.

Proof is in Appendix D.5.

Comment 141 *The statement of Theorem 140 is physically clear. System observability signifies the system property that every system initial state vector $\mathbf{s}_0 = \mathbf{y}_0^{\nu-1}$ can be determined from the system response $\mathbf{y}(t; \mathbf{s}_0; \mathbf{0}_{(\mu+1)M})$,*

$$\mathbf{y}(t; \mathbf{s}_0; \mathbf{0}_{(\mu+1)M}) = Ce^{At}\mathbf{s}_0 = Ce^{At}\mathbf{y}_0^{\nu-1}, \ \forall t \in \mathfrak{T}_0.$$

The knowledge of the system response $\mathbf{y}(t; \mathbf{s}_0; \mathbf{0}_{(\mu+1)M})$ determines all its derivatives

$$\mathbf{y}^{(k)}(t; \mathbf{s}_0; \mathbf{0}_{(\mu+1)M}), \ \forall k = 1, 2, ..., \ \forall t \in \mathfrak{T}_0,$$

hence, the system response $\mathbf{y}(t; \mathbf{s}_0; \mathbf{0}_{(\mu+1)M})$ determines

$$\mathbf{y}^{\nu-1}(t; \mathbf{s}_0; \mathbf{0}_{(\mu+1)M}), \ \forall t \in \mathfrak{T}_0,$$

and the initial state vector $\mathbf{y}_0^{\nu-1} = \mathbf{s}_0$ fully independently of all its matrices A_k, $k = 0, 1, ..., \nu$, and B_k, $k = 0, 1, ..., \mu$, i.e., independently of its matrices $A^{(\nu)}$ and $B^{(\mu)}$. The result is that every IO system is invariably observable.

Conclusion 142 *Every IO system (2.15), equivalently, (9.2), and (9.4), is (invariably) observable.*

The problem of the observability test does not exist in the framework of the IO systems (2.15), equivalently, (9.2), and (9.4).

Note 143 *It is to remind ourselves that Definition 129, Definition 130, and Theorem 140 concern the real, physical, initial state total vector $\mathbf{S}_0 = \mathbf{Y}_0^{\nu-1}$, i.e., its deviation $\mathbf{s}_0 = \mathbf{y}_0^{\nu-1}$, and the system real, physical, output response $\mathbf{Y}(t; \mathbf{Y}_0^{\nu-1}; \mathbf{0}_M)$, i.e., its deviation $\mathbf{y}(t; \mathbf{y}_0^{\nu-1}; \mathbf{0}_M)$ in the free regime.*

Comment 144 *The formal mathematical algorithm defined by Equations (C.4)-(C.9) transforms (see Section C.1 herein or, for more details, see Appendix B in [40]) the IO system (2.1) into the ISO system (3.1), (3.2), i.e., into*

$$\frac{d\mathbf{X}(t)}{dt} = A\mathbf{X}(t) + H\mathbf{I}(t), \ \forall t \in \mathfrak{T}_0, \tag{9.6}$$

$$\mathbf{Y}(t) = C\mathbf{X}(t) + Q\mathbf{I}(t), \ \forall t \in \mathfrak{T}_0, \tag{9.7}$$

in which the state vector $\mathbf{X}$ *is determined by (C.4)-(C.10) (Section C.1):*

$$\mathbf{X} = \begin{bmatrix} \mathbf{X}_1 \\ \mathbf{X}_2 \\ : \\ \mathbf{X}_{\nu-1} \\ \mathbf{X}_\nu \end{bmatrix} = \begin{bmatrix} \mathbf{Y} - H_\nu\mathbf{I} \\ \overset{\bullet}{\mathbf{X}}_1 + A_{\nu-1}\ \mathbf{Y} - H_{\nu-1}\mathbf{I} \\ \cdots \\ \overset{\bullet}{\mathbf{X}}_{\nu-2} + A_2\ \mathbf{Y} - H_2\mathbf{I} \\ \overset{\bullet}{\mathbf{X}}_{\nu-1} + A_1\mathbf{Y} - H_1\mathbf{I} \end{bmatrix} ., \tag{9.8}$$

which is different from the state vector (4.5) (Section 4.1):

$$\mathbf{X} = \begin{bmatrix} \mathbf{X}_1 \\ \mathbf{X}_2 \\ : \\ \mathbf{X}_{\nu-1} \\ \mathbf{X}_\nu \end{bmatrix} = \begin{bmatrix} \mathbf{Y} \\ \mathbf{Y}^{(1)} \\ \cdots \\ \mathbf{Y}^{(\nu-2)} \\ \mathbf{Y}^{(\nu-1)^T} \end{bmatrix}.$$

The state vector $\mathbf{X}$ *(9.8) depends on the system output vector* $\mathbf{Y}$ *and on the system input vector* $\mathbf{I}$ *in spite their physical natures are most often very different so that the state vector* $\mathbf{X}$ *and its subvectors* $\mathbf{X}_1$, $\mathbf{X}_2$, ... , $\mathbf{X}_\nu$ *are physically meaningless if there exists* $k \in \{1, 2, ... , \nu\}$ *such that* $H_k \neq O_{N,M}$*. Equations (C.4)-(C.10) together with (2.15) define the matrices* A *(C.11),* $B = H$ *(C.12),* C *(C.13) and* Q *(C.14) (Section C.1). They determine* $(n + N) \times n$ *complex matrix* $O_O(s)$ *(D.62),*

$$O_O(s) = \begin{bmatrix} sI_n - A \\ C \end{bmatrix} \in \mathbb{C}^{(n+N)\times n}, \tag{9.9}$$

which takes the following form:

$$O_O(s) = \begin{bmatrix} sI_N + A_{\nu-1} & -I_N & O_N & \dots & O_N & O_N & O_N \\ A_{\nu-2} & sI_N & -I_N & \dots & O_N & O_N & O_N \\ \dots & \dots & \dots & \dots & \dots & \dots & \dots \\ A_3 & O_N & O_N & \dots & -I_N & O_N & O_N \\ A_2 & O_N & O_N & \dots & sI_N & -I_N & O_N \\ A_1 & O_N & O_N & \dots & O_N & sI_N & -I_N \\ A_0 & O_N & O_N & \dots & O_N & O_N & sI_N \\ I_N & O_N & O_N & \dots & O_N & O_N & O_N \end{bmatrix}$$

Its rank is obtained from:

$$rank O_O(s) = rank \begin{bmatrix} sI_n - A \\ C \end{bmatrix} =$$

$$= rank \begin{bmatrix} sI_N + A_{\nu-1} & -I_N & O_N & \dots & O_N & O_N & O_N \\ A_{\nu-2} & sI_N & -I_N & \dots & O_N & O_N & O_N \\ \dots & \dots & \dots & \dots & \dots & \dots & \dots \\ A_3 & O_N & O_N & \dots & -I_N & O_N & O_N \\ A_2 & O_N & O_N & \dots & sI_N & -I_N & O_N \\ A_1 & O_N & O_N & \dots & O_N & sI_N & -I_N \\ I_N & O_N & O_N & \dots & O_N & O_N & O_N \end{bmatrix} =$$

$$= rank \begin{bmatrix} -I_N & O_N & \dots & O_N & O_N & O_N \\ sI_N & -I_N & \dots & O_N & O_N & O_N \\ \dots & \dots & \dots & \dots & \dots & \dots \\ O_N & O_N & \dots & -I_N & O_N & O_N \\ O_N & O_N & \dots & sI_N & -I_N & O_N \\ O_N & O_N & \dots & O_N & sI_N & -I_N \end{bmatrix} = \nu N = n \implies$$

$$O_O(s) = \begin{bmatrix} sI_n - A \\ C \end{bmatrix} = n, \ \forall (s, A) \in \mathbb{C} \times \mathfrak{R}^{n \times n}.$$

The ISO system (9.6), (9.7), (9.8) is also observable (Theorem 131, Section 8.3) independently of its matrices A_j and B_k, i.e., it is invariably observable. The result seems the same as expressed in Theorem 140, but it is not as explained in what follows. The explanation is in the fact that the obtained ISO system (9.6), (9.7), (9.8) is also considered in the free regime, i.e., for the zero input vector,

$$\mathbf{I}(t) = \mathbf{0}_M, \ \forall t \in \mathfrak{T}_0,$$

which implies

$$\mathbf{I}^{(k)}(t) = \mathbf{0}_M, \ \forall k = 0, 1, 2, .., \nu, \ i.e., \ \mathbf{I}^{\nu}(t) = \mathbf{0}_{(\nu+1)M}, \ \forall t \in \mathfrak{T}_0.$$

The state subvectors then become due to the free regime:

$$\mathbf{X}_1 = \mathbf{Y}, \tag{9.10}$$

$$\mathbf{X}_2 = \mathbf{Y}^{(1)} + A_{\nu-1}\,\mathbf{Y}, \tag{9.11}$$

$$\mathbf{X}_3 = \mathbf{Y}^{(2)} + A_{\nu-1}\,\mathbf{Y}^{(1)} + A_{\nu-2}\,\mathbf{Y}, \tag{9.12}$$

$$.... \tag{9.13}$$

$$\mathbf{X}_{\nu-2} = \mathbf{Y}^{(\nu-3)} + A_{\nu-1}\,\mathbf{Y}^{(\nu-2)} + ... + A_3\,\mathbf{Y}, \tag{9.14}$$

$$\mathbf{X}_{\nu-1} = \mathbf{Y}^{(\nu-2)} + A_{\nu-1}\,\mathbf{Y}^{(\nu-3)} + ... + A_2\,\mathbf{Y}, \tag{9.15}$$

$$\mathbf{X}_\nu = \mathbf{Y}^{(\nu-1)} + A_{\nu-1}\,\mathbf{Y}^{(\nu-2)} + ... + A_1\,\mathbf{Y}. \tag{9.16}$$

and the state vector of the transformed IO system (2.15) into the ISO system (3.1), (3.2) is the vector $\mathbf{X}$ defined by (9.8) and by (9.10)-(9.16). It is not the physical state vector $\mathbf{S}_{IO} = \mathbf{Y}^{\nu-1}$ of the original IO system (2.1), $\mathbf{X} \neq \mathbf{Y}^{\nu-1} = \mathbf{S}_{IO}$.

It is very rare for the mathematical subvectors $\mathbf{X}_k$ (9.10)-(9.16) of the state vector $\mathbf{X}$ to have a physical sense, i.e., for the initial state vector $\mathbf{X}_0$ to be physically meaningful. The formal mathematical approach left open the problem of the observability of the real, physical, initial state vector $\mathbf{S}_0 = \mathbf{Y}_0^{\nu-1}$ of the original IO system (2.1). This illustrates the advantage of the usage of the new notation (Section 1.2) and of the generalization of the state concept (Section 1.4, Definition 14), which enabled Theorem 140, i.e., the complete solution of the physical state observability problem for all IO systems (2.1).

Exercise 145 *Test the observability of the selected IO physical plant in Exercise 40, Section 2.3.*

9.2 *ISO* systems observability

Definition 129 and Definition 130 hold for the *ISO* systems.

The observability criterion for the *ISO* systems (3.1), (3.2) equivalently for the *ISO* systems (3.39), (3.40), is the same for both of them because they have the same matrices, orders, and structures. The following theorem is well known:

Theorem 146 *Observability criterion for the ISO system (3.1), (3.2)*

For the ISO system (3.1), (3.2) to be observable it is necessary and sufficient that any of the following equivalent conditions holds:

1. *The observability Gram matrix* $G_{OB}(t)$,

$$G_{OB}(t) = \int_0^t \Phi^T(t)\, C^T C \Phi(t)\, dt, \qquad (9.17)$$

is nonsingular for any $t \in In\mathfrak{T}_0$.

2. *All columns of* $C\Phi(t)$ *are linearly independent on* $[0,t]$ *for any* $t \in In\mathfrak{T}_0$.

3. *All columns of* $C(sI_N - A)^{-1}$ *are linearly independent on* $\mathbb{C}$.

4. *The* $nN \times n$ *observability matrix* O_{OBS},

$$O_{OBS} = \begin{bmatrix} C \\ CA \\ CA^2 \\ \cdots \\ CA^{n-1} \end{bmatrix} \in \mathfrak{R}^{nN \times n}, \qquad (9.18)$$

has the full rank n,

$$rank O_{OBS} = n. \qquad (9.19)$$

5. *For every complex number* $s \in \mathbb{C}$, *equivalently for every eigenvalue* $s_i(A)$ *of the matrix* A, $\forall i = 1, 2, ..., n$, *the* $(n + N) \times n$ *complex matrix* $O_O(s)$,

$$O_O(s) = \begin{bmatrix} sI_n - A \\ C \end{bmatrix} \in \mathfrak{R}^{(n+N) \times n}, \qquad (9.20)$$

has the full rank n,

$$rank O_O(s) = n, \ \forall s \in \mathbb{C}. \qquad (9.21)$$

Proof. The state vector $\mathbf{S}$ of the *ISO* system (3.1), (3.2) is its vector $\mathbf{X}$, and the state vector $\mathbf{s}$ of the *ISO* system (3.39), (3.40) is its vector $\mathbf{x}$. The input vectors of these systems are $\mathbf{I}$ and $\mathbf{i}$, respectively. When we replace $\mathbf{S}$ by $\mathbf{X}$, and set $\mu = 0$ that reduces $\mathbf{I}^\mu$ to $\mathbf{I}^0 = \mathbf{I}$ in (8.1), (8.2) then they become (3.1), (3.2). When we replace $\mathbf{s}$ by $\mathbf{x}$ in (8.11), (8.2) for $\mathbf{v} = \mathbf{0}_m$ then they become (3.39), (3.40). This proves that the Theorem 131 holds also for *ISO* systems (3.1), (3.2) equivalently for the *ISO* systems (3.39), (3.40). Theorem 146 is Theorem 131, which ends the proof. ∎

Differently than the observability test for the *IO* systems, which is trivial, the observability test for the *ISO* systems is not trivial at all. The observability property of the latter depend essentially on the properties of the system matrices A and C, as shown by the above theorem, while it is independent of them for the *IO* systems (Theorem 140, Comment 141 and Conclusion 142).

Exercise 147 *Test the output controllability of the selected ISO physical plant in Exercise 48, Section 3.3.*

9.3 *EISO* systems observability

Definition 129 and Definition 130 are directly applicable in this framework.

The observability criterion for the *EISO* system (4.1), (4.2), equivalently of the *EISO* system (4.71), (4.72), is the same as for the linear dynamical system (8.1), (8.2):

Theorem 148 *Observability criterion for the EISO system (4.1), (4.2), equivalently for the EISO system (4.71), (4.72)*

For the EISO system (4.1), (4.2), equivalently for the EISO system (4.71), (4.72), to be observable it is necessary and sufficient that any of the following equivalent conditions holds:

1. The observability Gram matrix $G_{OB}(t)$,

$$G_{OB}(t) = \int\limits_0^t \Phi^T(t) C^T C \Phi(t) \, dt, \tag{9.22}$$

is nonsingular for any $t \in In\mathfrak{T}_0$.

2. All columns of $C\Phi(t)$ are linearly independent on $[0,t]$ for any $t \in In\mathfrak{T}_0$.

3. All columns of $C(sI_N - A)^{-1}$ are linearly independent on $\mathbb{C}$.

4. The $nN \times n$ observability matrix O_{OBS},

$$O_{OBS} = \begin{bmatrix} C \\ CA \\ CA^2 \\ \cdots \\ CA^{n-1} \end{bmatrix} \in \mathfrak{R}^{nN \times n}, \tag{9.23}$$

has the full rank n,

$$rank O_{OBS} = n. \tag{9.24}$$

5. For every complex number $s \in \mathbb{C}$, equivalently for every eigenvalue $s_i(A)$ of the matrix A, $\forall i = 1, 2, \cdots, n$, the $(n + N) \times n$ complex matrix $O_O(s)$,

$$O_O(s) = \begin{bmatrix} sI_n - A \\ C \end{bmatrix} \in \mathfrak{R}^{(n+N) \times n}, \tag{9.25}$$

has the full rank n,

$$rank O_O(s) = n, \quad \forall s \in \mathbb{C}. \tag{9.26}$$

Proof. The *EISO* system (4.1), (4.2) is the system (8.1), (8.2). Theorem 148 is Theorem 131 that is valid for the system (8.1), (8.2). ∎

Exercise 149 *Test the observability of the selected EISO physical plant in Exercise 61, Section 4.3.*

9.4   *HISO* systems observability

The state vector $\mathbf{S}$ (5.3) of the *HISO* system (5.1), (5.2) (Section 5.1) is the extended vector $\mathbf{R}^{\alpha-1} \in \mathfrak{R}^n$ for $n = \alpha\rho$, and the state vector $\mathbf{s}$ of the *HISO* system (5.1), (5.2) (Section 5.1) is the extended vector $\mathbf{r}^{\alpha-1} \in \mathfrak{R}^n$,

$$\mathbf{S} = \mathbf{R}^{\alpha-1} \in \mathfrak{R}^n, \ n = \alpha\rho; \ \mathbf{s} = \mathbf{r}^{\alpha-1} \in \mathfrak{R}^n. \tag{9.27}$$

Let us remind that the *HISO* system description (5.1), (5.2) is in the compact form.

Definition 129 and Definition 130, (Section 8.2) are valid for the *HISO* system (5.1), (5.2), i.e., (5.44), (5.45), respectively.

Definition 150 *Observability of the HISO systems (5.1), (5.2), i.e., of the HISO systems (5.44), (5.45)*

An initial state $\mathbf{S}_0 \in \mathfrak{R}^n$ at $t_0 = 0$ of the HISO system (5.1), (5.2) is observable if and only if the output system response $\mathcal{Y}(.; \mathbf{S}_0; \mathbf{I}^\mu)$ on $\mathfrak{T}_0$ to the known extended input vector $\mathbf{I}^\mu(t)$ on $\mathfrak{T}_0$ uniquely determines $\mathbf{S}_0$.

An initial state $\mathbf{s}_0 \in \mathfrak{R}^n$ at $t_0 = 0$ of the HISO system (5.44), (5.45) is observable if and only if the output system response $\mathcal{Y}(.; \mathbf{s}_0; \mathbf{0}_{(\mu+1)M})$ in the free regime on $\mathfrak{T}_0$ uniquely determines the initial state vector $\mathbf{s}_0$.

The system is observable if and only if every its state is observable.

Let

$$A = \begin{bmatrix} O_\rho & I_\rho & O_\rho & \dots & O_\rho \\ O_\rho & O_\rho & I_\rho & \dots & O_\rho \\ \vdots & \vdots & \vdots & \vdots & \vdots \\ O_\rho & O_\rho & O_\rho & \dots & I_\rho \\ -A_\alpha^{-1}A_1 & -A_\alpha^{-1}A_2 & -A_\alpha^{-1}A_3 & \dots & -A_\alpha^{-1}A_{\alpha-1} \end{bmatrix} \tag{9.28}$$

$$P^{(\mu)} = \begin{bmatrix} O_{\rho,\rho\mu} \\ O_{\rho,\rho\mu} \\ \vdots \\ O_{\rho,\rho\mu} \\ A_\alpha^{-1} H^{(\mu)}, \end{bmatrix} \in \mathfrak{R}^{n \times \rho(\mu+1)}, \tag{9.29}$$

$$H^{(\mu)} = \begin{bmatrix} H_0 \vdots H_1 \vdots \dots \vdots H_\mu \end{bmatrix} \in \mathfrak{R}^{\rho \times (\mu+1)M}, \tag{9.30}$$

$$\mathbf{I}^\mu = \begin{bmatrix} \mathbf{I} \\ \mathbf{I}^{(1)} \\ \vdots \\ \mathbf{I}^{(\mu-1)} \\ \mathbf{I}^{(\mu)}, \end{bmatrix} \in \mathfrak{R}^{(\mu+1)M}, \tag{9.31}$$

and

$$C = R_y^{\alpha-1} = \begin{bmatrix} R_{y0} \vdots R_{y1} \vdots \dots \vdots R_{y(\alpha-1)} \end{bmatrix} \in \mathfrak{R}^{N \times n}. \tag{9.32}$$

Theorem 151 *Observability criterion for the HISO system (5.1),*
(5.2), equivalently for the HISO system (5.44), (5.45)

For the HISO system (5.1), (5.2), equivalently for the HISO system
(5.44), (5.45), to be observable it is necessary and sufficient that any of the
following equivalent conditions holds for the matrices A, $P^{(\mu)}$, $H^{(\mu)}$, C, and
the vector $\mathbf{I}^\mu$ defined by Equations (9.28)-(11.146), respectively:

1. The observability Gram matrix $G_{OB}(t)$,

$$G_{OB}(t) = \int_0^t \Phi^T(t) C^T C \Phi(t)\, dt, \tag{9.33}$$

is nonsingular for any $t \in In\mathfrak{T}_0$.

2. All columns of $C\Phi(t)$ are linearly independent on $[0,t]$ for any $t \in$
$In\mathfrak{T}_0$.

3. All columns of $C(sI_N - A)^{-1}$ are linearly independent on $\mathbb{C}$.

4. The $nN \times n$ observability matrix O_{OBS},

$$O_{OBS} = \begin{bmatrix} C \\ CA \\ CA^2 \\ \dots \\ CA^{n-1} \end{bmatrix} \in \mathfrak{R}^{nN \times n}, \tag{9.34}$$

has the full rank n,

$$rank O_{OBS} = n. \qquad (9.35)$$

5. For every complex number $s \in \mathbb{C}$, equivalently for every eigenvalue $s_i (A)$ of the matrix A, $\forall i = 1, 2, \cdots, n$, the $(n + N) \times n$ complex matrix $O_O (s)$,

$$O_O (s) = \left[\begin{array}{c} sI_n - A \\ C \end{array} \right] \in \mathfrak{R}^{(n+N) \times n}, \qquad (9.36)$$

has the full rank n,

$$rank O_O (s) = n, \ \forall s \in \mathbb{C}. \qquad (9.37)$$

For the proof of the theorem see the Appendix D.6.

Note 152 *In view of Equation (9.27) we conclude that Definition 150 and Theorem 151 deal with the physical initial state total vector $\mathbf{S}_0 = \mathbf{R}_0^{\alpha-1}$, i.e., with its deviation $\mathbf{s}_0 = \mathbf{y}_0^{\nu-1}$, and with the system real, physical, output response $\mathbf{Y}(t; \mathbf{Y}_0^{\alpha-1}; \mathbf{I}^\mu)$, i.e., its deviation $\mathbf{y}(t; \mathbf{y}_0^{\alpha-1}; \mathbf{0}_{(\mu+1)M})$ in the free regime.*

Exercise 153 *Test the output controllability of the selected HISO physical plant in Exercise 68, Section 5.3.*

9.5   *IIO* systems observability

Differently than all previously treated (*IO, ISO, EISO,* and *HISO*) systems, the *IIO* systems (6.1), (6.2) (Section 5.1), i.e., the *IIO* systems (6.65), (6.66) (Section 5.2), have the internal state vector $\mathbf{S}_I = \mathbf{R}^{\alpha-1}$ (1.25)(6.1) and the output state vector $\mathbf{S}_O = \mathbf{Y}^{\nu-1}$ (1.26) (Section 1.4). Their full state vector $\mathbf{S}_f$,

$$\mathbf{S}_f = \left[\begin{array}{c} \mathbf{S}_I \\ \mathbf{S}_O \end{array} \right] = \mathbf{S} = [\mathbf{S}_1^T \ \mathbf{S}_2^T \ ... \ \mathbf{S}_\alpha^T \ \mathbf{S}_{\alpha+1}^T \ ... \ \mathbf{S}_{\alpha+\nu}^T]^T \in R^{\alpha\rho+\nu N}, \qquad (9.38)$$

is their state vector $\mathbf{S}$ (1.27),

$$\mathbf{S} = \left[\begin{array}{c} \mathbf{R}^{\alpha-1} \\ \mathbf{Y}^{\nu-1} \end{array} \right] = \left[\begin{array}{c} \mathbf{S}_I \\ \mathbf{S}_O \end{array} \right] = \mathbf{S}_f, \ \mathbf{S}_I = \left[\begin{array}{c} \mathbf{S}_1 \\ \vdots \\ \mathbf{S}_\alpha \end{array} \right], \ \mathbf{S}_O = \left[\begin{array}{c} \mathbf{S}_{\alpha+1} \\ \vdots \\ \mathbf{S}_{\alpha+\nu} \end{array} \right]. \qquad (9.39)$$

Its deviation $\mathbf{s}$ from the nominal state vector $\mathbf{S}_N$,

$$\mathbf{s} = \mathbf{S} - \mathbf{S}_N, \qquad (9.40)$$

is composed of the internal state deviation vector $\mathbf{r}^{\alpha-1}$ and of the output state deviation vector $\mathbf{y}^{\nu-1}$:

$$\mathbf{s} = \begin{bmatrix} \mathbf{s}_I \\ \mathbf{s}_O \end{bmatrix} = \begin{bmatrix} \mathbf{s}_1^T & \mathbf{s}_2^T & \cdots & \mathbf{s}_\alpha^T & \mathbf{s}_{\alpha+1}^T & \cdots & \mathbf{s}_{\alpha+\nu}^T \end{bmatrix}^T = \begin{bmatrix} \mathbf{r}^{\alpha-1} \\ \mathbf{y}^{\nu-1} \end{bmatrix}. \tag{9.41}$$

Definition 154 *Full state observability of the IIO systems (6.1), (6.2), i.e., of the IIO systems (6.65), (6.66)*

An initial total state vector $\mathbf{S}_0 \in \mathfrak{R}^n$ at $t_0 = 0$ of the IIO system (6.1), (6.2) is observable if and only if the output system response $\mathcal{Y}(.; \mathbf{S}_0; \mathbf{I}^\mu)$ on $\mathfrak{T}_0$ to the known extended input vector $\mathbf{I}^\mu(t)$ on $\mathfrak{T}_0$ uniquely determines $\mathbf{S}_0$.

An initial state deviation vector $\mathbf{s}_0 \in \mathfrak{R}^n$ at $t_0 = 0$ of the IIO system (6.65), (6.66), is observable if and only if the output system response $\mathcal{Y}(.; \mathbf{s}_0; \mathbf{0}_{(\mu+1)M})$ in the free regime on $\mathfrak{T}_0$ uniquely determines $\mathbf{s}_0$.

The system is observable if and only if every its state is observable.

*This is the system initial **full state** observability.*

This definition determines the full state, i.e., the state, observability of the *IIO* systems (6.1), (6.2), i.e., of the *IIO* systems (6.65), (6.66). An essential characteristic of them is to have also the internal state $\mathbf{S}_I = \mathbf{R}^{\alpha-1}$ and the output state $\mathbf{S}_O = \mathbf{Y}^{\nu-1}$. This poses the question whether it is meaningful to think of their observability and if yes, in which sense. The following definitions reply affirmatively to that question. They open the problem of the criteria for their observability.

Definition 155 *Internal state observability of the IIO systems (6.1), (6.2), i.e., of the IIO systems (6.65), (6.66)*

An initial total internal state vector $\mathbf{S}_{I0} \in \mathfrak{R}^{\alpha\rho}$ at $t_0 = 0$ of the IIO system (6.1), (6.2) is observable if and only if the output system response $\mathcal{Y}(.; \mathbf{S}_{I0}; \mathbf{S}_{O0}; \mathbf{I}^\mu)$ on $\mathfrak{T}_0$ to the known both initial total output state vector $\mathbf{S}_{O0}$ and the extended input vector $\mathbf{I}^\mu(t)$ on $\mathfrak{T}_0$ uniquely determines $\mathbf{S}_{I0}$.

An initial internal state deviation vector $\mathbf{s}_{I0} \in \mathfrak{R}^n$ at $t_0 = 0$ of the IIO system (6.65), (6.66) is observable if and only if the output system response $\mathcal{Y}(.; \mathbf{s}_{I0}; \mathbf{s}_{O0}; \mathbf{0}_{(\mu+1)M})$ in the free regime on $\mathfrak{T}_0$ to the known initial output state deviation vector $\mathbf{s}_{O0}$ uniquely determines $\mathbf{s}_{I0}$.

The system is internal state observable if and only if every its internal state is observable.

Analogously:

Definition 156 *Output state observability of the IIO systems (6.1),*
(6.2), i.e., of the IIO systems (6.65), (6.66)

An initial total output state vector $\mathbf{S}_{OO} \in \mathfrak{R}^{\alpha\rho}$ at $t_0 = 0$ of the IIO
system (6.1), (6.2) is observable if and only if the output system response
$\mathcal{Y}(.; \mathbf{S}_{IO}; \mathbf{S}_{OO}; \mathbf{I}^{\mu})$ on $\mathfrak{T}_0$ to the known both the total internal state vector
$\mathbf{S}_I(t)$ and the extended input vector $\mathbf{I}^{\mu}(t)$ on $\mathfrak{T}_0$ uniquely determines $\mathbf{S}_{OO}$.

An initial output state deviation vector $\mathbf{s}_{OO} \in \mathfrak{R}^n$ at $t_0 = 0$ of the IIO
system (6.65), (6.66) is observable if and only if the output system response
$\mathcal{Y}(.; \mathbf{0}_{\alpha\rho}; \mathbf{s}_{OO}; \mathbf{0}_{(\mu+1)M})$ in the free regime on $\mathfrak{T}_0$ to the known internal state
vector deviation $\mathbf{s}_I(t)$ uniquely determines $\mathbf{s}_{OO}$.

The system is output state observable if and only if every its output state
is observable.

The observability criterion for the *IIO* systems (6.1), (6.2), and for the
IIO systems (6.65), (6.66), is the same for both of them because they have
the same matrices, orders and structures.

Theorem 157 *Observability criterion for the IIO system defined*
by (6.1), (6.2), equivalently for the IIO system (6.65), (6.66)

The IIO system (6.1), (6.2), equivalently the IIO system (6.65), (6.66),
is not full state observable

Appendix D.7 contains the proof of this theorem.

We have solved the problem of the initial full state $\mathbf{S}_0$, equivalently $\mathbf{s}_0$,
observability.

What are the conditions for their internal state $\mathbf{S}_{IO}$ observability? The
following theorem replies to this question:

Theorem 158 *Internal state observability criterion for the IIO*
system (6.1), (6.2), equivalently for the IIO system (6.65), (6.66)

The IIO system (6.1), (6.2), equivalently the IIO system (6.65), (6.66),
is internal state unobservable.

For the proof see Appendix D.8

Since the system is internal state unobservable (Theorem 158) it is also
full state unobservable (Theorem 157).

We have solved the problems of the initial full state $\mathbf{s}_0$ observability in
Theorem 157 and the problem of the initial internal state $\mathbf{s}_{IO}$ observability in
Theorem 158. These theorems discover that every *IIO* system (6.1), (6.2),
equivalently the *IIO* system (6.65), (6.66), is both internal state and full
state unobservable.

The criterion for the output state observability has very simple form.

Theorem 159 *Output state observability criterion for the IIO system (6.1), (6.2), equivalently for the IIO system (6.65), (6.66)*

The IIO system (6.1), (6.2), equivalently the IIO system (6.65), (6.66), is invariably output state observable, i.e., output state observable independently of the system matrices.

Appendix D.9 presents the proof of this theorem. It signifies that every *IIO* system (6.1), (6.2), equivalently the *IIO* system (6.65), (6.66), is output observable independently of its matrices.

Exercise 160 *Test the output controllability of the selected IIO physical plant in Exercise 78, Section 6.3.*

Part III

CONTROLLABILITY

Chapter 10

Controllability fundamentals

10.1 Controllability and system regime

Kalman's concept of *the state controllability* has become a fundamental control concept, [58]-[63]. E. G. Gilbert [32] generalized it to the *MIMO* systems. M. L. J. Hautus [48] established for them the simple form of the controllability criterion in the complex domain.

J. E. Bertram and P. E. Sarachik [8] broadened *the state controllability* concept to *the output controllability concept.*

Both the state controllability concept and the output controllability concept consider the system possibility of steering a state or an output from any initial state or from any initial output to another state or another output, in general, or to the zero state or to the zero output, in particular, respectively.

R. W. Brockett and M. D. Mesarović (Mesarovitch) [12] introduced *the concept of functional (output) reproducibility*, called also *the output function controllability* [2, page 313], [18, page 216], [97, pages 72 and 164], in which the target is not a particular output (e.g., the zero output) but a given function representing a reference (desired) output response.

All these concepts *concern the systems free of any external disturbance action:* $\mathbf{D}(t) \equiv \mathbf{0}_d$. They assume the nonexistence of any external perturbation acting on the system. The only external influences on the system are control actions. This explains why they are mainly important for plants, i.e., objects (Definition 9 in Section 1.4).

10.2 Controllability concepts

There are two main controllability concepts of linear dynamical systems:

159

- *The concept of the system state controllability.*

- *The concept of the system output controllability.*

The state controllability concept treats the vector state regardless of its physical sense whether its entries are physical or mathematical variables. If the *ISO* mathematical model of a physical system results directly from the system physical properties then the state variables (the entries of the state vector) are physical variables. Their state vector is also physical vector rather than only mathematical one. Then the state controllability concept has both mathematical and physical sense. This controllability concept can have only purely mathematical sense and meaning. It permits for an *ISO* mathematical model of a physical system to be expressed in terms of either mathematical or physical variables.

If the mathematical model obtained directly from the physical properties of the physical system is in the form of an *IO* system then it is used to introduce purely mathematical, without any physical sense, state variables, hence their state vector without any physical sense (Section C.1).

These facts open the need to distinguish in the framework of the state controllability

- *The physical state controllability*

from

- *The mathematical state controllability.*

The former demands for a system mathematical model to be expressed in terms of physical variables only. The latter concerns system mathematical models determined mainly in terms of mathematical variables that can be without any physical sense.

The following controllability definitions clearly and precisely explain and determine the corresponding controllability concept. They are valid for both total coordinates and for their deviations; i.e., for the total state vector **S** and the total output vector **Y** as well as for their deviation vectors **s** and **y**, respectively.

10.3   Controllability definitions in general

Definition 161 *State controllability of dynamical systems*

A dynamical system is the **mathematical state or physical state controllable** *if and only if for every initial mathematical state or physical state vector* $\mathbf{S}_0 \in \mathfrak{R}^n$ *at* $t_0 \in \mathfrak{T}$ *and for any final mathematical state or physical state vector* $\mathbf{S}_1 \in \mathfrak{R}^n$, *respectively, there exist a moment* $t_1 \in In\mathfrak{T}_0$ *and an extended control* $\mathbf{U}^\mu_{[t_0,t_1]}$ *on the time interval* $[t_0, t_1]$ *such that*

$$\mathbf{S}(t_1; t_0; \mathbf{S}_0; \mathbf{0}_d; \mathbf{U}^\mu_{[t_0,t_1]}) = \mathbf{S}_1, \qquad (10.1)$$

where μ *is the order of the highest control vector derivative* $\mathbf{U}^{(\mu)}(t)$ *acting on the system,* $\mu \geq 0$.

Note 162 *The zero vector* $\mathbf{0}_d \in \mathfrak{R}^d$ *in the notation* $\mathbf{S}(t_1; t_0; \mathbf{S}_0; \mathbf{0}_d; \mathbf{U}^\mu_{[t_0,t_1]})$ *(10.1) signifies that the system is unperturbed, i.e., there is not an external disturbance acting on the system:* $\mathbf{D}(t) \equiv \mathbf{0}_d$. *For the sake of the simplicity we will write* $\mathbf{S}(t; t_0; \mathbf{S}_0; \mathbf{U}^\mu_{[t_0,t_1]})$ *for* $\mathbf{S}(t; t_0; \mathbf{S}_0; \mathbf{0}_d; \mathbf{U}^\mu_{[t_0,t_1]})$ *in the sequel.*

We accept to consider the controllability of the unperturbed system, i.e. for $\mathbf{D}(t) \equiv \mathbf{0}_d$, *so that the input vector* $\mathbf{I} = \begin{bmatrix} \mathbf{D}^T & \vdots & \mathbf{U}^T \end{bmatrix}^T$ *reduces to* $\mathbf{I} = \begin{bmatrix} \mathbf{0}_d^T & \vdots & \mathbf{U}^T \end{bmatrix}^T$.

Comment 163 *On the notions: mathematical state controllability and physical state controllability*

We use explicitly adjectives "mathematical" or "physical" before the term "state controllability" only when it is necessary to emphasize the single state nature. Otherwise, i.e., **when the state is both mathematical and physical** *we use simply the therm* **"state controllability" without being preceded by the adjective "mathematical" or "physical"** *(see Conclusion 229, Section C.1).*

Comment 164 *In a special case the zero state vector* $\mathbf{S} = \mathbf{0}_n$ *can be accepted for the final state vector* $\mathbf{S}_1$.

By the definition we are interested primarily in the system output response, which emphasizes the importance of the system output controllability.

Definition 165 *Output controllability of linear dynamical systems*

A linear dynamical system is the output controllable *if and only if for every initial output vector* $\mathbf{Y}_0 \in \mathfrak{R}^N$ *at* $t_0 \in \mathfrak{T}$ *and for any final output vector* $\mathbf{Y}_1 \in \mathfrak{R}^N$ *there exist a moment* $t_1 \in In\mathfrak{T}_0$ *and an extended control* $\mathbf{U}^{\mu}_{[t_0,t_1]}$ *on the time interval* $[t_0, t_1]$ *such that*

$$\mathbf{Y}(t_1; t_0; \mathbf{Y}_0; \mathbf{U}^{\mu}_{[t_0,t_1]}) = \mathbf{Y}_1, \qquad (10.2)$$

where μ *is the order of the highest control vector derivative* $\mathbf{U}^{(\mu)}(t)$ *acting on the system,* $\mu \geq 0$.

Claim 166 *Output controllability versus physical output controllability*

    By the definition the output variables and the output vector are physical variables and physical vector, respectively. Therefore, the output controllability of a dynamical system is simultaneously its physical output controllability.

Note 167 *Output controllability versus state controllability*

    The state controllability and output controllability are independent system qualitative properties in general.

Comment 168 *In a special case the zero output vector* $\mathbf{Y} = \mathbf{0}_N$ *can be accepted for the final output vector* $\mathbf{Y}_1$: $\mathbf{Y}_1 = \mathbf{0}_N$.

Comment 169 *The controllability concept has the full sense and the greatest importance for the plants (Definition 9 in Section 1.4).*

Note 170 *Controllability and the control law*

    Definition 161 and Definition 165 do not specify the control law. They demand only the existence of an extended control that can steer an arbitrary initial state or output to an arbitrary final state or output, respectively, over a finite time interval $[t_0, t_1] \subset \mathfrak{T}_0$. *They leave the selection of the control law, control algorithm, open and free.*

10.4   General state controllability criteria

Let the system (8.1), (8.2), (Section 8.1), be unperturbed, i.e., $\mathbf{D}(t) \equiv \mathbf{0}_d$ so that it takes the following form:

$$\frac{d\mathbf{S}(t)}{dt} = A\mathbf{S}(t) + B^{(\mu)}\mathbf{U}^{\mu}(t), \mathbf{U}^{\mu}(t) = \begin{bmatrix} \mathbf{U}(t) \\ \mathbf{U}^{(1)}(t) \\ \dots \\ \mathbf{U}^{(\mu)}(t) \end{bmatrix} \in \mathfrak{R}^{(\mu+1)r}, \qquad (10.3)$$

$$\mathbf{Y}(t) = C\mathbf{S}(t) + Q\mathbf{U}(t). \qquad (10.4)$$

For the justification of allowing the existence of the input vector function derivatives in the state Equation (10.3) see Note 15 (Section 1.4) and for its physical origin see Note 50 (Section 4.1).

The fundamental matrix function $\Phi(.,t_0) : \mathfrak{T} \longrightarrow \mathfrak{R}^{n \times n}$ of the system (10.3), (10.4) is the same as of the *ISO* system (3.1), (3.2). It is defined in Equation (3.4). Equations (3.5)-(3.7) express its properties.

The solution of the state equation (10.3) obtained after its integration is given by

$$\mathbf{S}(t;t_0;\mathbf{U}^\mu) = \Phi(t,t_0)\,\mathbf{S}_0 + \int_{t_0}^{t} \Phi(t,\tau)\,\mathrm{B}^{(\mu)}\mathbf{U}^\mu(\tau)\,d\tau =$$

$$= \Phi(t,t_0)\left[\mathbf{S}_0 + \int_{t_0}^{t} \Phi(t_0,\tau)\,\mathrm{B}^{(\mu)}\mathbf{U}^\mu(\tau)\,d\tau\right]. \qquad (10.5)$$

This and the system output Equation (10.4) determine the system response:

$$\mathbf{Y}(t;t_0;\mathbf{S}_0;\mathbf{U}^\mu) = C\Phi(t,t_0)\,\mathbf{S}_0 + \int_{t_0}^{t} C\Phi(t,\tau)\,\mathrm{B}^{(\mu)}\mathbf{U}^\mu(\tau)\,d\tau + Q\mathbf{U}(t) =$$

$$= C\Phi(t,t_0)\left[\mathbf{S}_0 + \int_{t_0}^{t} \Phi(t_0,\tau)\,\mathrm{B}^{(\mu)}\mathbf{U}^\mu(\tau)\,d\tau\right] + Q\mathbf{U}(t). \qquad (10.6)$$

The general state controllability criteria follow:

Theorem 171 *Conditions for the state controllability of the system (10.3), (10.4)*

For the system (10.3), (10.4) to be state controllable it is necessary and sufficient that:

a) any of the following equivalent conditions 1. through 6.a) holds if $\mu = 0$,

b) any of the following equivalent conditions 1. through 5, 6.b) or 6.c) holds if $\mu > 0$:

1. All rows of both matrices $\Phi(t_1,t)B^{(\mu)}$ and $\Phi(t_0,t)B^{(\mu)}$ are linearly independent on $[t_0,t_1]$ for any $(t_0,t_1 > t_0) \in In\mathfrak{T}_0 \times In\mathfrak{T}_0$.

2. All rows of $\Phi(s)B^{(\mu)} = (sI - A)^{-1}B^{(\mu)}$ are linearly independent on $\mathbb{C}$.

3. The Gram matrix $G_{\Phi V}(t_1, t_0)$ (10.7) of $\Phi(t_1, t) B^{(\mu)}$,

$$G_{\Phi B}(t_1, t_0) = \int_{t_0}^{t_1} \Phi(t_1, \tau) B^{(\mu)} \left(B^{(\mu)}\right)^T \Phi^T(t_1, \tau)\, d\tau \qquad (10.7)$$

is nonsingular for any $(t_0, t_1 > t_0) \in In\mathfrak{T}_0 \times In\mathfrak{T}_0$; i.e.,

$$rank G_{\Phi B}(t_1, t_0) = n, \quad for \ any \ (t_0, t_1 > t_0) \in In\mathfrak{T}_0 \times In\mathfrak{T}_0. \qquad (10.8)$$

4. The $n \times n(\mu + 1)r$ controllability matrix $\mathcal{C}$,

$$\mathcal{C} = \left[B^{(\mu)} \vdots AB^{(\mu)} \vdots A^2 B^{(\mu)} \vdots ... \vdots A^{n-1} B^{(\mu)}\right] \in \mathfrak{R}^{n \times n(\mu+1)r}, \qquad (10.9)$$

has the full rank n,

$$rank \mathcal{C} = n. \qquad (10.10)$$

5. For every eigenvalue $s_i(A)$ of the matrix A, equivalently for every complex number $s \in \mathbb{C}$, the $n \times (n + (\mu + 1) r)$ matrix $\left[sI - A \vdots B^{(\mu)}\right]$ has the full rank n,

$$rank \left[sI - A \vdots B^{(\mu)}\right] = n,$$

$$\forall s = s_i(A) \in \mathbb{C}, \ \forall i = 1, 2, ...n, \ i.e., \ \forall s \in \mathbb{C}. \qquad (10.11)$$

6. a) If $\mu = 0$ the control vector function $\mathbf{U}(.)$ obeys

$$\mathbf{U}(t) = (\Phi(t_1, t) B_0)^T G_{\Phi B}^{-1}(t_1, t_0) [\mathbf{S}_1 - \Phi(t_1, t_0) \mathbf{S}_0],$$

$$\forall t \in [t_0, t_1], \ t_1 \in In\mathfrak{T}_0. \qquad (10.12)$$

b) If $\mu > 0$ the control vector function $\mathbf{U}(.)$ obeys either

$$T^{-1} \mathbf{U}^{\mu}(t) = T^T \left(B^{(\mu)}\right)^T \Phi^T(t_1, t) G_{\Phi BT}^{-1}(t_1, t_0) [\mathbf{S}_1 - \Phi(t_1, t_0) \mathbf{S}_0], \ \mu > 0, \qquad (10.13)$$

where $T \in \mathfrak{R}^{(\mu+1)r \times (\mu+1)r}$ is any nonsingular matrix, $detT \neq 0$, and the matrix $G_{\Phi BT}(t_1, t_0)$ is the Gram matrix of $\Phi(t_1, t) B^{(\mu)} T$,

$$G_{\Phi BT}(t_1, t_0) = \int_{t_0}^{t_1} \Phi(t_1, \tau) B^{(\mu)} T T^T \left(B^{(\mu)}\right)^T \Phi^T(t_1, \tau)\, d\tau, \qquad (10.14)$$

c) or the control vector function $\mathbf{U}(.)$ obeys

$$B^{(\mu)} \mathbf{U}^{\mu}(t) = \Phi^T(t_1, t) G_{\Phi}^{-1}(t_1, t_0) [\mathbf{S}_1 - \Phi(t_1, t_0) \mathbf{S}_0], \ \mu > 0, \qquad (10.15)$$

where $G_\Phi(t_1, t_0)$ *is the Gram matrix of* $\Phi(t_1, t)$,

$$G_\Phi(t_1, t_0) = \int_{t_0}^{t_1} \Phi(t_1, \tau) \Phi^T(t_1, \tau) \, d\tau. \qquad (10.16)$$

Appendix D.10 contains the proof of Theorem 171.

Comment 172 *It seems that the conditions 1. - 5. of Theorem 171 have been known. However, in the existing state controllability conditions the matrix* $B \in \mathfrak{R}^{n \times r}$ *describes the transmition of the action of the control vector* **U** *on the system. In Theorem 171 the extended matrix* $B^{(\mu)} \in \mathfrak{R}^{n \times (\mu+1)r}$ *describes the action transmission of the whole extended control vector* $\mathbf{U}^\mu$, *i.e., of the control vector and of its derivatives derivatives up to the order* μ, *on the system. For* $\mu = 0$ *the conditions 1. - 5. of Theorem 171 become those well known so far for the ISO systems and for the IO systems formally transformed into the ISO systems by loosing completely the physical meaning if* $\mu > 0$.

10.5 General output controllability criteria

Equation (10.6) is the basis for the study of the system output controllability (via the combined state-space approach and the output space approach).

The integrability of the Dirac impulse $\delta(.)$ (see [2, Lemma 16.1, pp. 72-75] and [40, Appendix B.2, p. 401]) enables the following:

$$\int_{t_0}^{t} \delta(t - \tau) Q\mathbf{U}(\tau) = Q \int_{t_0}^{t} \delta(t, \tau) \mathbf{U}(\tau) = Q\mathbf{U}(t), \ \forall t \in \mathfrak{T}_0. \qquad (10.17)$$

This leads to the following equivalent form of Equation (10.6) if $\mu = 0$:

$$\mathbf{Y}(t; t_0; \mathbf{S}_0; \mathbf{U}) = C\Phi(t, t_0) \mathbf{S}_0 + \int_{t_0}^{t} [C\Phi(t, \tau) + \delta(t, \tau) Q] \mathbf{U}(\tau) \, d\tau. \qquad (10.18)$$

The following facts are valid if $\mu > 0$:

$$Q\mathbf{U}(t) = \int_{t_0}^{t} Q\mathbf{U}^{(1)}(t)dt + Q\mathbf{U}(t_0) = \int_{t_0}^{t} \widetilde{Q}\mathbf{U}^\mu(t)dt + Q^{(\mu-1)}\mathbf{U}^{\mu-1}(t_0),$$

$$\widetilde{Q} = \left[O_{N,r} \vdots Q \vdots O_{N,r} \vdots .... \vdots O_{N,r} \right] \in \mathfrak{R}^{N \times (\mu+1)r},$$

$$Q^{(\mu-1)} = \left[Q \vdots O_{N,r} \vdots O_{N,r} \vdots .... \vdots O_{N,r} \right] \in \mathfrak{R}^{N \times \mu r}, \qquad (10.19)$$

Equation (10.18) and Equations (10.19) permit us to present the system response, Equation (10.6), in new forms valid for $\mu \geq 0$:

$$\mathbf{Y}(t; t_0; \mathbf{S}_0; \mathbf{U}^\mu) = C\Phi(t, t_0)\,\mathbf{S}_0 +$$

$$+ \left\{ \begin{array}{l} \displaystyle\int_{t_0}^{t} \left[C\Phi(t, \tau)\,\mathrm{B}^{(\mu)} + \widetilde{Q} \right] \mathbf{U}^\mu(\tau)\,d\tau + Q^{(\mu-1)}\mathbf{U}_0^{\mu-1},\ \mu > 0, \\[2ex] \displaystyle\int_{t_0}^{t} \left[C\Phi(t, \tau)\,\mathrm{B}_0 + \delta(t, \tau)\,Q \right] \mathbf{U}(\tau)\,d\tau,\ \mu = 0. \end{array} \right\}, \quad (10.20)$$

Equation (10.20) suggests us to introduce a new system matrix function $H_S(., t_0) : \mathfrak{T} \longrightarrow Cl\mathfrak{R}^{N \times (\mu+1)r}$, for any $t_0 \in \mathfrak{T}$, defined by the following:

$$H_S(t, t_0) =$$

$$= \left\{ \begin{array}{l} \left(C\Phi(t, t_0)\mathrm{B}^{(\mu)} + \widetilde{Q} \right) \in \mathfrak{R}^{N \times (\mu+1)r}, \mu > 0, \\[1.5ex] (C\Phi(t, t_0)\mathrm{B}_0 + \delta(t, t_0)\,Q) \in \mathfrak{R}^{N \times r},\ \mu = 0. \end{array} \right\} \quad (10.21)$$

Its Laplace transform $H_S(s)$ reads:

$$H_S(s) = \mathcal{L}\{H_S(t, t_0)\} =$$

$$= \left\{ \begin{array}{l} C(sI - A)^{-1}\mathrm{B}^{(\mu)} + s^{-1}\widetilde{Q} \in \mathbb{C}^{N \times (\mu+1)r}, \mu > 0, \\[1.5ex] C(sI - A)^{-1}\mathrm{B}_0 + Q \in \mathbb{C}^{N \times r},\ \mu = 0. \end{array} \right\} \quad (10.22)$$

$H_S(t, t_0)$ and its Laplace transform $H_S(s)$ should be distinguished, respectively, from $G_S(t, t_0)$ if $\mu > 0$,

$$G_S(t, t_0) = \mathcal{L}^{-1}\{G_S(s)\} =$$

$$= \mathcal{L}^{-1} \left\{ \begin{array}{l} \left(C(sI - A)^{-1}\mathrm{B}^{(\mu)} + s^{-1}\widetilde{Q} \right) S_r^{(\mu)}(s),\ \mu > 0, \\[1.5ex] \left(C(sI - A)^{-1}\mathrm{B}_0 + Q \right),\ \mu = 0, \end{array} \right\}, \quad (10.23)$$

and from its Laplace transform $G_S(s)$,

$$G_S(s) = \left\{ \begin{array}{l} \left(C(sI - A)^{-1}\mathrm{B}^{(\mu)} + s^{-1}\widetilde{Q} \right) S_r^{(\mu)}(s), \mu > 0, \\[1.5ex] \left(C(sI - A)^{-1}\mathrm{B}_0 + Q \right),\ \mu = 0, \end{array} \right\}, \quad (10.24)$$

which is the system transfer function matrix (relative to the control vector $\mathbf{U}$). Thus,

$$G_S(t,t_0) \begin{cases} \neq H_S(t,t_0) \ if \ \mu > 0, \\ = H_S(t,t_0) \ if \ \mu = 0, \end{cases}, \tag{10.25}$$

$$H_S(s) = \mathcal{L}\{H_S(t,t_0)\} =$$

$$= \begin{cases} C\,(sI - A)^{-1}\,\mathrm{B}^{(\mu)} + s^{-1}\tilde{Q} \neq G_S(s), \ \mu > 0, \\ C\,(sI - A)^{-1}\,\mathrm{B}_0 + Q = G_S(s), \ \mu = 0, \end{cases}, \tag{10.26}$$

$$G_S(s) = \begin{cases} H_S(s)S_r^{(\mu)}(s), \ \mu > 0, \\ H_S(s), \ \mu = 0. \end{cases}. \tag{10.27}$$

The introduction of $H_S(t,t_0)$ sets the system response (10.20) in a more compact form (10.28),

$$\mathbf{Y}\,(t;t_0;\mathbf{S}_0;\mathbf{U}) = C\Phi\,(t,t_0)\,\mathbf{S}_0+$$

$$+ \begin{cases} \displaystyle\int_{t_0}^{t} H_S\,(t,\tau)\,\mathbf{U}^{\mu}\,(\tau)\,d\tau \; + Q^{(\mu-1)}\mathbf{U}_0^{\mu-1}, \ \mu > 0, \\ \displaystyle\int_{t_0}^{t} H_S\,(t,\tau)\,\mathbf{U}\,(\tau)\,d\tau, \ \mu = 0. \end{cases}. \tag{10.28}$$

The general output controllability conditions for the system (10.3), (10.4) follow:

Theorem 173 *Conditions for the output controllability of the system (10.3), (10.4)*

For the system (10.3), (10.4) to be output controllable it is necessary that

$$rank\left[C \vdots Q\right] = N \tag{10.29}$$

and if this condition is satisfied then:

A) it is necessary that any of the following conditions 1.-6.a) holds and sufficient that any of the following conditions 1.-4., 6.a) holds if $\mu = 0$, where the conditions 1.-4.,6.a) are equivalent,

B) it is necessary that any of the following conditions 1.-5., 6.b) holds and sufficient that any of the following conditions 1.-4., 6.b) holds if $\mu > 0$, where the conditions 1.-4.,6.b) are equivalent:

1. All rows of the system matrix $H_S(t,t_0)$ (10.21) are linearly independent on $[t_0,t_1]$ for any $(t_0,t_1 > t_0) \in In\mathfrak{T}_0 \times In\mathfrak{T}_0$.

2. *All rows of the system matrix $H_S(s)$ (10.22) are linearly independent on $\mathbb{C}$.*

3. *The Gram matrix $G_{H_S}(t_1, t_0)$ of $H_S(t, t_0)$ (10.21),*

$$G_{H_S}(t_1, t_0) = \int_{t_0}^{t_1} H_S(t_1, \tau) H_S^T(t_1, \tau) d\tau,$$

$$\text{for any } (t_0, t_1 > t_0) \in In\mathfrak{T}_0 \times In\mathfrak{T}_0. \tag{10.30}$$

is nonsingular, i.e.,

$$rank G_{H_S}(t_1, t_0) = N, \text{ any } (t_0, t_1 > t_0) \in In\mathfrak{T}_0 \times In\mathfrak{T}_0. \tag{10.31}$$

4. *The output $N \times (n+1)(\mu+1)r$ controllability matrix $\mathcal{C}_{Sout}$,*

$$\mathcal{C}_{Sout} = \left\{ \begin{array}{l} \left[CB^{(\mu)} \vdots CAB^{(\mu)} \vdots ... \vdots CA^{n-1}B^{(\mu)} \vdots \widetilde{Q} \right], \ \mu > 0, \\[2mm] \left[CB_0 \vdots CAB_0 \vdots ... \vdots CA^{n-1}B_0 \vdots Q \right], \ \mu = 0, \end{array} \right\} \tag{10.32}$$

has the full rank N,

$$rank \mathcal{C}_{Sout} = N. \tag{10.33}$$

5. *For every eigenvalue $s_i(A)$ of the matrix A, equivalently for every complex number $s \in \mathbb{C}$, the $N \times (n+2(\mu+1)r)$ matrix $\mathcal{C}_{outS}$,*

$$\mathcal{C}_{outS} = \left\{ \begin{array}{l} \left[C(sI-A) \vdots CB^{(\mu)} \vdots \widetilde{Q} \right], \ \mu > 0, \\[2mm] \left[C(sI-A) \vdots CB_0 \vdots Q \right], \ \mu = 0, \end{array} \right\} \tag{10.34}$$

has the full rank N,

$$rank \mathcal{C}_{outS} = rank \left\{ \begin{array}{l} \left[C(sI-A) \vdots CB^{(\mu)} \vdots \widetilde{Q} \right], \ \mu > 0, \\[2mm] \left[C(sI-A) \vdots CB_0 \vdots Q \right], \ \mu = 0, \end{array} \right\} = N,$$

$$\forall s = s_i(A) \in \mathbb{C}, \ \forall i = 1, 2, ...n, \text{ i.e., } \forall s \in \mathbb{C}. \tag{10.35}$$

6. *The control vector function $\mathbf{U}(.)$ obeys:*
   a) The following equation if $\mu = 0$:

$$\mathbf{U}(t) = (H_S(t_1, \tau))^T G_{H_S}^{-1}(t_1, t_0) \left[\mathbf{Y}_1 - C\Phi(t_1, t_0) \mathbf{S}_0 \right], \ \mu = 0, \tag{10.36}$$

b) The following equations for any $(\mu + 1) r \times (\mu + 1) r$ nonsingular matrix R:

$$R^{-1}\mathbf{U}^{\mu}(t) = \left\{ \begin{array}{c} (H_S(t_1,t)\,R)^T \bullet G^{-1}_{H_S R}(t_1,t_0) \bullet \\ \bullet \left[\mathbf{Y_1} - C\Phi(t_1,t_0)\,\mathbf{S_0} - Q^{(\mu-1)}\mathbf{U}_0^{\mu-1}\right] \end{array} \right\}, \quad \mu > 0, \quad (10.37)$$

where $G_{H_S R}(t_1,t_0)$ is the Gram matrix of $H_S(t_1,\tau)\,R$,

$$G_{H_S R}(t_1,t_0) = \int_{t_0}^{t_1} H_S(t_1,\tau)\,RR^T H_S^T(t_1,\tau)\,d\tau. \quad (10.38)$$

For the proof consult Appendix D.11.

Theorem 174 *On the necessary output controllability condition of Theorem 173*

If the matrix $\left[C \vdots Q\right] = C$ for $Q = O_{N,r}$ does not have the full rank N, i.e., if Equation (10.29) is not satisfied, then the system (10.3), (10.4) is not output controllable in view of Definition 165.

Note 175 *The existing literature treats the system state controllability and output controllability only for the case $\mu = 0$.*

Chapter 11

Various systems controllability

11.1   *IO* system state controllability

11.1.1   Definition

The general Definition 161 of the state controllability takes the following form for the *IO* system (2.1):

Definition 176 *State controllability of the IO system (2.1), i.e., (2.15)*

*The IO system (2.1), i.e., (2.15), is **mathematical state or physical state controllable** if and only if for every initial mathematical state or physical state vector $\mathbf{Y}_0^{\nu-1} \in \mathfrak{R}^{\nu N}$ at $t_0 \in \mathfrak{T}$ and for any final mathematical state or physical state vector $\mathbf{Y}_1^{\nu-1} \in \mathfrak{R}^{\nu N}$, respectively, there exist a moment $t_1 \in In\mathfrak{T}_0$ and an extended control $\mathbf{U}_{[t_0,t_1]}^{\mu}$ on the time interval $[t_0, t_1]$ such that*

$$\mathbf{Y}^{\nu-1}(t_1; t_0; \mathbf{Y}_0^{\nu-1}; \mathbf{0}_d; \mathbf{U}_{[t_0,t_1]}^{\mu}) = \mathbf{Y}_1^{\nu-1}, \qquad (11.1)$$

where μ is the order of the highest control vector derivative $\mathbf{U}^{(\mu)}(t)$ acting on the system, $\mu \geq 0$.

Note 177 *The zero vector $\mathbf{0}_d \in \mathfrak{R}^d$ in $\mathbf{Y}^{\nu-1}(t_1; t_0; \mathbf{Y}_0^{\nu-1}; \mathbf{0}_d; \mathbf{U}_{[t_0,t_1]}^{\mu})$ (11.1) denotes that the system is unperturbed, i.e., there is not an external disturbance acting on the system: $\mathbf{D}(t) \equiv \mathbf{0}_d$. For the sake of the simplicity we will write $\mathbf{Y}^{\nu-1}(t; t_0; \mathbf{Y}_0^{\nu-1}; \mathbf{U}_{[t_0,t_1]}^{\mu})$ for $\mathbf{Y}^{\nu-1}(t; t_0; \mathbf{Y}_0^{\nu-1}; \mathbf{0}_d; \mathbf{U}_{[t_0,t_1]}^{\mu})$ in the sequel.*

This permits us to accept $D^{(\mu)} = O_{N,d(\mu+1)}$.

Comment 178 *The existence of the control* $\mathbf{U}_{[t_0,t_1]}$ *on the time interval* $[t_0,t_1]$ *means the existence of the extended control* $\mathbf{U}^{\mu}_{[t_0,t_1]}$ *on the time interval* $[t_0,t_1]$ *in view of (2.1), i.e., (2.15).*

Note 179 *The equivalence among (2.1), (2.15) and (2.57) permits us to refer in the sequel to anyone of them.*

11.1.2   Criteria

What follows resolves completely the problem of the state controllability; i.e., it resolves the problem of the necessary and sufficient conditions for the state controllability, of the *IO* system (2.15).

We recall $det A_\nu \neq 0$ due to (2.1) so that also $det A_\nu^{-1} \neq 0$.

Comment 144 (Section 9.1) explains the difference between the physical state $\mathbf{X} = \mathbf{Y}^{\nu-1}$ (2.16) (Section 2.1) and the formal mathematical state $\widehat{\mathbf{X}}$ (C.4)-(C.10) (Appendix C.1) of the *IO* system (2.15) . We are interested in the physical state controllability. The criteria for the formal mathematical state controllability of the *IO* system (2.15) are well known in the control theory.

The equivalent description of the *IO* system (2.15), which preserves the physical sense of the state is its *EISO* form (4.1), (4.2) specified by (4.4)-(4.11) (Section 4.1). The matrix A is defined by Equation (4.6) (Section 4.1), which is repeated as (11.2):

$$\nu > 1 \Longrightarrow A =$$

$$\begin{bmatrix} O_N & I_N & \dots & O_N & O_N \\ O_N & O_N & \dots & O_N & O_N \\ \dots & \dots & \dots & \dots & \dots \\ O_N & O_N & \dots & I_N & O_N \\ O_N & O_N & \dots & O_N & I_N \\ -A_\nu^{-1}A_0 & -A_\nu^{-1}A_1 & \dots & -A_\nu^{-1}A_{\nu-2} & -A_\nu^{-1}A_{\nu-1} \end{bmatrix} \in \mathfrak{R}^{\nu N \times \nu N},$$

$$n = \nu N,$$

$$\nu = 1 \Longrightarrow A = -A_1^{-1}A_0 \in \mathfrak{R}^{N \times N},$$

$$(11.2)$$

In view of Note 162, the input vector $\mathbf{I} = \begin{bmatrix} \mathbf{D}^T & \vdots & \mathbf{U}^T \end{bmatrix}^T$ reduces to $\mathbf{I} = \begin{bmatrix} \mathbf{0}_d^T & \vdots & \mathbf{U}^T \end{bmatrix}^T$ and simplifies the product $H^{(\mu)}\mathbf{I}^\mu$ (Equation (2.15) in Section 2.1) to $B^{(\mu)}\mathbf{U}^\mu$. These explanations enable us to use the following compact form of the unperturbed *IO* system (2.15):

$$A^{(\nu)}\mathbf{Y}^\nu(t) = B^{(\mu)}\mathbf{U}^\mu(t), \ \forall t \in \mathfrak{T}_0. \tag{11.3}$$

The reduction of $H^{(\mu)}\mathbf{I}^\mu$ to $B^{(\mu)}\mathbf{U}^\mu$ reduces $P^{(\mu)}\mathbf{I}^\mu$ in Equation (4.1) (Section 4.1) to $\mathrm{B}^{(\mu)}\mathbf{U}^\mu$, where the matrices B_{inv},

$$\mathrm{B}_{inv} = \left\{ \begin{bmatrix} O_{(\nu-1)N,N} \\ I_N \end{bmatrix} \in \mathfrak{R}^{n \times N}, \ \nu > 1, \\ I_N \in \mathfrak{R}^{N \times N}, \ \nu = 1, \right\}, \tag{11.4}$$

A_ν^{-1} and $B^{(\mu)}$,

$$B^{(\mu)} = \begin{bmatrix} B_0 & \vdots & B_1 & \vdots & ... & \vdots & B_\mu \end{bmatrix} \in \mathfrak{R}^{N \times (\mu+1)r}, \ B^{(0)} = B_0 \tag{11.5}$$

compose the matrix $\mathrm{B}^{(\mu)}$,

$$\mathrm{B}^{(\mu)} = \left\{ \begin{bmatrix} O_{(\nu-1)N,(\mu+1)r} \\ A_\nu^{-1}B^{(\mu)} \end{bmatrix}, \ 0 \le \mu \le \nu, \ \nu > 1, \\ I_N A_1^{-1}B^{(\mu)}, \ \nu = 1, \ 0 \le \mu \le 1, \right\} =$$
$$= \left\{ \begin{bmatrix} O_{(\nu-1)N,N} \\ I_N \end{bmatrix} A_\nu^{-1}B^{(\mu)}, \ 0 \le \mu \le \nu > 1, \\ I_N A_1^{-1}B^{(\mu)}, \ \nu = 1, \ 0 \le \mu \le 1, \right\} = \mathrm{B}_{inv}A_\nu^{-1}B^{(\mu)}. \tag{11.6}$$

The output matrices C and Q are given by (4.11) repeated as (11.7):

$$C = [I_N \ O_N \ O_N \ O_N \ ... \ O_N] \in \mathfrak{R}^{N \times n}, \ RankC \equiv N,$$
$$Q = O_{N,M} \in \mathfrak{R}^{N \times M}. \tag{11.7}$$

The matrix B_{inv} is constant, invariant, independent of the system matrices. The order ν of the system and the dimension N of the system output vector $\mathbf{Y}$ determine its structure as shown in (11.4).

In view of (11.2) - (11.6), the following unperturbed form (11.8), (11.9) of the *EISO* system (4.1), (4.2), is both physically and mathematically equivalent to the *IO* system (2.15):

$$\frac{d\mathbf{X}(t)}{dt} = A\mathbf{X}(t) + \mathrm{B}^{(\mu)}\mathbf{U}^\mu(t), \ \forall t \in \mathfrak{T}_0, \ 1 \le \mu, \ \mathrm{B}^{(\mu)} \in \mathfrak{R}^{n \times (\mu+1)r}, \tag{11.8}$$

$$\mathbf{Y}(t) = C\mathbf{X}(t), \ \forall t \in \mathfrak{T}_0. \tag{11.9}$$

In view of Equations (4.19) and (4.22) (Section 4.1), the motion and response of this system are determined by the following equations:

$$\mathbf{X}(t; t_0; \mathbf{X}_0; \mathbf{U}^\mu) = \Phi(t, t_0)\mathbf{X}_0 + \int_{t_0}^{t} \Phi(t, \tau) \mathbf{B}^{(\mu)} \mathbf{U}^\mu(\tau)\, d\tau =$$

$$= \Phi(t, t_0)\left[\mathbf{X}_0 + \int_{t_0}^{t} \Phi(t_0, \tau) \mathbf{B}^{(\mu)} \mathbf{U}^\mu(\tau)\, d\tau\right], \ \forall t \in \mathfrak{T}_0, \qquad (11.10)$$

$$\mathbf{Y}(t; t_0; \mathbf{X}_0; \mathbf{U}^\mu) = C\Phi(t, t_0) + \int_{t_0}^{t} C\Phi(t, \tau) \mathbf{B}^{(\mu)} \mathbf{U}^\mu(\tau)\, d\tau =$$

$$= C\Phi(t, t_0)\left[\mathbf{X}_0 + \int_{t_0}^{t} \Phi(t_0, \tau) \mathbf{B}^{(\mu)} \mathbf{U}^\mu(\tau)\, d\tau\right], \ \forall t \in \mathfrak{T}_0, \qquad (11.11)$$

where (see Equations (3.3)-(3.6), Section 3.1):

$$\Phi(t, t_0) = [\phi_{ij}(t, t_0)] = e^{A(t-t_0)} \in \mathfrak{R}^{n\times n}, \ \Phi(t_0, t_0) \equiv I. \qquad (11.12)$$

Theorem 180 *State controllability criteria for the IO system (2.15)*

For the IO system (2.15) to be state controllable it is necessary and sufficient that:

a) Any of the following equivalent conditions 1. through 6.a) holds if $\mu = 0$,

b) Any of the following equivalent conditions 1. through 5, 6.b) or 6.c) holds if $\mu > 0$,

in which Equations (11.2)-(11.6) induced by (2.15) determine the matrices:

1) All rows of matrices $\Phi(t_1, t)\mathbf{B}^{(\mu)}$, $\Phi(t_1, t)\mathbf{B}_{inv}$ and $\Phi(t_0, t)\mathbf{B}^{(\mu)}$ are linearly independent on $[t_0, t_1]$ for any $(t_0, t_1 > t_0) \in In\mathfrak{T}_0 \times In\mathfrak{T}_0$.

2) All rows of both $\Phi(s)\mathbf{B}^{(\mu)} = (sI - A)^{-1}\mathbf{B}^{(\mu)}$ and $(sI_n - A)^{-1}\mathbf{B}_{inv}$ are linearly independent on $\mathbb{C}$.

3) The Gram matrices $G_{ctB}(t_1, t_0)$ of $\Phi(t_1, t)\mathbf{B}^{(\mu)}$ and $G_{ctBinv}(t_1, t_0)$ of $\Phi(t_1, t)\mathbf{B}_{inv}$,

$$G_{ctB}(t_1, t_0) = \int_{t_0}^{t_1} \Phi(t_1, \tau) \mathbf{B}^{(\mu)}\left(\mathbf{B}^{(\mu)}\right)^T \Phi^T(t_1, \tau)\, d\tau, \ \forall t \in \mathfrak{T}_0,$$

$$G_{ctBinv}(t_1, t_0) = \int_{t_0}^{t_1} \Phi(t_1, \tau) \mathbf{B}_{inv}\mathbf{B}_{inv}^T \Phi^T(t_1, \tau)\, d\tau, \ \forall t \in \mathfrak{T}_0, \qquad (11.13)$$

are nonsingular for any $(t_0, t_1 > t_0) \in In\mathfrak{T}_0 \times In\mathfrak{T}_0$; *i.e.,*

$$rankG_{ctB}(t_1, t_0) = rank\ G_{ctBinv}(t_1, t_0) = n,$$
$$for\ any\ (t_0, t_1 > t_0) \in In\mathfrak{T}_0 \times In\mathfrak{T}_0. \tag{11.14}$$

4) The $n \times n(\mu + 1)r$ *controllability matrix* $\mathcal{C}_B$ *and* $n \times nN$ *controllability matrix* $\mathcal{C}_{Binv}$,

$$\mathcal{C}_B = \left[B^{(\mu)} \vdots AB^{(\mu)} \vdots A^2 B^{(\mu)} \vdots ... \vdots A^{n-1}B^{(\mu)} \right],$$

$$\mathcal{C}_{Binv} = \left[B_{inv} \vdots AB_{inv} \vdots A^2 B_{inv} \vdots ... \vdots A^{n-1}B_{inv} \right], \tag{11.15}$$

have the full rank n,

$$rank\ \mathcal{C}_B = rank\ \mathcal{C}_{Binv} = n = \nu N. \tag{11.16}$$

5) For every eigenvalue $s_i(A)$ *of the matrix* A, *equivalently for every complex number* $s \in \mathbb{C}$, *the* $n \times (n + (\mu + 1)r)$ *matrix* $\left[sI - A \vdots B^{(\mu)} \right]$ *and the* $n \times (n + N)$ *matrix* $\left[sI - A \vdots B_{inv} \right]$ *have the full rank* n,

$$rank \left[sI - A \vdots B^{(\mu)} \right] = rank \left[sI - A \vdots B_{inv} \right] = n,$$
$$\forall s = s_i(A) \in \mathbb{C},\ \forall i = 1, 2, ..., n,\ \forall s \in \mathbb{C}. \tag{11.17}$$

6. a) If $\mu = 0$ *the control vector function* $\mathbf{U}(.)$ *obeys*

$$\mathbf{U}(t) = (\Phi(t_1, t) B_0)^T G_{\Phi B}^{-1}(t_1, t_0) [\mathbf{X}_1 - \Phi(t_1, t_0) \mathbf{X}_0],$$
$$\forall t \in [t_0, t_1],\ t_1 \in In\mathfrak{T}_0. \tag{11.18}$$

b) If $\mu > 0$ *the control vector function* $\mathbf{U}(.)$ *obeys either*

$$T^{-1}\mathbf{U}^\mu(t) = T^T \left(B^{(\mu)} \right)^T \Phi^T(t_1, t) G_{\Phi BT}^{-1}(t_1, t_0) [\mathbf{X}_1 - \Phi(t_1, t_0) \mathbf{X}_0],\ \mu > 0, \tag{11.19}$$

where $T \in \mathfrak{R}^{(\mu+1)r \times (\mu+1)r}$ *is any nonsingular matrix and* $G_{\Phi BT}(t_1, t_0)$ *is the Gram matrix of* $\Phi(t_1, t) B^{(\mu)}T$,

$$G_{\Phi BT}(t_1, t_0) = \int_{t_0}^{t_1} \Phi(t_1, \tau) B^{(\mu)}TT^T \left(B^{(\mu)} \right)^T \Phi^T(t_1, \tau) d\tau. \tag{11.20}$$

c) or the control vector function $\mathbf{U}(.)$ *obeys*

$$\mathbf{B}^{(\mu)}\mathbf{U}^{\mu}(t) = \Phi^{T}(t_1, t)\, G_{\Phi}^{-1}(t_1, t_0)\left[\mathbf{X}_1 - \Phi(t_1, t_0)\mathbf{X}_0\right], \quad \mu > 0, \qquad (11.21)$$

where $G_{\Phi}(t_1, t_0)$ *is the Gram matrix of* $\Phi(t_1, t)$,

$$G_{\Phi}(t_1, t_0) = \int_{t_0}^{t_1} \Phi(t_1, \tau)\, \Phi^{T}(t_1, \tau)\, d\tau. \qquad (11.22)$$

Proof. This is Theorem 171 in which $\mathbf{B}^{(\mu)}$ can be appropriately replaced by $\mathbf{B}_{inv}$ and $Q = O_{N,r}$ by noting that the linear independence of all rows of $\Phi(t, t_0)\,\mathbf{B}^{(\mu)}$ on $]t_0, t_1]$ for $t_1 \in In\mathfrak{T}_0$, together with $\mathbf{B}^{(\mu)} = \mathbf{B}_{inv}A_{\nu}^{-1}B^{(\mu)}$ (11.6), implies the linear independence of the rows of the matrix product $\Phi(t, t_0)\,\mathbf{B}_{inv}$ and of $\Phi(t, t_0)$ on $[t_0, t_1]$ for $t_1 \in In\mathfrak{T}_0$ due to the statement under 1) of Lemma 107 in Section 7.4. ∎

Example 181 *The following example illustrates Theorem 180*
In this example the control is scalar variable U:

$$2Y_1 + 2Y_1^{(1)} + 2Y_2 = U,$$
$$-Y_1 + 3Y_2 + Y_2^{(1)} = 2U,$$

i.e.,

$$\nu = 1, \; n = \nu N = N = 2, \; \mu = 0, \; M = r = 1, \; B^{(\mu)} = B^{(0)},$$

$$\underbrace{\begin{bmatrix} 2 & 0 \\ 0 & 1 \end{bmatrix}}_{A_1}\mathbf{Y}^{(1)} + \underbrace{\begin{bmatrix} 2 & 2 \\ -1 & 3 \end{bmatrix}}_{A_0}\mathbf{Y} = \underbrace{\begin{bmatrix} 1 \\ 2 \end{bmatrix}}_{H^{(0)}=B^{(0)}=B_0}U.$$

The preceding data specify the following:

$$\mathbf{Y} = \begin{bmatrix} Y_1 \\ Y_2 \end{bmatrix} \in \mathfrak{R}^2,$$

$$\begin{bmatrix} 2 & 0 \\ 0 & 1 \end{bmatrix}\mathbf{Y}^{(1)} + \begin{bmatrix} 2 & 2 \\ -1 & 3 \end{bmatrix}\mathbf{Y} = \begin{bmatrix} 1 \\ 2 \end{bmatrix}U$$

$$A_1 = \begin{bmatrix} 2 & 0 \\ 0 & 1 \end{bmatrix}, \; A_0 = \begin{bmatrix} 2 & 2 \\ -1 & 3 \end{bmatrix}, \; B^{(0)} = \begin{bmatrix} 1 \\ 2 \end{bmatrix}, \; \mathbf{U} = [U],$$

$$\mathbf{X} = \mathbf{Y} = \begin{bmatrix} Y_1 \\ Y_2 \end{bmatrix} = \begin{bmatrix} X_1 \\ X_2 \end{bmatrix} \in \mathfrak{R}^2,$$

$$A_1^{-1} = \begin{bmatrix} 0,5 & 0 \\ 0 & 1 \end{bmatrix} \Longrightarrow A = -A_1^{-1}A_0 = \begin{bmatrix} -1 & -1 \\ 1 & -3 \end{bmatrix},$$

$$B_0 = A_1^{-1}B^{(0)} = \begin{bmatrix} 0,5 \\ 2 \end{bmatrix} \Longrightarrow rankB_0 = 1 < 2 = n,$$

$$\mathbf{Y}^{(1)} = \begin{bmatrix} -1 & -1 \\ 1 & -3 \end{bmatrix} \mathbf{Y} + \begin{bmatrix} 0.5 \\ 2 \end{bmatrix} U \Longrightarrow$$

$$\mathbf{X}^{(1)} = \begin{bmatrix} -1 & -1 \\ 1 & -3 \end{bmatrix} \mathbf{X} + \begin{bmatrix} 0.5 \\ 2 \end{bmatrix} U$$

$$sI_2 - A = \begin{bmatrix} s+1 & 1 \\ -1 & s+3 \end{bmatrix} \Longrightarrow s^2 + 4s + 4, \; s_1(A) = s_2(A) = -2,$$

$$\begin{bmatrix} sI_2 - A \vdots B_0 \end{bmatrix} = \begin{bmatrix} s+1 & 1 & 0,5 \\ -1 & s+3 & 2 \end{bmatrix} \Longrightarrow$$

$$rank \begin{bmatrix} s_i(A) I_2 - A \vdots B_0 \end{bmatrix} = rank \begin{bmatrix} -1 & 1 & 0,5 \\ -1 & 1 & 2 \end{bmatrix} =$$

$$= rank \begin{bmatrix} 1 & 0,5 \\ 1 & 2 \end{bmatrix} = 2, \; i = 1,2.$$

The matrix $\begin{bmatrix} sI_2 - A \vdots B_0 \end{bmatrix}$ *has the invariant full rank* $n = 2$. *The matrix* $\begin{bmatrix} sI_2 - A \vdots B_0 \end{bmatrix}$ *has the full rank on* $\mathbb{C}$ *in spite* $rankB_0 = 1 < 2 = n = N$.

Example 182 *Let*

$$Y_1 + 2Y_1^{(1)} = 4U_1 + 12U_2,$$
$$Y_2 + Y_2^{(1)} = U_1 + 3U_2,$$

i.e.,

$$\mathbf{Y} = \begin{bmatrix} Y_1 \\ Y_2 \end{bmatrix}, \; \mathbf{U} = \begin{bmatrix} U_1 \\ U_2 \end{bmatrix},$$

$$\begin{bmatrix} 2 & 0 \\ 0 & 1 \end{bmatrix} \mathbf{Y}^{(1)} + \begin{bmatrix} 1 & 0 \\ 0 & 1 \end{bmatrix} \mathbf{Y} = \begin{bmatrix} 4 & 12 \\ 1 & 3 \end{bmatrix} \mathbf{U} \Longrightarrow$$

$$A_1 = \begin{bmatrix} 2 & 0 \\ 0 & 1 \end{bmatrix}, \; A_0 = \begin{bmatrix} 1 & 0 \\ 0 & 1 \end{bmatrix}, \; B^{(0)} = \begin{bmatrix} 4 & 12 \\ 1 & 3 \end{bmatrix}$$

For this system $\nu = 1$ and $\mu = 0$, The state vector form of the system reads

$$\frac{d\mathbf{X}}{dt} = \begin{bmatrix} -0,5 & 0 \\ 0 & -1 \end{bmatrix} \mathbf{X} + \begin{bmatrix} 2 & 6 \\ 1 & 3 \end{bmatrix} \mathbf{U},$$

$$\mathbf{X} = \begin{bmatrix} X_1 \\ X_2 \end{bmatrix} = \begin{bmatrix} Y_1 \\ Y_2 \end{bmatrix}, \quad A = -A_1^{-1} = \begin{bmatrix} -0,5 & 0 \\ 0 & -1 \end{bmatrix},$$

$$\mathbf{B}_0 = .A_1^{-1} B^{(0)} = \begin{bmatrix} 0,5 & 0 \\ 0 & 1 \end{bmatrix} \begin{bmatrix} 4 & 12 \\ 1 & 3 \end{bmatrix} = \begin{bmatrix} 2 & 6 \\ 1 & 3 \end{bmatrix}$$

The matrix A is nonsingular. The eigenvalues $s_i(A)$ of A are $s_1(A) = -0,5$ and $s_2(A) = -1$. The controllability test follows:

$$rank\begin{bmatrix} sI_2 - A \vdots \mathbf{B}_0 \end{bmatrix} = \begin{bmatrix} s+0,5 & 0 & 2 & 6 \\ 0 & s+1 & 1 & 3 \end{bmatrix},$$

$$rank\begin{bmatrix} s_1(A) I_2 - A \vdots \mathbf{B}_0 \end{bmatrix} = \begin{bmatrix} 0 & 0 & 2 & 6 \\ 0 & 0,5 & 1 & 3 \end{bmatrix},$$

its submatrix

$$\begin{bmatrix} 0 & 2 \\ 0,5 & 1 \end{bmatrix}$$

is nonsingular, and

$$rank\begin{bmatrix} s_2(A) I_2 - A \vdots \mathbf{B}_0 \end{bmatrix} = \begin{bmatrix} -0,5 & 0 & 2 & 6 \\ 0 & 0 & 1 & 3 \end{bmatrix},$$

yields its nonsingular submatrix

$$\begin{bmatrix} -0,5 & 2 \\ 0 & 1 \end{bmatrix}.$$

The system is state controllable. The system state controllability ensures the existence of a control $\mathbf{U}$ that forces the system from any initial state at the initial moment to an arbitrarily chosen state at some finite moment after the initial one.

Let us consider the case of the *IO* system (2.15) with $A_0 = O_N$ to be state controllable for arbitrary matrices A_i, $i = 0, 1, ..., \nu - 1$. This means *the system state controllability robustness relative to the matrices A_i, $i = 0, 1, ..., \nu - 1$.* It is a kind of *the robust state controllability.*

Theorem 183 *Robust state controllability criteria for the IO system (2.15) with $A_0 = O_N$ relative to arbitrary matrices A_i, $i = 1, ..., \nu - 1$*

For the IO system (2.15) with $A_0 = O_N$ to be invariantly (i.e., robust) state controllable relative to arbitrary matrices A_i, $i = 1, ..., \nu - 1$, it is necessary and sufficient that any of the above equivalent conditions 1. -5. of Theorem 180 holds, or equivalently, that the matrix $B^{(\mu)}$ has the full rank N,

$$rank B^{(\mu)} = N, \tag{11.23}$$

where Equations (11.2)-(11.6) induced by (2.15) determine the matrices.

Appendix D.12 contains the proof of Theorem 183.

Example 184 *The following example illustrates Theorem 183.*

$$Y_1^{(2)} + 2Y_1^{(1)} + Y_1 = U,$$
$$Y_2^{(2)} + Y_2^{(1)} = U^{(1)} \tag{11.24}$$

$$\mathbf{Y} = \begin{bmatrix} Y_1 \\ Y_2 \end{bmatrix},$$

$$\begin{bmatrix} 1 & 0 \\ 0 & 1 \end{bmatrix} \mathbf{Y}^{(2)} + \begin{bmatrix} 2 & 0 \\ 0 & 1 \end{bmatrix} \mathbf{Y}^{(1)} + \begin{bmatrix} 1 & 0 \\ 0 & 0 \end{bmatrix} \mathbf{Y} =$$

$$= \underbrace{\begin{bmatrix} 1 & 0 \\ 0 & 1 \end{bmatrix}}_{B^{(1)}} \underbrace{\begin{bmatrix} U \\ U^{(1)} \end{bmatrix}}_{\mathbf{U}^1},$$

$$A_2 = I_2, \; A_1 = \begin{bmatrix} 2 & 0 \\ 0 & 1 \end{bmatrix}, \; A_0 = \begin{bmatrix} 1 & 0 \\ 0 & 0 \end{bmatrix},$$

$$B^{(1)} = \begin{bmatrix} 1 & 0 \\ 0 & 1 \end{bmatrix} = I_2, \; rank B^{(1)} = 2, \; \mathbf{U}^1 = \begin{bmatrix} U \\ U^{(1)} \end{bmatrix}$$

We apply Equations (11.2)-(11.7):

$$\mathbf{X} = \begin{bmatrix} Y_1 \\ Y_2 \\ Y_1^{(1)} \\ Y_2^{(1)} \end{bmatrix} = \begin{bmatrix} X_1 \\ X_2 \\ X_3 \\ X_4 \end{bmatrix} = \begin{bmatrix} \mathbf{Y} \\ \mathbf{Y}^{(1)} \end{bmatrix} = \begin{bmatrix} \mathbf{X}_1 \\ \mathbf{X}_2 \end{bmatrix}$$

$$\mathbf{X}^{(1)} = \begin{bmatrix} Y_1^{(1)} \\ Y_2^{(1)} \\ Y_1^{(2)} \\ Y_2^{(2)} \end{bmatrix} = \begin{bmatrix} X_3 \\ X_4 \\ X_3^{(1)} \\ X_4^{(1)} \end{bmatrix} = \begin{bmatrix} \mathbf{Y}^{(1)} \\ \mathbf{Y}^{(2)} \end{bmatrix} =$$

$$= \underbrace{\begin{bmatrix} 0 & 0 & 1 & 0 \\ 0 & 0 & 0 & 1 \\ -1 & 0 & -2 & 0 \\ 0 & 0 & 0 & -1 \end{bmatrix}}_{A} \underbrace{\begin{bmatrix} \mathbf{X}_1 \\ \mathbf{X}_2 \end{bmatrix}}_{\mathbf{X}} + \underbrace{\begin{bmatrix} 0 & 0 \\ 0 & 0 \\ 1 & 0 \\ 0 & 1 \end{bmatrix}}_{B^{(1)}} \mathbf{U}^1,$$

$$\mathbf{Y} = \underbrace{\begin{bmatrix} 1 & 0 & 0 & 0 \\ 0 & 1 & 0 & 0 \end{bmatrix}}_{C} \mathbf{X},$$

$$A = \begin{bmatrix} O_2 & I_2 \\ -A_2^{-1}A_0 & -A_2^{-1}A_1 \end{bmatrix} = \begin{bmatrix} 0 & 0 & 1 & 0 \\ 0 & 0 & 0 & 1 \\ -1 & 0 & -2 & 0 \\ 0 & 0 & 0 & -1 \end{bmatrix},$$

$$B^{(1)} = \begin{bmatrix} O_2 \\ A_2^{-1}B^{(1)} \end{bmatrix} = \begin{bmatrix} 0 & 0 \\ 0 & 0 \\ 1 & 0 \\ 0 & 1 \end{bmatrix}, \quad C = \begin{bmatrix} 1 & 0 & 0 & 0 \\ 0 & 1 & 0 & 0 \end{bmatrix}.$$

The rank of the matrix A is 3. It does not have the full rank 4. It is singular. The matrix $B^{(1)}$ has the full rank 2.

$$sI_4 - A = \begin{bmatrix} s & -I_2 \\ A_0 & sI_2 + A_1 \end{bmatrix} = \begin{bmatrix} s & 0 & -1 & 0 \\ 0 & s & 0 & -1 \\ 1 & 0 & s+2 & 0 \\ 0 & 0 & 0 & s+1 \end{bmatrix}$$

$$\left[sI_4 - A \vdots B^{(1)} \right] = \left[\begin{array}{cc:c} sI_2 & -I_2 & O_2 \\ A_0 & sI_2 + A_1 & I_2 \end{array} \right] =$$

$$= \left[\begin{array}{cccc:cc} s & 0 & -1 & 0 & 0 & 0 \\ 0 & s & 0 & -1 & 0 & 0 \\ 1 & 0 & s+2 & 0 & 1 & 0 \\ 0 & 0 & 0 & s+1 & 0 & 1 \end{array} \right] =$$

$$= \left[\begin{array}{cccccc} s & 0 & -1 & 0 & 0 & 0 \\ 0 & s & 0 & -1 & 0 & 0 \\ 1 & 0 & s+2 & 0 & 1 & 0 \\ 0 & 0 & 0 & s+1 & 0 & 1 \end{array} \right].$$

The eigenvalues $s_i(A)$ of the matrix A are $s_1(A) = 0$, and $s_i(A) = -1$, $i = 1, 2, 3$. They imply:

$$s_i(A) = 0 \Longrightarrow \left[sI_4 - A \vdots B^{(1)} \right] = \left[-I_4 - A \vdots B^{(1)} \right] =$$

$$= \left[\begin{array}{cccccc} 0 & 0 & -1 & 0 & 0 & 0 \\ 0 & 0 & 0 & -1 & 0 & 0 \\ 1 & 0 & 2 & 0 & 1 & 0 \\ 0 & 0 & 0 & 1 & 0 & 1 \end{array} \right].$$

Its submatrix

$$\left[\begin{array}{cccc} -1 & 0 & 0 & 0 \\ 0 & -1 & 0 & 0 \\ 2 & 0 & 1 & 0 \\ 0 & 1 & 0 & 1 \end{array} \right]$$

has the full rank. For the eigenvalue $s_i(A) = -1$, $i = 1, 2, 3$, the matrix $\left[sI_4 - A \vdots B^{(1)} \right]$ becomes $\left[-I_4 - A \vdots B^{(1)} \right]$:

$$\left[\begin{array}{cccccc} -1 & 0 & -1 & 0 & 0 & 0 \\ 0 & -1 & 0 & -1 & 0 & 0 \\ 1 & 0 & 1 & 0 & 1 & 0 \\ 0 & 0 & 0 & 0 & 0 & 1 \end{array} \right].$$

Its submatrix

$$\left[\begin{array}{cccc} -1 & 0 & 0 & 0 \\ 0 & -1 & 0 & 0 \\ 1 & 0 & 1 & 0 \\ 0 & 0 & 0 & 1 \end{array} \right]$$

has the full rank and the matrix $\left[sI_4 - A \,\vdots\, B^{(1)} \right]$ *has the full rank on* $\mathbb{C}$. *The system is physical state controllable.*

Comment 185 *Comparison with the classical controllability theory for* $\mu > 0$

    We will apply the existing, i.e., the classical, state controllability condition to the original system (11.24),

$$\left(\begin{array}{c} Y_1^{(2)} + 2Y_1^{(1)} + Y_1 \\ Y_2^{(2)} + Y_2^{(1)} \end{array} \right) = \left(\begin{array}{c} U \\ U^{(1)} \end{array} \right) \implies$$

$$\underbrace{I_2}_{A_2} \mathbf{Y}^{(2)} + \underbrace{\left[\begin{array}{cc} 2 & 0 \\ 0 & 1 \end{array} \right]}_{A_1} \mathbf{Y}^{(1)} + \underbrace{\left[\begin{array}{cc} 1 & 0 \\ 0 & 0 \end{array} \right]}_{A_0} \mathbf{Y} = \underbrace{\left[\begin{array}{c} 1 \\ 0 \end{array} \right]}_{B_0} U + \underbrace{\left[\begin{array}{c} 0 \\ 1 \end{array} \right]}_{B_1} U^{(1)}, \quad (11.25)$$

The physical state vector $\mathbf{X}$ *of this system is* $\mathbf{Y}^1$,

$$\mathbf{X} = \left[\begin{array}{c} X_1 \\ X_2 \\ X_3 \\ X_4 \end{array} \right] = \mathbf{Y}^1 = \left[\begin{array}{c} \mathbf{Y} \\ \mathbf{Y}^{(1)} \end{array} \right] = \left[\begin{array}{c} Y_1 \\ Y_2 \\ Y_1^{(1)} \\ Y_2^{(1)} \end{array} \right] \quad (11.26)$$

    We transform now the system (11.25) by applying the transformations (C.4), (C.5). They imply Equations (C.10)-(C.13) in which $\nu = 2$, $\mu = 1$, $N = 2$, $n = \nu N = 4$, $H_k = B_k$, $k = 0, 1$:

$$\widehat{\mathbf{X}} = \left[\begin{array}{c} \widehat{\mathbf{X}}_1 \\ \widehat{\mathbf{X}}_2 \end{array} \right] \in \mathfrak{R}^4, \ \widehat{\mathbf{X}}_1 = \left[\begin{array}{c} \widehat{X}_1 \\ \widehat{X}_2 \end{array} \right] = \mathbf{Y} = \left[\begin{array}{c} Y_1 \\ Y_2 \end{array} \right], \quad (11.27)$$

$$\widehat{\mathbf{X}}_2 = \left[\begin{array}{c} \widehat{X}_3 \\ \widehat{X}_4 \end{array} \right] = \widehat{\mathbf{X}}_1^{(1)} + A_1 \, \mathbf{Y} - B_1 U =$$

$$= \left[\begin{array}{c} Y_1^{(1)} \\ Y_2^{(1)} \end{array} \right] + \left[\begin{array}{cc} 2 & 0 \\ 0 & 1 \end{array} \right] \left[\begin{array}{c} Y_1 \\ Y_2 \end{array} \right] - \left[\begin{array}{c} 0 \\ 1 \end{array} \right] U, \quad (11.28)$$

so that Equations (C.10)-(C.13) become, respectively:

$$\widehat{\mathbf{X}} = \left[\begin{array}{c} \widehat{X}_1 \\ \widehat{X}_2 \\ \widehat{X}_3 \\ \widehat{X}_4 \end{array} \right] = \left[\begin{array}{c} Y_1 \\ Y_2 \\ Y_1^{(1)} + 2Y_1 \\ Y_2^{(1)} + Y_2 - U \end{array} \right] \neq \mathbf{X} = \left[\begin{array}{c} X_1 \\ X_2 \\ X_3 \\ X_4 \end{array} \right] = \left[\begin{array}{c} Y_1 \\ Y_2 \\ Y_1^{(1)} \\ Y_2^{(1)} \end{array} \right], \quad (11.29)$$

$$\widehat{A} = \begin{bmatrix} -A_1 & I_2 \\ -A_0 & O_2 \end{bmatrix}, \ \widehat{B} = \begin{bmatrix} B_1 \\ B_0 \end{bmatrix}, \ \widehat{C} = \begin{bmatrix} I_2 \vdots O_2 \end{bmatrix} \Longrightarrow$$

$$\widehat{A} = \begin{bmatrix} -2 & 0 & 1 & 0 \\ 0 & -1 & 0 & 1 \\ -1 & 0 & 0 & 0 \\ 0 & 0 & 0 & 0 \end{bmatrix}, \ \widehat{B} = \begin{bmatrix} 0 \\ 1 \\ 1 \\ 0 \end{bmatrix}, \ .$$

$\widehat{\mathbf{X}}$ *is the mathematical, but not physical, state vector of the system (11.24).*
From (11.29) follows that the fourth mathematical state variable $\widehat{X}_4$ *depends*
on the input control variable U, $\widehat{X}_4 = Y_2^{(1)} + Y_2 - U$. *It is meaningless. For*
example,

$$\mathbf{X} = \mathbf{0}_4 \Longleftrightarrow Y_1 = Y_2 = 0, \ Y_1^{(1)} = Y_2^{(1)} = 0,$$

but

$$\widehat{\mathbf{X}} = \mathbf{0}_4 \Longleftrightarrow Y_1 = Y_2 = 0, \ Y_1^{(1)} = 0, \ Y_2^{(1)} = U.$$

The zero value of the fourth mathematical state variable $\widehat{X}_4$,

$$\widehat{X}_4 = 0 = Y_2^{(1)} - U,$$

is not the zero value of the fourth physical state variable $X_4 = Y_2^{(1)} = 0$.
This imposes the following:

Claim 186 *On the physical sense of the mathematical state con-*
trollability of the IO system (2.15) with $\mu > 0$
   The mathematical state controllability of the IO system (2.15) with $\mu > 0$
does not have a physical sense.

Notice that for the existence of the control vector function $\mathbf{U}(.)$ that
obeys the condition 6.c) of Theorem 180 it is necessary and sufficient that
the rank condition (11.23) holds. Theorem 183 proves that Theorem 180
guarantees the robust state controllability of the *IO* system (2.15) with
$\mu \geq 0$ and $A_0 = O_N$ relative to arbitrary matrices $A_i, \ i = 1, ..., \nu - 1$.
   The above results illustrate the existence of various control algorithms
that satisfy the state controllability definition.
   Let us present the counter Example 187 to the existing controllability
criterion for the *IO* systems:

Comment 187 *Counter example to the existing controllability cri-*
terion for the IO systems

The original system (11.24) of Example 184 is physical state control-lable. It can be formally mathematically transformed into the following sys-tem (Section C.1):

$$
\begin{bmatrix} \widehat{X}_1^{(1)} \\ \widehat{X}_2^{(1)} \\ \widehat{X}_3^{(1)} \\ \widehat{X}_4^{(1)} \end{bmatrix} = \begin{bmatrix} -2 & 0 & 1 & 0 \\ 0 & -1 & 0 & 1 \\ -1 & 0 & 0 & 0 \\ 0 & 0 & 0 & 0 \end{bmatrix} \begin{bmatrix} \widehat{X}_1 \\ \widehat{X}_2 \\ \widehat{X}_3 \\ \widehat{X}_4 \end{bmatrix} + \begin{bmatrix} 0 \\ 1 \\ 1 \\ 0 \end{bmatrix} [U],
$$

$$
\begin{bmatrix} Y_1 \\ Y_2 \end{bmatrix} = \begin{bmatrix} 1 & 0 & 0 & 0 \\ 0 & 1 & 0 & 0 \end{bmatrix} \begin{bmatrix} \widehat{X}_1 \\ \widehat{X}_2 \\ \widehat{X}_3 \\ \widehat{X}_4 \end{bmatrix}. \tag{11.30}
$$

The characteristic determinant $det\,(sI_4 - A)$ of the matrix A and its eigenvalues $s_i\,(A)$ are:

$$
det\,(sI_4 - A) = s^4 + 3s^3 + 3s^2 + s = s\,(s+1)^3,
$$
$$
s_1\,(A) = 0, \; s_i\,(A) = -1, \; i = 2,3,4.
$$

We test the mathematical state controllability condition 5. for the system (11.30):

$$
rank\left[sI_4 - A \vdots B\right] =
$$

$$
= rank \begin{bmatrix} s+2 & 0 & -1 & 0 & 0 \\ 0 & s+1 & 0 & -1 & -1 \\ 1 & 0 & s & 0 & -1 \\ 0 & 0 & 0 & s & 0 \end{bmatrix} = 4, \; \forall s_i\,(A), \; i = 1,2,3,4?
$$

For the eigenvalue $s_1\,(A) = 0$ follows:

$$
rank\left[s_1\,(A)\,I_4 - A \vdots B\right] =
$$

$$
= rank \begin{bmatrix} 2 & 0 & -1 & 0 & 0 \\ 0 & 1 & 0 & -1 & -1 \\ 1 & 0 & 0 & 0 & -1 \\ 0 & 0 & 0 & 0 & 0 \end{bmatrix} = 3 < 4.
$$

The system (11.30) is not (mathematical) state controllable. This contradicts the (physical) state controllability of the original system (11.24) despite the mathematical equivalence between the original system (11.24) and the for-mally mathematically transformed system (11.30).

Conclusion 188 *On the mathematical state controllability of the IO system (2.15) with $\mu > 0$*

The formal mathematical state $\widehat{\mathbf{X}}$ controllability of the IO system (2.15) is physically meaningless if $\mu > 0$.

Exercise 189 *Test the physical state controllability of the selected IO physical plant in Exercise 40, Section 2.3.*

11.2 *IO* system output controllability

11.2.1 Definition

The general output controllability Definition 165 holds unchanged for the *IO* system (2.15).

11.2.2 Criteria

The output controllability of the *IO* system (2.15) can be studied via its *state space-output space product* or directly in its *output space*.

State space-output space product approach

The approach via the state space-output space product uses the *IO* system (2.15) equivalent *EISO* form (4.1), (4.2), in which Equations (11.2)-(11.6) induced by the *IO* system (2.15) determine the matrices (Section 11.1).

Equations (10.6) (Section 10.4) determine the system output response:

$$
\mathbf{Y}(t; t_0; \mathbf{X}_0; \mathbf{U}^\mu) = C\Phi(t, t_0)\mathbf{X}_0 + \int_{t_0}^{t} C\Phi(t, \tau)B^{(\mu)}\mathbf{U}^\mu(\tau)\, d\tau =
$$

$$
= C\Phi(t, t_0)\left[\mathbf{X}_0 + \int_{t_0}^{t} \Phi(t_0, \tau)B^{(\mu)}\mathbf{U}^\mu(\tau)\, d\tau\right] \tag{11.31}
$$

The system matrix $H_S(t, t_0)$ (10.21) (Section 10.5) takes the following form for the *IO* system (2.15):

$$
H_{IO}(t, t_0) = C\Phi(t, t_0)B^{(\mu)} = C\Phi(t, t_0)B_{inv}A_\nu^{-1}B^{(\mu)} \in \mathfrak{R}^{N \times (\mu+1)r}. \tag{11.32}
$$

Its Laplace transform reads due to $Q = O_{N,r}$:

$$
H_{IO}(s) = C\Phi(s)B^{(\mu)} = C\Phi(s)B_{inv}A_\nu^{-1}B^{(\mu)} \in \mathfrak{C}^{N \times (\mu+1)r} \tag{11.33}
$$

Theorem 173 (Section 10.5) forms the basis for the following results. At first we present it adjusted to the *IO* system (2.15) by noting once more that $Q = O_{N,r}$:

Theorem 190 *Equivalent conditions for the output controllability of the IO system (2.15): state space-output space product approach*

The IO system (2.15) obeys invariantly the output controllability necessary rank condition (10.29) (Section 10.5), i.e.,

$$rankC = N. \tag{11.34}$$

For the IO system (2.15) to be output controllable:

A) It is necessary that any of the following conditions 1.-6.a) holds and sufficient that any of the following conditions 1.-4.,6.a) holds if $\mu = 0$, where the conditions 1.-4.,6.a) are equivalent,

B) It is necessary that any of the following conditions 1.-5., 6.b) holds and sufficient that any of the following conditions 1.-4., 6.b) holds if $\mu > 0$, where the conditions 1.-4.,6.b) are equivalent,

In A) and B) Equations (11.2)-(11.6) induced by (2.15) determine the matrices:

1. All rows of the system matrix $H_{IO}(t, t_0)$ (11.32) are linearly independent on $[t_0, t_1]$ for any $(t_0, t_1 > t_0) \in In\mathfrak{T}_0 \times In\mathfrak{T}_0$.

2. All rows of the system matrix $H_{IO}(s)$ (11.33) are linearly independent on $\mathbb{C}$.

3. The Gram matrix $G_{HIO}(t, t_0)$ of $H_{IO}(t, t_0)$ (11.32),

$$G_{HIO}(t, t_0) = \int_{t_0}^{t} H_{IO}(t, \tau) H_{IO}^{T}(t, \tau) d\tau,$$
$$for\ any\ (t_0, t_1 > t_0) \in In\mathfrak{T}_0 \times In\mathfrak{T}_0, \tag{11.35}$$

is nonsingular; i.e.,

$$rankG_{HIO}(t_1, t_0) = N,\ any\ (t_0, t_1 > t_0) \in In\mathfrak{T}_0 \times In\mathfrak{T}_0. \tag{11.36}$$

4. The output $N \times n(\mu + 1)r$ controllability matrix $\mathcal{C}_{IOout}$,

$$\mathcal{C}_{IOout} = \begin{bmatrix} CB^{(\mu)} \vdots CAB^{(\mu)} \vdots CA^2B^{(\mu)} \vdots ... \vdots CA^{n-1}B^{(\mu)} \end{bmatrix}, \tag{11.37}$$

has the full rank N,

$$rank\mathcal{C}_{IOout} = N. \tag{11.38}$$

5. For every eigenvalue $s_i(A)$ of the matrix A, equivalently for every complex number $s \in \mathbb{C}$, the $N \times (n + (\mu + 1)r)$ matrix $\mathcal{C}_{outIO}$,

$$\mathcal{C}_{outIO}(s) = \left[C(sI_n - A) \;\vdots\; CB^{(\mu)} \right] \in \mathfrak{R}^{N \times (n + (\mu+1)r)}, \quad \mu \geq 0,$$

has the full rank N,

$$rank\mathcal{C}_{outIO}(s) = N,$$
$$\forall s = s_i(A) \in \mathbb{C}, \;\; \forall i = 1, 2, ...n, \; i.e., \; \forall s \in \mathbb{C}. \tag{11.39}$$

6. The control vector function $\mathbf{U}(.)$ obeys:
  a) The following equation if $\mu = 0$:

$$\mathbf{U}(t) = (H_{IO}(t_1, t))^T G_{HIO}^{-1}(t_1, t_0)[\mathbf{Y}_1 - C\Phi(t_1, t_0)\mathbf{X}_0], \quad \mu = 0, \tag{11.40}$$

  b) The following equations for any $(\mu + 1)r \times (\mu + 1)r$ nonsingular matrix R:

$$R^{-1}\mathbf{U}^{\mu}(t) = \left\{ \begin{array}{c} (H_{IO}(t_1, t)R)^T \bullet G_{HIOR}^{-1}(t_1, t_0) \bullet \\ \bullet [\mathbf{Y}_1 - C\Phi(t_1, t_0)\mathbf{X}_0] \end{array} \right\}, \quad \mu > 0, \tag{11.41}$$

where $G_{HIOR}(t_1, t_0)$ is the Gram matrix of $H_{IO}(t, t_0)R$,

$$G_{HIOR}(t_1, t_0) = \int_{t_0}^{t_1} H_{IO}(t_1, \tau)RR^T H_{IO}^T(t_1, \tau) d\tau. \tag{11.42}$$

Proof. After replacing H_S by H_{IO} in Theorem 173 it is adjusted to the *IO* system (2.15) and becomes Theorem 190, by noting that the conditions 2. and 3. are equivalent because they are related through the Laplace transform or its inverse, which are linear transformations, and that the matrix C (11.7) has the invariant full rank N:

$$C = [I_N \;\; O_N \;\; O_N \;\; O_N \; ... \; O_N] \in \mathfrak{R}^{N \times n}, \;\; RankC = N. \tag{11.43}$$

∎

Let $G_{C\Phi Binv}(t_1, t_0)$ be the Gram matrix of $C\Phi(t_1, \tau)\mathbf{B}_{inv}$,

$$G_{C\Phi Binv}(t_1, t_0) = \int_{t_0}^{t_1} C\Phi(t_1, \tau)\mathbf{B}_{inv}(C\Phi(t_1, \tau)\mathbf{B}_{inv})^T d\tau. \tag{11.44}$$

Theorem 191 *Alternative sufficient condition and control algorithm for the output controllability of the IO system (2.15): state space-output space product approach*

For the IO system (2.15) to be output controllable it is sufficient that the rank condition (11.45) holds,

$$rank B^{(\mu)} = N, \qquad (11.45)$$

and that the control vector function is the solution of the following equation:

$$B^{(\mu)} \mathbf{U}^\mu (t) = A_\nu \left\{ \begin{array}{c} (C\Phi(t_1,t)\mathbf{B}_{inv})^T \, G^{-1}_{C\Phi Binv} (t_1,t_0) \bullet \\ \bullet [\mathbf{Y}_1 - C\Phi(t_1,t_0)\mathbf{X}_0] \end{array} \right\}. \qquad (11.46)$$

The proof of this theorem is in Appendix D.13.

Comment 192 *On the insufficiency of the condition 5. of Theorem 173, i.e., of the condition 5. of Theorem 190*

Let $\nu > 1$ and $\mu \geq 0$.

We apply the condition 5. of Theorem 190 to the IO system (2.15):

$$rank \left[C(sI_n - A) \vdots CB \right] = rank C \left[(sI_n - A) \vdots B \right] =$$

$$rank \left[\bullet \left[\left[\begin{array}{ccccc} [I_N & O_N & O_N & O_N ... O_N] \bullet & \\ sI_N & -I_N & ... & O_N & O_N \\ O_N & sI_N & ... & O_N & O_N \\ ... & ... & ... & ... & ... \\ O_N & O_N & ... & -I_N & O_N \\ O_N & O_N & ... & sI_N & -I_N \\ A_\nu^{-1}A_0 & A_\nu^{-1}A_1 & ... & A_\nu^{-1}A_{\nu-2} & sI_N + A_\nu^{-1}A_{\nu-1} \end{array} \right] \vdots \left[\begin{array}{c} O_{(\nu-1)N,(\mu+1)r} \\ A_\nu^{-1}B^{(\mu)} \end{array} \right] \right] \vdots \right] = $$

$$= rank \left[\begin{array}{ccccc} sI_N & -I_N & O_N & ... & O_N \end{array} \right] = N, \; \forall s \in \mathbb{C}.$$

This shows that the condition (11.39) is invariantly satisfied, i.e., independently of the system matrices A_i, $i = 0, 1, ..., \nu - 1$, and of the control matrix $B^{(\mu)}$,

$$B^{(\mu)} = \left[B_0 \vdots B_1 \vdots ... \vdots B_\mu \right] \neq O_{N,(\mu+1)r}.$$

The rank of $\left[C(sI_n - A) \vdots CB\right]$ *is independent of* $B^{(\mu)}$ *if* $\nu > 1$ *because for* $\nu > 1$ *due to Equations (11.2) (Section 11.1):*

$$CB = [I_N \quad O_N \quad O_N \quad O_N \quad ... \quad O_N] \begin{bmatrix} O_{N,(\mu+1)r} \\ O_{N,(\mu+1)r} \\ ... \\ O_{N,(\mu+1)r} \\ A_\nu^{-1} B^{(\mu)} \end{bmatrix} = O_{N,(\mu+1)r}. \quad (11.47)$$

If $\nu > 1$ *then the rank condition (11.39) is satisfied for any* $B^{(\mu)}$ *including* $B^{(\mu)} = O_{N,(\mu+1)r}$ *that means the disconnection of the system from its controller, i.e., the control vector does not act on the system. This explains why the rank condition (11.39) is not sufficient (despite it is necessary) for the output controllability of the IO system (2.15).*

Output space approach

The direct output controllability study in the output space starts with the system response fully described by Equation (2.39) or by its equivalent form in Equation (2.40) repeated as follows:

$$\mathbf{Y}(t; \mathbf{Y}_{0-}^{\nu-1}; \mathbf{U}) = \int_{0-}^t [\Gamma_{IOU}(\tau)\mathbf{U}(t-\tau)d\tau] + \Gamma_{IOU_0}(t)\mathbf{U}_{0-}^{\mu-1} + \Gamma_{IOY_0}(t)\mathbf{Y}_{0-}^{\nu-1},$$

$$\int_{0-}^t [\Gamma_{IOU}(\tau)\mathbf{U}(t-\tau)d\tau] = \int_{0-}^t [\Gamma_{IOU}(\tau)\mathbf{U}(t,\tau)d\tau] =$$

$$= \int_{0-}^t [\Gamma_{IOU}(t-\tau)\mathbf{U}(\tau)d\tau] = \int_{0-}^t [\Gamma_{IOU}(t,\tau)\mathbf{U}(\tau)d\tau], \quad (11.48)$$

and by the first Equation (2.46) that reads for $t_0 = 0$:

$$\Gamma_{IOU}(t) = \mathcal{L}^{-1}\{G_{IOU}(s)\} = \mathcal{L}^{-1}\left\{\Phi_{IO}(s)\left(B^\mu S_r^{(\mu)}(s)\right)\right\}, \quad (11.49)$$

the Laplace transform of which is $G_{IOU}(s)$:

$$G_{IOU}(s) = \left(A^{(\nu)}S_N^{(\nu)}(s)\right)^{-1} B^\mu S_r^{(\mu)}(s) = \mathcal{L}\{\Gamma_{IOU}(t)\}, \quad (11.50)$$

by Equation (2.47):

$$\Phi_{IO}(s) = \mathcal{L}^-\{\Phi_{IO}(t)\}, \quad \Phi_{IO}(t) = \mathcal{L}^{-1}\{\Phi_{IO}(s)\}, \quad (11.51)$$

and by Equation (11.52):

$$\Phi_{IO}(s) = \left(A^{(\nu)} S_N^{(\nu)}(s)\right)^{-1}, \quad \Phi_{IO}(t) = \mathcal{L}^{-1}\left\{\left(A^{(\nu)} S_N^{(\nu)}(s)\right)^{-1}\right\}. \quad (11.52)$$

(all from Section 2.1). The preceding formulae come out from the Laplace transform $\mathbf{Y}(s; \mathbf{Y}_{0-}^{\nu-1}; \mathbf{U}) = F_{IO}(s)\, \mathbf{U}_{IO}(s)$, Equation (2.24) for $\mathbf{D}(t) \equiv \mathbf{0}_d$, of the system response $\mathbf{Y}(t; \mathbf{Y}_{0-}^{\nu-1}; \mathbf{U})$ (11.48) where:
- The system IO full transfer function matrix $F_{IO}(s)$,

$$F_{IO}(s) =$$

$$= \Phi_{IO}(s) \left[B^{(\mu)} S_r^{(\mu)}(s) \;\vdots\; - B^{(\mu)} Z_r^{(\mu-1)}(s) \;\vdots\; A^{(\nu)} Z_N^{(\nu-1)}(s) \right] =$$

$$= \left[G_{IOU}(s) \;\vdots\; G_{IOU_0}(s) \;\vdots\; G_{IOY_0}(s) \right], \quad (11.53)$$

- The Laplace transform $\mathbf{V}_{IO}(s)$ of the system action vector $\mathbf{V}_{IO}(t)$,

$$\mathbf{V}_{IO}(s) = \begin{bmatrix} \mathbf{U}(s) \\ \mathbf{U}_0^{\mu-1} \\ \mathbf{Y}_0^{\nu-1} \end{bmatrix}, \quad \mathbf{V}_{IO}(t) = \begin{bmatrix} \mathbf{U}(t) \\ \delta(t)\, \mathbf{U}_0^{\mu-1} \\ \delta(t)\, \mathbf{Y}_0^{\nu-1} \end{bmatrix}, \quad (11.54)$$

where $\delta(t)$ is Dirac impulse (for full details see [36, Section E.2, pp. 411-426], [40, Section B.2, pp. 401-416]).

The leading coefficient of the denominator polynomial $det\left(A^{(\nu)} S_N^{(\nu)}(s)\right)$ from the denominator of $\left(A^{(\nu)} S_N^{(\nu)}(s)\right)^{-1}$ in general is not equal to 1, which happens if the matrix A_ν is not the unity matrix I_N : if $A_\nu \neq I_N$. In order to make that coefficient equal to 1 we introduce the *IO system normalized fundamental matrix function* $\Theta_{IO}(.) : \mathfrak{T} \longrightarrow \mathfrak{R}^{N \times N}$,

$$\Theta_{IO}(t) = \mathcal{L}^{-1}\{\Theta_{IO}(s)\} \in \mathfrak{R}^{N \times N}, \quad \Theta_{IO}(s) = \mathcal{L}^-\{\Theta_{IO}(t)\} \in \mathbb{C}^{N \times N}, \quad (11.55)$$

and its left Laplace transform $\Theta_{IO}(s)$,

$$\Theta_{IO}(s) = \left(A_\nu^{-1} A^{(\nu)} S_N^{(\nu)}(s)\right)^{-1} = \Phi_{IO}(s) A_\nu, \quad \Phi_{IO}(s) = \Theta_{IO}(s) A_\nu^{-1},$$

$$\Theta_{IO}(t) = \mathcal{L}^{-1}\left\{\left(A_\nu^{-1} A^{(\nu)} S_N^{(\nu)}(s)\right)^{-1}\right\} = \Phi_{IO}(t) A_\nu. \quad (11.56)$$

The system IO full transfer function matrix $F_{IO}(s)$ (2.45) (also Equations (2.41), (2.46), Section 2.1), has then the following equivalent form:

$$F_{IO}(s) =$$

$$= \Theta_{IO}(s) \left[A_\nu^{-1} B^{(\mu)} S_r^{(\mu)}(s) \vdots - A_\nu^{-1} B^{(\mu)} Z_r^{(\mu-1)}(s) \vdots A_\nu^{-1} A^{(\nu)} Z_N^{(\nu-1)}(s) \right] =$$

$$= \left[G_{IOU}(s) \vdots G_{IOU_0}(s) \vdots G_{IOY_0}(s) \right], \tag{11.57}$$

The system response $\mathbf{Y}(t; \mathbf{Y}_{0-}^{\nu-1}; \mathbf{U})$ (11.48) can be then set into the following form obtained after the application of the inverse Laplace transform to $\mathbf{Y}(s; \mathbf{Y}_{0-}^{\nu-1}; \mathbf{U}) = F_{IO}(s) \mathbf{V}_{IO}(s)$ (11.54), (11.53), and (11.54):

$$\mathbf{Y}(t; \mathbf{Y}_{0-}^{\nu-1}; \mathbf{U}) =$$

$$= \int_0^t \left\{ \bullet \begin{bmatrix} \Theta_{IO}(t,\tau) \bullet \\ A_\nu^{-1} B^{(\mu)} \mathcal{L}^{-1} \left\{ S_r^{(\mu)}(s) \mathbf{U}(s) - Z_r^{(\mu-1)}(s) \mathbf{U}_{0\mp}^{\mu-1} \right\} + \\ + A_\nu^{-1} A^{(\nu)} \mathcal{L}^{-1} \left\{ Z_N^{(\nu-1)}(s) \right\} \mathbf{Y}_{0\mp}^{\nu-1} \end{bmatrix} d\tau \right\} \Longrightarrow$$

$$\mathbf{Y}(t; \mathbf{Y}_{0-}^{\nu-1}; \mathbf{U}) =$$

$$= \int_0^t \left\{ \bullet \left[A_\nu^{-1} B^{(\mu)} \mathbf{U}^\mu(\tau) + A_\nu^{-1} A^{(\nu)} \mathcal{L}^{-1} \left\{ Z_N^{(\nu-1)}(s) \right\} \mathbf{Y}_{0\mp}^{\nu-1} \right] d\tau \right\} \Longrightarrow$$

$$\tag{11.58}$$

$$\mathbf{Y}(t; \mathbf{Y}_{0-}^{\nu-1}; \mathbf{U}) =$$

$$= \int_0^t \left\{ \bullet \left[B^{(\mu)} \mathbf{U}^\mu(\tau) + A^{(\nu)} \mathcal{L}^{-1} \left\{ Z_N^{(\nu-1)}(s) \right\} \mathbf{Y}_{0\mp}^{\nu-1} \right] d\tau \right\}. \tag{11.59}$$

Equations (11.55), Equations (11.56), and Equation (11.49) permit us to express the output fundamental matrix $\Gamma_{IOU}(t)$ of the IO system (2.15) in terms of $\Theta_{IO}(s)$:

$$\Gamma_{IOU}(t) = \mathcal{L}^{-1} \left\{ G_{IOU}(s) \right\} = \mathcal{L}^{-1} \left\{ \Theta_{IO}(s) \left(A_\nu^{-1} B^\mu S_r^{(\mu)}(s) \right) \right\}, \tag{11.60}$$

as well as the following:

-The system transfer function $G_{IOU}(s)$ relating the output $\mathbf{Y}$ to the control vector $\mathbf{U}$:

$$G_{IOU}(s) = \Phi_{IO}(s) B^{(\mu)} S_r^{(\mu)}(s) = \Theta_{IO}(s) A_\nu^{-1} B^{(\mu)} S_r^{(\mu)}(s), \tag{11.61}$$

- The system transfer function $G_{IOU_0}(s) = \mathcal{L}\{\Gamma_{IOU_0}(t)\}$ relating the output to the extended initial input vector $\mathbf{U}_{0\mp}^{\mu-1}$:

$$G_{IOU_0}(s) = -\Theta_{IO}(s)A_\nu^{-1}B^{(\mu)}Z_r^{(\mu-1)}(s), \qquad (11.62)$$

- The system transfer function $G_{IOY_0}(s) = \mathcal{L}\{\Gamma_{IOY_0}(t)\}$ relating the output to the extended initial output vector $\mathbf{Y}_{0\mp}^{\nu-1}$:

$$G_{IOY_0}(s) = \Theta_{IO}(s)A_\nu^{-1}A^{(\nu)}Z_N^{(\nu-1)}(s). \qquad (11.63)$$

Equations (11.49), (11.50), and (11.52) imply the following:

Claim 193 *The output fundamental matrix $\Gamma_{IOU}(t)$ of the IO system (2.15) is:*

1. Determined, via the IO system EISO model (4.1), (4.2) (Section 11.1), in the time domain by the inverse Laplace transform of the product of the Laplace transform of the system state fundamental matrix $\Phi(t) = e^{At}$ induced by the matrix A (11.2) (Section 11.1) premultiplied by the matrix C (11.7), and of the matrix B (11.4)-(11.6) (all from Section 11.1) postmultiplied by $S_r^{(\mu)}(s)$:

$$\Gamma_{IOU}(t) = \mathcal{L}^{-1}\{G_{IOU}(s)\} = \mathcal{L}^{-1}\left\{C\,(sI_n - A)^{-1}\,BS_r^{(\mu)}(s)\right\} =$$
$$= \mathcal{L}^{-1}\left\{C\,(sI_n - A)^{-1}\,B_{inv}A_\nu^{-1}B^{(\mu)}S_r^{(\mu)}(s)\right\}, \qquad (11.64)$$

2. Determined in the time domain by the inverse Laplace transform of the system transfer function matrix $G_{IOU}(s)$ due to Equation (2.30) (Section 2.1):

$$\Gamma_{IOU}(t) = \mathcal{L}^{-1}\{G_{IOU}(s)\} =$$
$$= \mathcal{L}^{-1}\left\{\left(A^{(\nu)}S_N^{(\nu)}(s)\right)^{-1}\left(B^{(\mu)}S_r^{(\mu)}(s)\right)\right\} =$$
$$= \mathcal{L}^{-1}\left\{\left(A_\nu^{-1}A^{(\nu)}S_N^{(\nu)}(s)\right)^{-1}\left(A_\nu^{-1}B^{(\mu)}S_r^{(\mu)}(s)\right)\right\}, \qquad (11.65)$$

and, due to Equation (11.64):

$$\Gamma_{IOU}(t) = \mathcal{L}^{-1}\left\{\left(C\,(sI_n - A)^{-1}\,B_{inv}A_\nu^{-1}\right)\left(B^{(\mu)}S_r^{(\mu)}(s)\right)\right\} =$$
$$= \mathcal{L}^{-1}\left\{\left(C\,(sI_n - A)^{-1}\,B_{inv}\right)\left(A_\nu^{-1}B^{(\mu)}S_r^{(\mu)}(s)\right)\right\}. \qquad (11.66)$$

Equations (11.65) and (11.66) lead to the following result:

$$\left(A_\nu^{-1}A^{(\nu)}S_N^{(\nu)}(s)\right)^{-1} = C(sI_n - A)^{-1}B_{inv}, \qquad (11.67)$$

which links the two approaches and expresses their equivalence.

Equation (11.48) determines the output response of the *IO* system (2.15). The Gram matrix $G_{\Gamma IO}(t_1)$ of $\Gamma_{IOU}(t, t_0) \in \mathfrak{R}^{N \times r}$ reads

$$G_{\Gamma IO}(t_1, t_0) = \int_{t_0}^{t_1} \Gamma_{IOU}(t_1, \tau)\Gamma_{IOU}^T(t_1, \tau)d\tau \in \mathfrak{R}^{N \times N}. \qquad (11.68)$$

Equation (11.56) and Equation (11.67) imply the following equivalent definition of $\Theta_{IO}(t)$:

$$\Theta_{IO}(t) = \mathcal{L}^{-1}\left\{C(sI_n - A)^{-1}B_{inv}\right\}, \ \det\Theta_{IO}(t) \neq 0, \ \forall t \in \mathfrak{T}_0. \qquad (11.69)$$

The Gram matrix $G_{\Theta IO}(t_1, t_0)$ of $\Theta_{IO}(t_1, \tau)$ reads:

$$G_{\Theta IO}(t_1, t_0) = \int_{t_0}^{t_1} \Theta_{IO}(t_1, \tau)\Theta_{IO}^T(t_1, \tau)d\tau \in \mathfrak{R}^{N \times N}. \qquad (11.70)$$

Lemma 194 *The relationship between $\Theta_{IO}(t_1, \tau)$ and $G_{\Theta IO}(t_1, t_0)$, between $G_{\Gamma IO}(t_1, t_0)$ and $G_{IOU}(s)$*

1) In order for the Gramm matrix $G_{\Theta IO}(t_1, t_0)$ (11.70) to be nonsingular:

$$\det G_{\Theta IO}(t_1, t_0) \neq 0, \ any \ (t_0, t_1 > t_0) \in In\mathfrak{T}_0 \times In\mathfrak{T}_0, \qquad (11.71)$$

it is necessary and sufficient that the rows of $\Theta_{IO}(t_1, \tau)$ are linearly independent on the time interval $[t_0, t_1]$, for any $(t_0, t_1 > t_0) \in In\mathfrak{T}_0 \times In\mathfrak{T}_0$.

In order for the Gramm matrix $G_{\Gamma IO}(t_1, t_0)$ (11.68) to be nonsingular it is necessary and sufficient that any of the following equivalent conditions holds:

2) The rows of $\Gamma_{IOU}(t, t_0)$ are linearly independent on the time interval $[t_0, t_1]$, for any $(t_0, t_1 > t_0) \in In\mathfrak{T}_0 \times In\mathfrak{T}_0$.

3) The rows of $G_{IOU}(s)$ are linearly independent on $\mathfrak{C}$.

Proof. Since the IO system is stationary and linear then its matrices $\Gamma_{IOU}(t,t_0)$ and $\Theta_{IO}(t,t_0)$ are integrable in $\tau \in \mathfrak{T}_0$, $t_1 \in In\mathfrak{T}_0$. They satisfy the condition of Theorem 114.

1) The statement under 1) follows directly from Theorem 114 (Section 7.4) applied to $\Theta_{IO}(t,t_0)$.

2) The statement under 2) follows directly from Theorem 114 (Section 7.4) applied to $\Gamma_{IOU}(t,t_0)$ and its Gramm matrix $G_{\Gamma IO}(t_1,t_0)$ (11.68).

3) The statement under 3) follows directly from Equations (11.64), (11.66), the linearity of the Laplace transform and condition 2). ∎

Equation (11.11) (Section 11.1) determines the output response of the IO system (2.15).

Theorem 195 *Output controllability criterion for the IO system (2.15)*

In order for the IO system (2.15) to be output controllable it is necessary and sufficient that any of the following equivalent conditions holds:

1) For any $(t_0, t_1 > t_0) \in In\mathfrak{T}_0 \times In\mathfrak{T}_0$ the rows of the system output fundamental matrix $\Gamma_{IOU}(t,t_0)$ (11.64) are linearly independent on the time interval $[t_0, t_1]$.

2) The rows of the system transfer function matrix $G_{IOU}(s)$ (11.61) are linearly independent on $\mathbb{C}$.

3) For any $(t_0, t_1 > t_0) \in In\mathfrak{T}_0 \times In\mathfrak{T}_0$ the Gram matrix $G_{\Gamma IO}(t_1, t_0)$ (11.68) is nonsingular, i.e., the condition (11.72),

$$det G_{\Gamma IO}(t_1,t_0) \neq 0, \ any \ (t_0, t_1 > t_0) \in In\mathfrak{T}_0 \times In\mathfrak{T}_0, \qquad (11.72)$$

holds.

4) For any $(t_0, t_1 > t_0) \in In\mathfrak{T}_0 \times In\mathfrak{T}_0$ the rows of the matrix product

$$\Theta_{IO}(t,t_0) A_\nu^{-1} B^{(\mu)}$$

and of the matrix $\Theta_{IO}(t,t_0)$ are linearly independent on the time interval $[t_0, t_1]$.

5) For any $(t_0, t_1 > t_0) \in In\mathfrak{T}_0 \times In\mathfrak{T}_0$ the Gram matrices $G_{\Theta IOAB}(t_1, t_0)$,

$$G_{\Theta IOAB}(t_1,t_0) = \int_{t_0}^{t_1} \Theta_{IO}(t_1,\tau) A_\nu^{-1} B^{(\mu)} \left(\Theta_{IO}(t_1,\tau) A_\nu^{-1} B^{(\mu)} \right)^T d\tau \quad (11.73)$$

and $G_{\Theta IO}(t_1, t_0)$ (11.70) are nonsingular, i.e., the conditions (11.74),

$$det G_{\Theta IOAB}(t_1,t_0) \neq 0, \ any \ (t_0, t_1 > t_0) \in In\mathfrak{T}_0 \times In\mathfrak{T}_0, \qquad (11.74)$$

and (11.71) hold.

6) The control vector function satisfies any of the following linear time-invariant [algebraic under a) or differential under b) and c)] vector equation:

a)

$$\mathbf{U}(t) = \Gamma_{IOU}(t_1, t) G_{\Gamma IO}^{-1}(t_1, t_0) \left[\mathbf{Y}_1 - \Gamma_{IOU_0}(t_1) \mathbf{U}_{0-}^{\mu-1} - \Gamma_{IOY_0}(t_1) \mathbf{Y}_{0-}^{\nu-1} \right].$$
(11.75)

b)

$$R^{-1}\mathbf{U}^{\mu}(t) = \left(\Theta_{IO}(t_1, t) A_{\nu}^{-1} B^{(\mu)} R \right)^T G_{\Theta IOABR}^{-1}(t_1, t_0) \bullet$$

$$\bullet \left[\mathbf{Y}_1 - A_{\nu}^{-1} A^{(\nu)} \mathcal{L}^{-1} \left\{ Z_N^{(\nu-1)}(s) \right\} \mathbf{Y}_0^{\nu-1} \right],$$
(11.76)

where $R \in \mathfrak{R}^{(\mu+1)r \times (\mu+1)r}$ *is any nonsingular square matrix, $\det R \neq 0$, and*

$$G_{\Theta IOABR}(t_1, t_0) = \int_{t_0}^{t_1} \Theta_{IO}(t_1, \tau) A_{\nu}^{-1} B^{(\mu)} R \left(\Theta_{IO}(t_1, \tau) A_{\nu}^{-1} B^{(\mu)} R \right)^T d\tau.$$

c)

$$B^{(\mu)}\mathbf{U}^{\mu}(t) = A_{\nu} \Theta_{IO}^T(t_1, t) G_{\Theta IO}^{-1}(t_1, t_0) \mathbf{Y}_1 + A^{(\nu)} \mathcal{L}^{-1} \left\{ Z_N^{(\nu-1)}(s) \right\} \mathbf{Y}_{0\mp}^{\nu-1}.$$
(11.77)

Appendix D.14 contains the proof of Theorem 195.

Exercise 196 *Test the output controllability of the selected IO physical plant in Exercise 40, Section 2.3.*

11.3 *ISO* system state controllability

11.3.1 Definition

The general Definition 161 of a dynamical system state controllability takes the following form for the unperturbed *ISO* system (3.1), (3.2), (Section 3.1), i.e., unperturbed *ISO* plant (3.1), (3.2), (Section 3.1), described herein by (11.78), (11.79),

$$\frac{d\mathbf{X}(t)}{dt} = A\mathbf{X}(t) + B\mathbf{U}(t), \ \forall t \in \mathfrak{T}_0, \ A \in \mathfrak{R}^{n \times n}, \ \mathbf{U} \in \mathfrak{R}^r, \ B \in \mathfrak{R}^{n \times r}, \ (11.78)$$

$$\mathbf{Y}(t) = C\mathbf{X}(t) + U\mathbf{U}(t), \forall t \in \mathfrak{T}_0, (C \neq O_{N,n}) \in \mathfrak{R}^{N \times n}, U \in \mathfrak{R}^{N \times r}, n \geq N :$$
(11.79)

Definition 197 *State controllability of the ISO system defined by (11.78), (11.79)*

The ISO system (11.78), (11.79) is **mathematical state or physical state controllable** if and only if for every initial mathematical state or physical state vector $\mathbf{X}_0 \in \mathfrak{R}^n$ at $t_0 \in \mathfrak{T}$ and for any final mathematical state or physical state vector $\mathbf{X}_1 \in \mathfrak{R}^n$, respectively, there exist a moment $t_1 \in In\mathfrak{T}_0$ and a control $\mathbf{U}_{[t_0,t_1]}$ on the time interval $[t_0, t_1]$ such that

$$\mathbf{X}(t_1; t_0; \mathbf{X}_0; \mathbf{0}_d; \mathbf{U}_{[t_0,t_1]}) = \mathbf{X}_1. \tag{11.80}$$

11.3.2   Criterion

Notice that $\mu = 0$ the *ISO* system (11.78), (11.79). Let the state vector $\mathbf{S}$ and the matrix $\mathbf{B}^{(\mu)} = \mathbf{B}^{(0)} = \mathbf{B}_0$ of the system (10.3), (10.4) (Section 10.4) be denoted, respectively, by $\mathbf{X}$, $\mathbf{S} = \mathbf{X}$, and by B, $\mathbf{B}_0 = B$. Then the system (10.3) becomes the *ISO* system (11.78), (11.79), and Theorem 171 becomes the following theorem that completes (due to the explicit statement of condition 6) the well-known conditions for the state controllability of the *ISO* system (11.78), (11.79) (e.g., [18, Theorem 5-7, pp. 183, 184]):

Theorem 198 *Equivalent conditions for the state controllability of the ISO system (11.78), (11.79)*

For the ISO system (11.78), (11.79) to be state controllable it is necessary and sufficient that

a) Any of the following equivalent conditions 1. through 6. holds:

1. All rows of $\Phi(t_1, .) B$ and $\Phi(t_0, .) B$ are linearly independent on $[t_0, t_1]$ for any $(t_0, t_1 > t_0) \in In\mathfrak{T}_0 \times In\mathfrak{T}_0$.

2. All rows of $\Phi(s) B = (sI - A)^{-1} B$ are linearly independent on $\mathbb{C}$.

3. The Gramm matrix $G_{\Phi B}(t, t_0)$,

$$G_{\Phi B}(t_1, t_0) = \int_{t_0}^{t_1} \Phi(t_1, \tau) BB^T \Phi^T(t_1, \tau) d\tau \tag{11.81}$$

is nonsingular for any $(t_0, t_1 > t_0) \in In\mathfrak{T}_0 \times In\mathfrak{T}_0$; i.e.,

$$rank G_{\Phi B}(t_1, t_0) = n, \; for \; any \; (t_0, t_1 > t_0) \in In\mathfrak{T}_0 \times In\mathfrak{T}_0. \tag{11.82}$$

4. The $n \times nr$ controllability matrix $\mathcal{C}$,

$$\mathcal{C} = \left[B \vdots AB \vdots A^2 B \vdots ... \vdots A^{n-1} B \right] \tag{11.83}$$

has the full rank n,

$$rank \mathcal{C} = n. \tag{11.84}$$

5. For every eigenvalue $s_i(A)$ of the matrix A, equivalently for every complex number $s \in \mathbb{C}$, the $n \times (n+r)$ matrix $\left[sI - A \vdots B\right]$ has the full rank n,

$$rank \left[sI - A \vdots B\right] = n,$$

$$\forall s = s_i(A) \in \mathbb{C}, \ \forall i = 1, 2, ...n, \ i.e., \ \forall s \in \mathbb{C}. \tag{11.85}$$

6. The control vector function $U(.)$ obeys

$$\mathbf{U}(t) = (\Phi(t_1, t)B)^T G_{\Phi B}^{-1}(t_1, t_0)[\mathbf{X}_1 - \Phi(t_1, t_0)\mathbf{X}_0],$$

$$\forall t \in [t_0, t_1], \ t_1 \in In\mathfrak{T}_0. \tag{11.86}$$

Exercise 199 *Test the state controllability of the selected ISO physical plant in Exercise 48, Section 3.3.*

11.4 *ISO* system output controllability

11.4.1 Definition

The general output controllability Definition 165 (Section 10.3) slightly simplifies for the unperturbed *ISO* system (11.78), (11.79).

Definition 200 *Output controllability of the ISO system (11.78), (11.79)*

The ISO system (11.78), (11.79) is output controllable if and only if for every initial output vector $\mathbf{Y}_0 \in \mathfrak{R}^N$ at $t_0 \in \mathfrak{T}$ and for any final output vector $\mathbf{Y}_1 \in \mathfrak{R}^N$ there exist a moment $t_1 \in In\mathfrak{T}_0$ and a control $\mathbf{U}_{[t_0,t_1]}$ on the time interval $[t_0, t_1]$ such that $\mathbf{Y}(t_1; t_0; \mathbf{Y}_0; \mathbf{U}_{[t_0,t_1]}) = \mathbf{Y}_1$.

11.4.2 Criteria

The matrix $G_{ISO}(t, t_0)$ is the matrix $G_S(t, t_0)$ (10.23), (Section 10.5), applied to the *ISO* system (11.78), (11.79)

$$G_{ISO}(t, t_0) = \mathcal{L}^{-1}\{G_{ISO}(s)\} = \mathcal{L}^{-1}\left\{C(sI - A)^{-1}B + U\right\} =$$

$$= C\Phi(t, t_0)B + \delta(t, t_0)U. \tag{11.87}$$

where $G_{ISO}(s)$ is the system transfer function matrix,

$$G_{ISO}(s) = C(sI - A)^{-1}B + U. \tag{11.88}$$

The matrix $H_S(t_1, \tau)$ (10.21), (Section 10.5), of the ISO system (11.78), (11.79) reads

$$H_{ISO}(t_1, t) = C\Phi(t_1, t)B + \delta(t_1, t)U = G_{ISO}(t_1, t), \tag{11.89}$$

$$H_{ISO}(s) = C(sI - A)^{-1}B + U = G_{ISO}(s). \tag{11.90}$$

State space-output space product approach

The following theorem is usually referred to in textbooks without its proof if $U \neq O_{N,r}$:

Theorem 201 *Equivalent conditions for the output controllability of the *ISO* system (11.78), (11.79)*

For the ISO system (11.78), (11.79) to be output controllable it is necessary that

$$rank \begin{bmatrix} C \vdots U \end{bmatrix} = N \tag{11.91}$$

is valid and if this condition is satisfied then it is necessary that any of the following conditions 1.-6. holds and sufficient that any of the following conditions 1.-4., 6. holds, where the conditions 1.-4.,6. are equivalent:

1. All rows of the system matrix $H_{ISO}(t_1, \tau)$ (11.159) are linearly independent on $[t_0, t_1]$, for any $(t_0, t_1 > t_0) \in In\mathfrak{T}_0 \times In\mathfrak{T}_0$.

2. All rows of the system transfer function matrix $G_{ISO}(s)$ (11.88) are linearly independent on $\mathbb{C}$.

3. The Gramm matrix $G_{GISO}(t_1, t_0)$ of $G_{ISO}(t_1, t)$,

$$G_{GISO}(t_1, t_0) = \int_{t_0}^{t_1} G_{ISO}(t_1, \tau) G_{ISO}^T(t_1, \tau) d\tau,$$
$$for \ any \ (t_0, t_1 > t_0) \in In\mathfrak{T}_0 \times In\mathfrak{T}_0, \tag{11.92}$$

is nonsingular; i.e.,

$$rank G_{GISO}(t_1, t_0) = n, \ any \ (t_0, t_1 > t_0) \in In\mathfrak{T}_0 \times In\mathfrak{T}_0.. \tag{11.93}$$

4. The output $N \times (n+1)r$ controllability matrix $\mathcal{C}_{ISO}$,

$$\mathcal{C}_{ISO} = \begin{bmatrix} CB \vdots CAB \vdots CA^2B \vdots ... \vdots CA^{n-1}B \vdots U \end{bmatrix} \tag{11.94}$$

has the full rank N,

$$rank\mathcal{C}_{ISO} = N. \tag{11.95}$$

5. For every eigenvalue $s_i(A)$ of the matrix A, equivalently for every complex number $s \in \mathbb{C}$, the $N \times (n + 2r)$ matrix $\left[C(sI_n - A) \vdots CB \vdots U\right]$ has the full rank N,

$$rank\left[C(sI_n - A) \vdots CB \vdots U\right] = N,$$

$$\forall s = s_i(A) \in \mathbb{C}, \ \forall i = 1, 2, ...n, \ i.e., \ \forall s \in \mathbb{C}. \tag{11.96}$$

6. The control vector function $\mathbf{U}(.)$ obeys the following equation:

$$\mathbf{U}(t) = (G_{ISO}(t_1, \tau))^T G_{GISO}^{-1}(t_1, t_0)[\mathbf{Y}_1 - C\Phi(t_1, t_0)\mathbf{X}_0], \ \mu = 0, \tag{11.97}$$

Proof. Let $\mu = 0$, the state vector $\mathbf{S}$, the matrix V and the control vector $\mathbf{V}$ of the system (10.3), (10.4) (Section 10.4) be denoted, respectively, by the vector $\mathbf{X}$, the matrix B and by the control vector $\mathbf{U}$. The system (10.3), (10.4) becomes the *ISO* system (11.78), (11.79) and Theorem 173 becomes Theorem 201 except for its condition 2., which results from its condition under 1. due to Equations (11.87)-(11.159) and the linearity of the Laplace transform. ■

Notice that Theorem 174 is valid for the *ISO* system (11.78), (11.79).

Output space approach

The inverse Laplace transform of Equation (3.13), (Section 3.1), leads to the following for $\mathbf{D}(t) \equiv \mathbf{0}_d$:

$$\mathbf{Y}(t; \mathbf{X}_0; \mathbf{U}) = \left\{ \begin{array}{c} \int_{0^-}^t [\Gamma_{ISOU}(\tau)\mathbf{U}(t - \tau)d\tau] + \\ +\Gamma_{ISOX_0}(t)\mathbf{X}_0, \end{array} \right\},$$

$$\int_{0^-}^t [\Gamma_{ISOU}(\tau)\mathbf{U}(t - \tau)d\tau] = \int_{0^-}^t [\Gamma_{ISOU}(\tau)\mathbf{U}(t, \tau)d\tau] =$$

$$= \int_{0^-}^t [\Gamma_{ISOU}(t - \tau)\mathbf{U}(\tau)d\tau] = \int_{0^-}^t [\Gamma_{ISOU}(t, \tau)\mathbf{U}(\tau)d\tau], \tag{11.98}$$

where

- $\Gamma_{ISO}(t)$ is the inverse Laplace transform of the *ISO* system (3.1), (3.2) *transfer function matrix $G_{ISO}(s)$ relating the output $\mathbf{Y}$ to the control vector* $\mathbf{U}$,

$$G_{ISO}(s) = C(sI - A)^{-1}B + U, \tag{11.99}$$

$$\Gamma_{ISOU}(t) = \mathcal{L}^- \{G_{ISOU}(s)\} = \mathcal{L}^- \left\{ C(sI - A)^{-1}B + U \right\}, \qquad (11.100)$$

and

- $\Gamma_{ISOX_0}(t)$ is the inverse Laplace transform of the *ISO* system (3.1), (3.2) *transfer function matrix* $G_{ISOX_0}(s)$ *relating the output* **Y** *to to the initial state* **X**$_0$,

$$\Gamma_{ISOX_0}(t) = \mathcal{L}^- \{G_{ISOX_0}(s)\}, \; G_{ISOX_0}(s) = C(sI - A)^{-1}. \qquad (11.101)$$

Theorem 202 *Output controllability criterion for the ISO system (3.1), (3.2)*

For the ISO system (3.1), (3.2) to be output controllable it is necessary that (11.91) is valid and if this condition is satisfied then it is necessary and sufficient that any of the following equivalent conditions holds:

1) For any $(t_0, t_1 > t_0) \in In\mathfrak{T}_0 \times In\mathfrak{T}_0$ *the rows of the system output fundamental matrix* $\Gamma_{ISO}(t, t_0)$ *(11.100) are linearly independent on the time interval* $[t_0, t_1]$.

2) The rows of the system transfer function matrix $G_{ISO}(s)$ *(11.99) are linearly independent on* $\mathbb{C}$.

3) For any $(t_0, t_1 > t_0) \in In\mathfrak{T}_0 \times In\mathfrak{T}_0$ *the Gram matrix* $G_{\Gamma ISO}(t_1, t_0)$ *(11.102),*

$$G_{\Gamma ISO}(t_1, t_0) = \int_{t_0}^{t_1} \Gamma_{ISO}(t_1, \tau) \Gamma_{ISO}^T(t_1, \tau) d\tau \in \mathfrak{R}^{N \times N}. \qquad (11.102)$$

is nonsingular, i.e., the condition (11.103),

$$det G_{\Gamma ISO}(t_1, t_0) \neq 0, \; any \; (t_0, t_1 > t_0) \in In\mathfrak{T}_0 \times In\mathfrak{T}_0, \qquad (11.103)$$

holds.

4) The control vector function satisfies the following linear time-invariant algebraic vector equation:

$$\mathbf{U}(t) = \Gamma_{ISO}^T(t, t_0) G_{\Gamma ISO}^{-1}(t_1, t_0) [\mathbf{Y}_1 - \Gamma_{ISOx_0}(t_1)\mathbf{X}_{0-}]. \qquad (11.104)$$

The proof results directly from the proof (Appendix D.14) of Theorem 195, Subsection 11.2.2.

Exercise 203 *Test the output controllability of the selected ISO physical plant in Exercise 48, Section 3.3.*

11.5 *EISO* system state controllability

11.5.1 Definition

The general state controllability Definition 161, (Section 10.3), reads as follows for the unperturbed *EISO* system (4.1), (4.2) (Section 4.1), which is described in the sequel by (11.105), (11.106)

$$\frac{d\mathbf{X}(t)}{dt} = A\mathbf{X}(t) + B^{(\mu)}\mathbf{U}^\mu(t), \ \forall t \in \mathfrak{T}_0, \ \mu \geq 1, \ B^{(\mu)} \in \mathfrak{R}^{n\times(\mu+1)r},$$
$$(11.105)$$

$$\mathbf{Y}(t) = C\mathbf{X}(t) + U\mathbf{U}(t), \ \forall t \in \mathfrak{T}_0. \tag{11.106}$$

Definition 204 *State controllability of the EISO system (11.105), (11.106)*

*The EISO system (11.105), (11.106) is **mathematical state or physical state controllable** if and only if for every initial mathematical state or physical state vector $\mathbf{X}_0 \in \mathfrak{R}^n$ at $t_0 \in \mathfrak{T}$ and for any final mathematical state or physical state vector $\mathbf{X}_1 \in \mathfrak{R}^n$, respectively, there exist a moment $t_1 \in In\mathfrak{T}_0$ and an extended control $\mathbf{U}^\mu_{[t_0,t_1]}$ on the time interval $[t_0, t_1]$ such that $\mathbf{X}(t_1; t_0; \mathbf{X}_0; \mathbf{U}^\mu_{[t_0,t_1]}) = \mathbf{X}_1$.*

11.5.2 Criterion

Let $\mathbf{S}$ be replaced by $\mathbf{X}$ in the system (10.3), (10.4) (Section 10.4), which then becomes the *EISO* system (11.105), (11.106) (Section 4.1) and Theorem 171 (Section 10.4) takes the following form:

Theorem 205 *State controliability criteria for the EISO system (11.105), (11.106)*

For the EISO system (11.105), (11.106) to be state controllable it is necessary and sufficient that any of the following equivalent conditions 1. through 5, 6.a) or 6.b) holds due to $\mu > 0$:

1. All rows of both matrices $\Phi(t_1, t) B^{(\mu)}$ and $\Phi(t_0, t) B^{(\mu)}$ are linearly independent on $[t_0, t_1]$ for any $(t_0, t_1 > t_0) \in In\mathfrak{T}_0 \times In\mathfrak{T}_0$.

2. All rows of $\Phi(s) B^{(\mu)} = (sI - A)^{-1} B^{(\mu)}$ are linearly independent on $\mathbb{C}$.

3. The Gram matrix $G_{\Phi B}(t_1, t_0)$ (11.107) of $\Phi(t_1, t) B^{(\mu)}$,

$$G_{\Phi B}(t_1, t_0) = \int_{t_0}^{t_1} \Phi(t_1, \tau) B^{(\mu)} \left(B^{(\mu)}\right)^T \Phi^T(t_1, \tau) d\tau, \tag{11.107}$$

is nonsingular for any $(t_0, t_1 > t_0) \in In\mathfrak{T}_0 \times In\mathfrak{T}_0$; *i.e.,*

$$rank\ G_{\Phi B}\ (t_1, t_0) = n,\ for\ any\ (t_0, t_1 > t_0) \in In\mathfrak{T}_0 \times In\mathfrak{T}_0. \quad (11.108)$$

4. The $n \times n\ (\mu + 1)\ r$ *controllability matrix* $\mathcal{C}$,

$$\mathcal{C} = \left[B^{(\mu)} \vdots AB^{(\mu)} \vdots A^2 B^{(\mu)} ... \vdots A^{n-1} B^{(\mu)} \right] \in \mathfrak{R}^{n \times n(\mu+1)r}, \quad (11.109)$$

has the full rank n,

$$rank\ \mathcal{C} = n. \quad (11.110)$$

5. For every eigenvalue $s_i\ (A)$ *of the matrix* A, *equivalently for every complex number* $s \in \mathbb{C}$, *the* $n \times (n + (\mu + 1)\ r)$ *matrix* $\left[sI - A \vdots B^{(\mu)} \right]$ *has the full rank* n,

$$rank\ \left[sI - A \vdots B^{(\mu)} \right] = n,\ \forall s = s_i\ (A) \in \mathbb{C},\ \forall i = 1, 2, ..., n,, \forall s \in \mathbb{C}.$$
$$(11.111)$$

6. a) If $\mu > 0$ *the control vector function* $\mathbf{U}\ (.)$ *obeys either*

$$T^{-1} \mathbf{U}^\mu\ (t) = T^T \left(B^{(\mu)} \right)^T \Phi^T\ (t_1, t)\ G_{\Phi BT}^{-1}\ (t_1, t_0)\ [\mathbf{X}_1 - \Phi\ (t_1, t_0)\ \mathbf{X}_0],\ \mu > 0,$$
$$(11.112)$$

where $T \in \mathfrak{R}^{(\mu+1)r \times (\mu+1)r}$ *is any nonsingular matrix,* $detT \neq 0$, *and the matrix* $G_{\Phi BT}\ (t_1, t_0)$ *is the Gram matrix of* $\Phi\ (t_1, t)\ B^{(\mu)} T$,

$$G_{\Phi BT}\ (t_1, t_0) = \int_{t_0}^{t_1} \Phi\ (t_1, \tau)\ B^{(\mu)} TT^T \left(B^{(\mu)} \right)^T \Phi^T\ (t_1, \tau)\ d\tau, \quad (11.113)$$

b) or the control vector function $\mathbf{U}\ (.)$ *obeys*

$$B^{(\mu)} \mathbf{U}^\mu\ (t) = \Phi^T\ (t_1, t)\ G_\Phi^{-1}\ (t_1, t_0)\ [\mathbf{X}_1 - \Phi\ (t_1, t_0)\ \mathbf{X}_0],\ \mu > 0, \quad (11.114)$$

where $G_\Phi\ (t_1, t_0)$ *is the Gram matrix of* $\Phi\ (t_1, t)$,

$$G_\Phi\ (t_1, t_0) = \int_{t_0}^{t_1} \Phi\ (t_1, \tau)\ \Phi^T\ (t_1, \tau)\ d\tau. \quad (11.115)$$

Exercise 206 *Test the state controllability of the selected EISO physical plant in Exercise 61, Section 4.3.*

11.6 *EISO* system output controllability

11.6.1 Definition

The general output controllability Definition 165 (Section 10.5) preserves its form in the framework of the unperturbed *EISO* system (11.105), (11.106), for which $\mu > 0$.

Definition 207 *Output controllability of the EISO system (11.105), (11.106)*

The *EISO* system (11.105), (11.106) is the output controllable if and only if for every initial output vector $\mathbf{Y}_0 \in \mathfrak{R}^N$ at $t_0 \in \mathfrak{T}$ and for any final output vector $\mathbf{Y}_1 \in \mathfrak{R}^N$ there exist a moment $t_1 \in In\mathfrak{T}_0$ and an extended control $\mathbf{U}^\mu_{[t_0,t_1]}$ on the time interval $[t_0, t_1]$ such that $\mathbf{Y}(t_1; t_0; \mathbf{Y}_0; \mathbf{U}^\mu_{[t_0,t_1]}) = \mathbf{Y}_1$.

11.6.2 Criteria

State space-output space product approach

The system (10.3), (10.4) with $\mu > 0$ becomes the *EISO* system (11.105), (11.106) when we replace in the former $\mathbf{S}$ by $\mathbf{X}$ and Q by U. This implies that $H_S(t, t_0)$ (10.21), (10.22) and Theorem 173 (Section 10.5) become, respectively, Equation (11.116),

$$H_{EISO}(t, t_0) = \left(C\Phi(t, t_0)\mathrm{B}^{(\mu)} + \widetilde{U} \right) \in \mathfrak{R}^{N \times (\mu+1)r},$$

$$\widetilde{U} = \left[O_{N,r} \vdots U \vdots O_{N,r} \vdots .... \vdots O_{N,r} \right] \in \mathfrak{R}^{N \times (\mu+1)r}, \qquad (11.116)$$

Equation (11.117),

$$H_{EISO}(s) = \mathcal{L}\left\{ H_{EISO}(t, t_0) \right\} =$$

$$= \left(C\left(sI - A\right)^{-1}\mathrm{B}^{(\mu)} + s^{-1}\widetilde{U} \right) \in \mathbb{C}^{N \times (\mu+1)r}, \qquad (11.117)$$

and the following theorem:

Theorem 208 *Output controllability of the EISO system (11.105), (11.106)*

For the *EISO* system (11.105), (11.106) to be output controllable it is necessary that

$$rank \left[C \vdots U \right] = N \qquad (11.118)$$

and if this condition is satisfied then it is necessary that any of the following conditions 1.-6. holds and sufficient that any of the following conditions 1.-4., 6. holds, where the conditions 1.-4., 6. are equivalent:

1. All rows of the system matrix $H_{EISO}(t, t_0)$ (11.116) are linearly independent on $[t_0, t_1]$ for any $(t_0, t_1 > t_0) \in In\mathfrak{T}_0 \times In\mathfrak{T}_0$.

2. All rows of the system matrix $H_{EISO}(s)$ (11.117) are linearly independent on $\mathbb{C}$.

3. The Gram matrix $G_{HEISO}(t_1, t_0)$ of the system matrix $H_{EISO}(t, t_0)$,

$$G_{HEISO}(t_1, t_0) = \int_{t_0}^{t_1} H_{EISO}(t_1, \tau) H_{EISO}^T(t_1, \tau) \, d\tau,$$

$$for \ any \ (t_0, t_1 > t_0) \in In\mathfrak{T}_0 \times In\mathfrak{T}_0, \qquad (11.119)$$

is nonsingular; i.e.,

$$rank G_{HEISO}(t_1, t_0) = n, \ any \ (t_0, t_1 > t_0) \in In\mathfrak{T}_0 \times In\mathfrak{T}_0. \qquad (11.120)$$

4. The output $N \times (n+1)(\mu+1) r$ controllability matrix $\mathcal{C}_{EISOout}$,

$$\mathcal{C}_{EISOout} = \left[CB^{(\mu)} \ \vdots \ CAB^{(\mu)} \ \vdots \ CA^2 B^{(\mu)} \ \vdots \ ... \vdots \ CA^{n-1} B^{(\mu)} \ \vdots \ \widetilde{U} \right] \quad (11.121)$$

has the full rank N,

$$rank \mathcal{C}_{EISOout} = N. \qquad (11.122)$$

5. For every eigenvalue $s_i(A)$ of the matrix A, equivalently for every complex number $s \in \mathbb{C}$, the $N \times (n + 2(\mu+1) r)$ matrix $\mathcal{C}_{outEISO}$,

$$\mathcal{C}_{outEISO} = \left[C(sI - A) \ \vdots \ CB^{(\mu)} \ \vdots \ \widetilde{U} \right]$$

has the full rank N,

$$rank \mathcal{C}_{outEISO} = rank \left[C(sI - A) \ \vdots \ CB^{(\mu)} \ \vdots \ \widetilde{U} \right] = N,$$

$$\forall s = s_i(A) \in \mathbb{C}, \ \forall i = 1, 2, ...n, \ i.e., \ \forall s \in \mathbb{C}. \qquad (11.123)$$

6. The control vector function $\mathbf{U}(.)$ obeys the following equation for any nonsingular matrix R, where the matrix $R \in \mathfrak{R}^{(\mu+1)r \times (\mu+1)r}$:

$$R^{-1} \mathbf{U}^\mu (t) = \left\{ \begin{array}{c} (H_{EISO}(t_1, t) R)^T \bullet G_{HEISOR}^{-1}(t_1, t_0) \bullet \\ \bullet \left[\mathbf{Y_1} - C\Phi(t_1, t_0) \mathbf{X}_0 - Q^{(\mu-1)} \mathbf{U}_0^{\mu-1} \right] \end{array} \right\}, \qquad (11.124)$$

where $G_{HEISOR}(t_1, t_0)$ is the Gram matrix of $H_{EISO}(t_1, t) R$,

$$G_{HEISOR}(t_1, t_0) = \int_{t_0}^{t_1} H_{EISO}(t_1, \tau) RR^T H_{EISO}^T(t_1, \tau) \, d\tau. \qquad (11.125)$$

Output space approach

The system response $\mathbf{Y}(t; \mathbf{X}_0; \mathbf{U})$ results as the inverse Laplace transform of the first Equation (4.27), (Section 4.1), applied to the undisturbed system (11.105), (11.106):

$$\mathbf{Y}(t; \mathbf{X}_0; \mathbf{U}) =$$

$$= \int_{0-}^{t} [\Gamma_{EISO}(\tau)\mathbf{U}(t - \tau)d\tau] + \Gamma_{EISOU_0}(t)\mathbf{U}_0^{\mu-1} + \Gamma_{EISOX_0}(t)\mathbf{X}_0,$$

$$\int_{0-}^{t} [\Gamma_{EISO}(\tau)\mathbf{U}(t - \tau)d\tau] = \int_{0-}^{t} [\Gamma_{EISO}(\tau)\mathbf{U}(t, \tau)d\tau] =$$

$$= \int_{0-}^{t} [\Gamma_{EISO}(t - \tau)\mathbf{U}(\tau)d\tau] = \int_{0-}^{t} [\Gamma_{EISO}(t, \tau)\mathbf{U}(\tau)d\tau], \qquad (11.126)$$

where

- The system output fundamental matrix $\Gamma_{EISO}(t)$ is the inverse Laplace transform of the system transfer function $G_U(s)$, Equation (4.34), (Subsection 4.1.2), relating the output $\mathbf{Y}$ to the control vector $\mathbf{U}$:

$$G_U(s) = C(sI - A)^{-1} B^{(\mu)} S_r^{(\mu)}(s) + U, \qquad (11.127)$$

$$\Gamma_{EISO}(t) = \mathcal{L}^{-1}\{G_{PU}(s)\} = \mathcal{L}^{-1}\left\{C(sI - A)^{-1} B^{(\mu)} S_r^{(\mu)}(s) + U\right\}, \qquad (11.128)$$

- $\Gamma_{EISOU_0}(t)$ is the inverse Laplace transform of the transfer function matrix $G_{EISOU_0}(s)$ relative to the extended initial control vector $\mathbf{U}_0^{\mu-1}$,

$$G_{EISOU_0}(s) = -C(sI - A)^{-1} B^{(\mu)} Z_r^{(\mu-1)}(s), \qquad (11.129)$$

$$\Gamma_{EISOU_0}(t) = \mathcal{L}^{-1}\{G_{EISOU_0}(s)\} = \mathcal{L}^{-1}\left\{-C(sI - A)^{-1} B^{(\mu)} Z_r^{(\mu-1)}(s)\right\}, \qquad (11.130)$$

- $\Gamma_{EISOX_0}(t)$ is the inverse Laplace transform of the transfer function matrix $G_{EISOX_0}(s)$ relative to the initial state vector $\mathbf{X}_0$,

$$G_{EISOX_0}(s) = C(sI - A)^{-1}, \qquad (11.131)$$

$$\Gamma_{EISOX_0}(t) = \mathcal{L}^{-1}\{G_{EISOX_0}(s)\} = \mathcal{L}^{-1}\left\{C(sI - A)^{-1}\right\}. \qquad (11.132)$$

Theorem 209 *Output controllability of the EISO system (11.105), (11.106)*

For the EISO system (11.105), (11.106) to be output controllable it is necessary that (11.118) is valid and if it is satisfied then it is necessary and sufficient that any of the following equivalent conditions holds:

1) For any $(t_0, t_1 > t_0) \in In\mathfrak{T}_0 \times In\mathfrak{T}_0$ the rows of the system output fundamental matrix $\Gamma_{EISO}(t_1, t)$ (11.128) are linearly independent on the time interval $[t_0, t_1]$.

2) The rows of the system transfer function matrix $G_{EISO}(s)$ (11.127) are linearly independent on $\mathbb{C}$.

3) For any $(t_0, t_1 > t_0) \in In\mathfrak{T}_0 \times In\mathfrak{T}_0$ the Gram matrix $G_{\Gamma EISO}(t_1, t_0)$ (11.133)

$$G_{\Gamma EISO}(t_1, t_0) = \int_{t_0}^{t_1} \Gamma_{EISO}(t_1, \tau)\Gamma_{EISO}^T(t_1, \tau)d\tau \in \mathfrak{R}^{N \times N}. \qquad (11.133)$$

is nonsingular, i.e., the condition (11.134),

$$detG_{\Gamma EISO}(t_1, t_0) \neq 0, \ \ any \ \ (t_0, t_1 > t_0) \in In\mathfrak{T}_0 \times In\mathfrak{T}_0, \qquad (11.134)$$

holds.

4) The control vector function satisfies the following linear time-invariant algebraic vector equation:

$$\mathbf{U}(t) = \Gamma_{EISO}^T(t_1, t)G_{\Gamma EISO}^{-1}(t_1, t_0) \bullet$$

$$\bullet \left[\mathbf{Y}_1 - \Gamma_{EISOU_0}(t_1)\mathbf{U}_{0-}^{\mu-1} - \Gamma_{EISOX_0}(t_1)\mathbf{X}_0\right]. \qquad (11.135)$$

The proof is essentially the same as the proof of Theorem 195, Subsection 11.2.2.

Exercise 210 *Test the output controllability of the selected EISO physical plant in Exercise 61, Section 4.3.*

11.7   *HISO* system state controllability

11.7.1   Definition

The vector $\mathbf{R}^{\alpha-1}$ (5.3) is the state vector of the *HISO* system (5.1), (5.2), (Section 5.1).

The following definition is the general Definition 161 of the state controllability adjusted to the *HISO* system (5.1), (5.2):

Definition 211 *State controllability of the HISO system (5.1), (5.2)*

The *HISO* system (5.1), (5.2) is **mathematical state or physical state controllable** *if and only if for every initial mathematical state or physical state vector* $\mathbf{R}_0^{\alpha-1} \in \mathfrak{R}^{\alpha\rho}$ *at* $t_0 \in \mathfrak{T}$ *and for any final mathematical state or physical state vector* $\mathbf{R}_1^{\alpha-1} \in \mathfrak{R}^{\alpha\rho}$, *respectively, there exist a moment* $t_1 \in In\mathfrak{T}_0$ *and an extended control* $\mathbf{U}_{[t_0,t_1]}^{\mu}$ *on the time interval* $[t_0, t_1]$ *such that*

$$\mathbf{R}^{\alpha-1}(t_1; t_0; \mathbf{R}_0^{\alpha-1}; \mathbf{0}_d; \mathbf{U}_{[t_0,t_1]}^{\mu}) = \mathbf{R}_1^{\alpha-1}. \tag{11.136}$$

11.7.2 Criterion

The unperturbed *HISO* system (5.1), (5.2) is described in the compact form by

$$A^{(\alpha)}\mathbf{R}^{\alpha}(t) = B^{(\mu)}\mathbf{U}^{\mu}(t), \ \forall t \in \mathfrak{T}_0, \tag{11.137}$$

$$\mathbf{Y}(t) = R_y^{(\alpha)}\mathbf{R}^{\alpha}(t) + U\mathbf{U}(t) = R_y^{(\alpha-1)}\mathbf{R}^{\alpha-1}(t) + U\mathbf{U}(t), \ \forall t \in \mathfrak{T}_0. \tag{11.138}$$

If we replace α by ν, ρ by N and $\mathbf{R}$ by $\mathbf{Y}$ then the matrix $A^{(\alpha)}$,

$$A^{(\alpha)} = \left[A_0 \vdots A_1 \vdots ... \vdots A_\alpha \right],$$

becomes the matrix $A^{(\nu)}$ (2.13) (Section 2.1),

$$\alpha = \nu \Longrightarrow A^{(\alpha)} = \left[A_0 \vdots A_1 \vdots ... \vdots A_\nu \right] = A^{(\nu)},$$

and Equation (11.137) becomes Equation (11.3) (Section 11.3). This explains that Equation (11.137) and Equation (11.138) can be set in the following equivalent *EISO* forms:

$$\frac{d\mathbf{X}(t)}{dt} = A\mathbf{X}(t) + B^{(\mu)}\mathbf{U}^{\mu}(t), \forall t \in \mathfrak{T}_0, \mu \geq 1, B^{(\mu)} \in \mathfrak{R}^{n\times(\mu+1)r}, \ n = \alpha\rho, \tag{11.139}$$

$$\mathbf{Y}(t) = C\mathbf{X}(t) + U\mathbf{U}(t), \ \forall t \in \mathfrak{T}_0. \tag{11.140}$$

where (for details see Theorem 49 in Section 4.1 and Subsection 11.1.2 of Section 11.1):

$$\left\{ \begin{array}{c} \alpha > 1 \Longrightarrow \mathbf{X}_i = \mathbf{R}^{(i-1)} \in \mathfrak{R}^{\rho}, \ \forall i = 1, 2, ..., \alpha, \ i.e., \\ \mathbf{X} = \left[\mathbf{X}_1^T \vdots \mathbf{X}_2^T \vdots ... \vdots \mathbf{X}_\alpha^T \right]^T = \left[\mathbf{R}^T \vdots \mathbf{R}^{(1)T} \vdots ... \vdots \mathbf{R}^{(\alpha-1)^T} \right]^T = \\ = \mathbf{R}^{\alpha-1} \in \mathfrak{R}^n, \ n = \alpha\rho, \\ \alpha = 1 \Longrightarrow \mathbf{X}_1 = \mathbf{X} = \mathbf{R} \in \mathfrak{R}^\rho, \end{array} \right\} \tag{11.141}$$

$$\alpha > 1 \Longrightarrow A =$$

$$\begin{bmatrix} O_\rho & I_\rho & \cdots & O_\rho & O_\rho \\ O_\rho & O_\rho & \cdots & O_\rho & O_\rho \\ O_\rho & O_\rho & \cdots & I_\rho & O_\rho \\ \cdots & \cdots & \cdots & \cdots & \cdots \\ O_\rho & O_\rho & \cdots & O_\rho & I_\rho \\ -A_\alpha^{-1}A_0 & -A_\alpha^{-1}A_1 & \cdots & -A_\alpha^{-1}A_{\alpha-2} & -A_\alpha^{-1}A_{\alpha-1} \end{bmatrix} \in \mathfrak{R}^{\alpha\rho\times\alpha\rho} \quad ,$$

$$\alpha = 1 \Longrightarrow A = -A_1^{-1}A_0 \in \mathfrak{R}^{\rho\times\rho},$$

$$(11.142)$$

$$B_{inv} = \left\{ \begin{array}{l} \begin{bmatrix} O_{(\alpha-1)\rho,\rho} \\ I_\rho \end{bmatrix} \in \mathfrak{R}^{n\times\rho}, \ \alpha > 1, \\ I_\rho \in \mathfrak{R}^{\rho\times\rho}, \ \alpha = 1, \end{array} \right\} \in \mathfrak{R}^{n\times\rho}, \qquad (11.143)$$

A_α^{-1} exists and $B^{(\mu)}$ has its general form:

$$B^{(\mu)} = \begin{bmatrix} B_0 \vdots B_1 \vdots \dots \vdots B_\mu \end{bmatrix} \in \mathfrak{R}^{\rho\times(\mu+1)r}, \ B^{(0)} = B_0 \qquad (11.144)$$

$$\mathrm{B}^{(\mu)} = \mathrm{B}_{inv}A_\alpha^{-1}B^{(\mu)} \in \mathfrak{R}^{n\times(\mu+1)r}, \qquad (11.145)$$

$$C = R_y^{(\alpha-1)} \in \mathfrak{R}^{N\times n}, \ Q = U. \qquad (11.146)$$

In view of the above explanations, Theorem 171 (Section 10.4) is applicable to the *HISO* system (5.1), (5.2) in the following form:

Theorem 212 *State controllability criteria for the HISO system (5.1), (5.2)*

For the HISO system (5.1), (5.2) to be state controllable it is necessary and sufficient that:

a) Any of the following equivalent conditions 1. through 6.a) holds if $\mu = 0$,

b) Any of the following equivalent conditions 1. through 5, 6.b) or 6.c) holds if $\mu > 0$.

In a) and b) Equations (11.142)-(11.145) induced by (11.139) determine the matrices:

1. All rows of both matrices $\Phi(t_1, t)\,\mathrm{B}^{(\mu)}$ *and* $\Phi(t_0, t)\,\mathrm{B}^{(\mu)}$ *are linearly independent on* $[t_0, t_1]$ *for any* $(t_0, t_1 > t_0) \in In\mathfrak{T}_0 \times In\mathfrak{T}_0$.

2. All rows of $\Phi(s)\,\mathrm{B}^{(\mu)} = (sI - A)^{-1}\,\mathrm{B}^{(\mu)}$ *and* $(sI - A)^{-1}\,\mathrm{B}_{inv}$ *are linearly independent on* $\mathbb{C}$.

3. The Gram matrix $G_{\Phi B}(t_1, t_0)$ *(11.147) of* $\Phi(t_1, t)\,\mathrm{B}^{(\mu)}$,

$$G_{\Phi B}(t_1, t_0) = \int_{t_0}^{t_1} \Phi(t_1, \tau)\,\mathrm{B}^{(\mu)}\left(\mathrm{B}^{(\mu)}\right)^T \Phi^T(t_1, \tau)\,d\tau \qquad (11.147)$$

is nonsingular for any $(t_0, t_1 > t_0) \in In\mathfrak{T}_0 \times In\mathfrak{T}_0$; *i.e.,*

$$rankG_{\Phi B}(t_1, t_0) = n, \ for \ any \ (t_0, t_1 > t_0) \in In\mathfrak{T}_0 \times In\mathfrak{T}_0. \qquad (11.148)$$

4. The $n \times n(\mu + 1)r$ *controllability matrix* $\mathcal{C}_{HISO}$,

$$\mathcal{C}_{HISO} = \left[B^{(\mu)} \vdots AB^{(\mu)} \vdots A^2 B^{(\mu)} \vdots ... \vdots A^{n-1} B^{(\mu)} \right] \in \mathfrak{R}^{n \times n(\mu+1)r}, \quad (11.149)$$

has the full rank n,

$$rank\mathcal{C}_{HISO} = n. \qquad (11.150)$$

5. For every eigenvalue $s_i(A)$ *of the matrix* A, *equivalently for every complex number* $s \in \mathbb{C}$, *the* $n \times (n + (\mu + 1)r)$ *matrix* $\left[sI - A \vdots B^{(\mu)} \right]$ *has the full rank* n,

$$rank \left[sI - A \vdots B^{(\mu)} \right] = n,$$

$$\forall s = s_i(A) \in \mathbb{C}, \ \forall i = 1, 2, ...n, \ i.e., \ \forall s \in \mathbb{C}. \qquad (11.151)$$

6. a) If $\mu = 0$ *the control vector function* $\mathbf{U}(.)$ *obeys*

$$\mathbf{U}(t) = (\Phi(t_1, t) B_0)^T G_{\Phi B}^{-1}(t_1, t_0) [\mathbf{X}_1 - \Phi(t_1, t_0) \mathbf{X}_0],$$

$$\forall t \in [t_0, t_1], \ for \ any \ (t_0, t_1 > t_0) \in In\mathfrak{T}_0 \times In\mathfrak{T}_0, \mu = 0. \qquad (11.152)$$

b) If $\mu > 0$ *the control vector function* $\mathbf{U}(.)$ *obeys either*

$$T^{-1}\mathbf{U}^\mu(t) = T^T \left(B^{(\mu)} \right)^T \Phi^T(t_1, t) G_{\Phi BT}^{-1}(t_1, t_0) [\mathbf{X}_1 - \Phi(t_1, t_0) \mathbf{X}_0], \ \mu > 0,$$
$$(11.153)$$

where $T \in \mathfrak{R}^{(\mu+1)r \times (\mu+1)r}$ *is any nonsingular matrix and* $G_{\Phi BT}(t_1, t_0)$ *is the Gram matrix of* $\Phi(t_1, t) B^{(\mu)} T$,

$$G_{\Phi BT}(t_1, t_0) = \int_{t_0}^{t_1} \Phi(t_1, \tau) B^{(\mu)} T T^T \left(B^{(\mu)} \right)^T \Phi^T(t_1, \tau) d\tau. \qquad (11.154)$$

c) or the control vector function $\mathbf{U}(.)$ *obeys*

$$B^{(\mu)} \mathbf{U}^\mu(t) = \Phi^T(t_1, t) G_\Phi^{-1}(t_1, t_0) [\mathbf{X}_1 - \Phi(t_1, t_0) \mathbf{X}_0], \ \mu > 0, \quad (11.155)$$

where $G_\Phi(t_1, t_0)$ *is the Gram matrix of* $\Phi(t_1, t)$,

$$G_\Phi(t_1, t_0) = \int_{t_0}^{t_1} \Phi(t_1, \tau) \Phi^T(t_1, \tau) d\tau. \qquad (11.156)$$

Exercise 213 *Test the state controllability of the selected HISO physical plant in Exercise 68, Section 5.3.*

11.8   *HISO* system output controllability

11.8.1   Definition

The general output controllability Definition 165, (Section 10.3), adjusted to the nondisturbed *HISO* system (11.137), (11.138) reads:

Definition 214 *Output controllability of the HISO system (11.137), (11.138)*
    The HISO system (11.137), (11.138) is output controllable if and only if for every initial output vector $\mathbf{Y}_0 \in \mathfrak{R}^N$ *at* $t_0 \in \mathfrak{T}$ *and for any final output vector* $\mathbf{Y}_1 \in \mathfrak{R}^N$ *there exist a moment* $t_1 \in In\mathfrak{T}_0$ *and an extended control* $\mathbf{U}^\mu_{[t_0,t_1]}$ *on the time interval* $[t_0, t_1]$ *such that*

$$\mathbf{Y}(t_1; t_0; \mathbf{Y}_0; \mathbf{U}^\mu_{[t_0,t_1]}) = \mathbf{Y}_1.$$

11.8.2   Criteria

State space-output space product approach

An equivalent form of the *HISO* system (11.137), (11.138) is its *EISO* form (11.139), (11.140), (Section 11.7), presented as follows:

$$\frac{d\mathbf{X}(t)}{dt} = A\mathbf{X}(t) + \mathbf{B}^{(\mu)}\mathbf{U}^\mu(t), \forall t \in \mathfrak{T}_0, \mu > 0, \ \mathbf{B}^{(\mu)} \in \mathfrak{R}^{n \times (\mu+1)r}, \ n = \alpha\rho,$$
$$(11.157)$$
$$\mathbf{Y}(t) = C\mathbf{X}(t) + U\mathbf{U}(t), \ \forall t \in \mathfrak{T}_0. \qquad (11.158)$$

Equations (11.141)-(11.146) link the *HISO* system (11.137), (11.138) with its *EISO* form (11.157), (11.158). This makes Theorem 208 directly applicable to the *HISO* system (11.137), (11.138) for which

$$H_{HISO}(t, t_0) = \left[C\Phi(t, t_0)B + \tilde{U} \right] \in \mathfrak{R}^{N \times (\mu+1)r}, \ \mathbf{B}^{(\mu)} = B_{inv}A_\alpha^{-1}B^{(\mu)},$$

$$\tilde{U} = \left[O_{N,r} \vdots U \vdots O_{N,r} \vdots .... \vdots O_{N,r} \right] \in \mathfrak{R}^{N \times (\mu+1)r}, \qquad (11.159)$$

$$H_{HISO}(s) = \mathcal{L}\{H_{HISO}(t, t_0)\} = C\left(sI - A\right)^{-1}\mathbf{B}^{(\mu)} + s^{-1}\tilde{U} =$$
$$= \left(C\Phi(s)\mathbf{B}^{(\mu)} + s^{-1}\tilde{U}\right) \in \mathbb{C}^{N \times (\mu+1)r}, \qquad (11.160)$$

where $\Phi(t, t_0) = e^{A(t-t_0)}$. The matrices A, B_{inv}, $B^{(\mu)}$, and C are defined, respectively, in (11.142), (11.143), (11.144), and (11.146), (Section 11.7).

Theorem 215 *Output controllability of the HISO system (11.137), (11.138)*

For the *HISO* system (11.137), (11.138) to be output controllable it is necessary that

$$rank \left[C \vdots U \right] = N \qquad (11.161)$$

and if this condition is satisfied then it is necessary that any of the following conditions 1.-6. holds and sufficient that any of the following conditions 1.-4., 6. holds, where the conditions 1.-4.,6. are equivalent:

1. All rows of the system matrix $H_{HISO}(t, t_0)$ (11.159) are linearly independent on $[t_0, t_1]$ for any $(t_0, t_1 > t_0) \in In\mathfrak{T}_0 \times In\mathfrak{T}_0$.

2. All rows of the system matrix $H_{HISO}(s)$ (11.160) are linearly independent on $\mathbb{C}$.

3. The Gram matrix $G_{HISO}(t_1, t_0)$ of $H_{HISO}(t, t_0)$,

$$G_{HISO}(t_1, t_0) = \int_{t_0}^{t_1} H_{HISO}(t_1, \tau) H_{HISO}^T(t_1, \tau) d\tau,$$

$$for \ any \ (t_0, t_1 > t_0) \in In\mathfrak{T}_0 \times In\mathfrak{T}_0, \qquad (11.162)$$

is nonsingular; i.e.,

$$rank G_{HISO}(t_1, t_0) = n, \ any \ (t_0, t_1 > t_0) \in In\mathfrak{T}_0 \times In\mathfrak{T}_0. \qquad (11.163)$$

4. The output $N \times (n + 1)(\mu + 1) r$ controllability matrix $\mathcal{C}_{HISOout}$,

$$\mathcal{C}_{HISOout} = \left[CB \vdots CAB \vdots CA^2B \vdots ...\vdots CA^{n-1}B \vdots \tilde{U} \right] \qquad (11.164)$$

has the full rank N,

$$rank \mathcal{C}_{HISOout} = N. \qquad (11.165)$$

5. For every eigenvalue $s_i(A)$ of the matrix A, equivalently for every complex number $s \in \mathbb{C}$, the $N \times (n + 2(\mu + 1) r)$ matrix

$$\left[C(sI - A) \vdots CB^{(\mu)} \vdots \tilde{U} \right]$$

has the full rank N,

$$rank \left[C(sI - A) \vdots CB^{(\mu)} \vdots \tilde{U} \right] = N,$$

$$\forall s = s_i(A) \in \mathbb{C}, \ \forall i = 1, 2, ...n, \ i.e., \ \forall s \in \mathbb{C}. \qquad (11.166)$$

6. The control vector function $\mathbf{U}(.)$ *obeys the following equation for any nonsingular matrix* R, *where the matrix* $R \in \mathfrak{R}^{(\mu+1)r \times (\mu+1)r}$:

$$R^{-1}\mathbf{U}^{\mu}(t) = \left\{ \begin{array}{c} (H_{HISO}(t_1,t)\,R)^T \bullet G_{HISOR}^{-1}(t_1,t_0) \bullet \\ \bullet \left[\mathbf{Y_1} - C\Phi(t_1,t_0)\,\mathbf{X_0} - U^{(\mu-1)}\mathbf{U}_0^{\mu-1} \right] \end{array} \right\}, \qquad (11.167)$$

where $G_{HISOR}(t_1,t_0)$ *is the Gram matrix of* $H_{HISO}(t_1,t)\,R$,

$$G_{HISOR}(t_1,t_0) = \int_{t_0}^{t_1} H_{HISO}(t_1,\tau)\,RR^T H_{HISO}^T(t_1,\tau)\,d\tau. \qquad (11.168)$$

Output space approach

The system response $\mathbf{Y}(t; \mathbf{R}_0^{\alpha-1}; \mathbf{U})$ results as the inverse Laplace transform of Equation (5.5), (Section 5.1), applied to the undisturbed system (5.1), (5.2):

$$\mathbf{Y}(t; \mathbf{R}_0^{\alpha-1}; \mathbf{U}) =$$

$$= \int_{0-}^{t} [\Gamma_{HISO}(\tau)\mathbf{U}(t-\tau)d\tau] + \Gamma_{HISOU_0}(t)\mathbf{U}_0^{\mu-1} + \Gamma_{HISOR_0}(t)\mathbf{R}_0^{\alpha-1},$$

$$\int_{0-}^{t} [\Gamma_{HISO}(\tau)\mathbf{U}(t-\tau)d\tau] = \int_{0-}^{t} [\Gamma_{HISO}(\tau)\mathbf{U}(t,\tau)d\tau] =$$

$$= \int_{0-}^{t} [\Gamma_{HISO}(t-\tau)\mathbf{U}(\tau)d\tau] = \int_{0-}^{t} [\Gamma_{HISO}(t,\tau)\mathbf{U}(\tau)d\tau], \qquad (11.169)$$

where

- The system output fundamental matrix $\Gamma_{HISO}(t)$ is the inverse Laplace transform of the system transfer function $G_{HISO}(s)$ relating the output $\mathbf{Y}$ to the control vector $\mathbf{U}$:

$$G_{HISO}(s) = R_y^{(\alpha)} S_\rho^{(\alpha)}(s) \left(A^{(\alpha)} S_\rho^{(\alpha)}(s) \right)^{-1} B^{(\mu)} S_r^{(\mu)}(s) + U, \qquad (11.170)$$

$$\Gamma_{HISO}(t) = \mathcal{L}^{-1}\{G_{HISO}(s)\} =$$

$$= \mathcal{L}^{-1}\left\{ R_y^{(\alpha)} S_\rho^{(\alpha)}(s) \left(A^{(\alpha)} S_\rho^{(\alpha)}(s) \right)^{-1} B^{(\mu)} S_r^{(\mu)}(s) + U \right\}, \qquad (11.171)$$

- $\Gamma_{HISOU_0}(t)$ is the inverse Laplace transform of the system transfer function matrix $G_{HISOU_0}(s)$ relative to the extended initial control vector $\mathbf{U}_0^{\mu-1}$,

$$G_{HISOU_0}(s) = -R_y^{(\alpha)} S_\rho^{(\alpha)}(s) \left(A^{(\alpha)} S_\rho^{(\alpha)}(s)\right)^{-1} B^{(\mu)} Z_r^{(\mu-1)}(s), \quad (11.172)$$

$$\Gamma_{HISOU_0}(t) = \mathcal{L}^{-1} \{G_{HISOU_0}(s)\} =$$
$$= \mathcal{L}^{-1} \left\{ -R_y^{(\alpha)} S_\rho^{(\alpha)}(s) \left(A^{(\alpha)} S_\rho^{(\alpha)}(s)\right)^{-1} B^{(\mu)} Z_r^{(\mu-1)}(s) \right\}, \quad (11.173)$$

- $\Gamma_{HISOR_0}(t)$ is the inverse Laplace transform of is the transfer function matrix $G_{HISOR_0}(s)$ relative to the initial state vector $\mathbf{R}_0^{\alpha-1}$,

$$G_{HISOR_0}(s) =$$
$$= R_y^{(\alpha)} S_\rho^{(\alpha)}(s) \left(A^{(\alpha)} S_\rho^{(\alpha)}(s)\right)^{-1} A^{(\alpha)} Z_\rho^{(\alpha-1)}(s) - R_y^{(\alpha)} Z_\rho^{(\alpha-1)}(s), \quad (11.174)$$

$$\Gamma_{HISOR_0}(t) = \mathcal{L}^{-1} \{G_{HISOR_0}(s)\} =$$
$$= \mathcal{L}^{-1} \left\{ R_y^{(\alpha)} S_\rho^{(\alpha)}(s) \left(A^{(\alpha)} S_\rho^{(\alpha)}(s)\right)^{-1} A^{(\alpha)} Z_\rho^{(\alpha-1)}(s) - R_y^{(\alpha)} Z_\rho^{(\alpha-1)}(s) \right\}.$$
$$(11.175)$$

Theorem 216 *Output controllability of the HISO system (11.137), (11.138)*

For the HISO system (11.137), (11.138) to be output controllable it is necessary that (11.161) is fulfilled and if it is satisfied then it is necessary and sufficient that any of the following equivalent conditions holds:

1) For any $(t_0, t_1 > t_0) \in In\mathfrak{T}_0 \times In\mathfrak{T}_0$ the rows of the system output fundamental matrix $\Gamma_{HISO}(t_1, t)$ (11.171) are linearly independent on the time interval $[t_0, t_1]$.

2) The rows of the system transfer function matrix $G_{HISO}(s)$ (11.170) are linearly independent on $\mathbb{C}$.

3) For any $(t_0, t_1 > t_0) \in In\mathfrak{T}_0 \times In\mathfrak{T}_0$ the Gram matrix $G_{\Gamma HISO}(t_1, t_0)$ (11.176)

$$G_{\Gamma HISO}(t_1, t_0) = \int_{t_0}^{t_1} \Gamma_{HISO}(t_1, \tau) \Gamma_{HISO}^T(t_1, \tau) d\tau \in \mathfrak{R}^{N \times N}. \quad (11.176)$$

is nonsingular, i.e., the condition (11.177),

$$det G_{\Gamma HISO}(t_1, t_0) \neq 0, \ any \ (t_0, t_1 > t_0) \in In\mathfrak{T}_0 \times In\mathfrak{T}_0, \qquad (11.177)$$

holds.

4) The control vector function satisfies the following linear time-invariant algebraic vector equation:

$$\mathbf{U}(t) = \Gamma_{HISO}^T(t_1, t) G_{\Gamma HISO}^{-1}(t_1, t_0) \bullet$$

$$\bullet \left[\mathbf{Y}_1 - \Gamma_{HISOU_0}(t_1) \mathbf{U}_{0-}^{\mu-1} - \Gamma_{HISOR_0}(t_1) \mathbf{R}_0^{\alpha-1} \right]. \qquad (11.178)$$

The proof repeats essentially the proof of Theorem 195, (Subsection 11.2.2).

Exercise 217 *Test the output controllability of the selected HISO physical plant in Exercise 68, Section 5.3.*

11.9   *IIO* system state controllability

11.9.1   Definition

The unperturbed *IIO* system (6.1), (6.2), (Section 6.1), has the following form:

$$A^{(\alpha)} \mathbf{R}^{\alpha}(t) = B^{(\mu)} \mathbf{U}^{\mu}(t), \ \forall t \in \mathfrak{T}_0, \qquad (11.179)$$

$$E^{(\nu)} \mathbf{Y}^{\nu}(t) = R_y^{(\alpha-1)} \mathbf{R}^{\alpha-1}(t) + Q^{(\mu)} \mathbf{U}^{\mu}(t), \ \forall t \in \mathfrak{T}_0, \ \nu > 0. \qquad (11.180)$$

The system has (see Note 72, Section 6.1):
- The internal state vector $\mathbf{S}_{IIOI}$,

$$\mathbf{S}_{IIOI} = \mathbf{R}^{\alpha-1} \in \mathfrak{R}^n, \ n = \alpha\rho, \qquad (11.181)$$

- The output state vector $\mathbf{S}_{IIOO}$,

$$\mathbf{S}_{IIOO} = \mathbf{Y}^{\nu-1} = \left[\mathbf{Y}^T \vdots \mathbf{Y}^{(1)^T} \vdots ... \vdots \mathbf{Y}^{(\nu-1)^T} \right]^T \in \mathfrak{R}^n, \ n = \nu N, \quad (11.182)$$

- And the full state vector $\mathbf{S}_{IIOf}$, which is its state vector $\mathbf{S}_{IIO}$. It is composed of the internal state vector $\mathbf{S}_{IIOI} = \mathbf{R}^{\alpha-1}$, Equation (11.181), and of the output state vector $\mathbf{S}_{IIOO} = \mathbf{Y}^{\nu-1}$, Equation (11.182),

$$\mathbf{S}_{IIOf} = \begin{bmatrix} \mathbf{S}_{IIOI} \\ \mathbf{S}_{IIOO} \end{bmatrix} = \begin{bmatrix} \mathbf{R}^{\alpha-1} \\ \mathbf{Y}^{\nu-1} \end{bmatrix} = \mathbf{S}_{IIO} \in \mathfrak{R}^n, \ n = \alpha\rho + \nu N. \quad (11.183)$$

The internal state vector $\mathbf{S}_{IIOI}$ is independent of the output state vector $\mathbf{S}_{IIOO}$ in view of Equation (11.179). However, the output state vector $\mathbf{S}_{IIOO}$ depends on the internal state vector $\mathbf{S}_{IIOI}$. They both determine the system (full) state vector $\mathbf{S}_{IIO}$. These facts lead to the following definitions (see Definition 161, Section 10.3):

Definition 218 *Internal state controllability of the IIO system defined by (11.179), (11.180)*

*The IIO system (11.179), (11.180) is the **mathematical internal state or physical internal state controllable** if and only if for every initial mathematical internal state or physical internal state vector $\mathbf{R}_0^{\alpha-1} \in \mathfrak{R}^{\alpha\rho}$ at $t_0 \in \mathfrak{T}$ and for any final mathematical internal state or physical internal state vector $\mathbf{R}_1^{\alpha-1} \in \mathfrak{R}^{\alpha\rho}$, respectively, there exist a moment $t_1 \in In\mathfrak{T}_0$ and an extended control $\mathbf{U}_{[t_0,t_1]}^{\mu}$ on the time interval $[t_0, t_1]$ such that*

$$\mathbf{R}^{\alpha-1}(t_1; t_0; \mathbf{R}_0^{\alpha-1}; \mathbf{0}_d; \mathbf{U}_{[t_0,t_1]}^{\mu}) = \mathbf{R}_1^{\alpha-1}. \tag{11.184}$$

Definition 219 *State controllability of the IIO system (11.179), (11.180)*

*The IIO system (11.179), (11.180) is the **mathematical state or physical state controllable** if and only if for every initial mathematical state or physical state vector*

$$\mathbf{S}_{IIO0} = \begin{bmatrix} \mathbf{R}_0^{\alpha-1} \\ \mathbf{Y}_0^{\nu-1} \end{bmatrix} \in \mathfrak{R}^n, \ n = \alpha\rho + \nu N,$$

at $t_0 \in \mathfrak{T}$ and for any final mathematical state or physical state vector

$$\mathbf{S}_{IIO1} = \begin{bmatrix} \mathbf{R}_1^{\alpha-1} \\ \mathbf{Y}_1^{\nu-1} \end{bmatrix} \in \mathfrak{R}^n,$$

respectively, there exist a moment $t_1 \in In\mathfrak{T}_0$ and an extended control $\mathbf{U}_{[t_0,t_1]}^{\mu}$ on the time interval $[t_0, t_1]$ such that

$$\mathbf{S}_{IIO}(t_1; t_0; \mathbf{S}_{IIO0}; \mathbf{0}_d; \mathbf{U}_{[t_0,t_1]}^{\mu}) = \begin{bmatrix} \mathbf{R}^{\alpha-1}(t_1; t_0; \mathbf{R}_0^{\alpha-1}; \mathbf{0}_d; \mathbf{U}_{[t_0,t_1]}^{\mu}) \\ \mathbf{Y}^{\nu-1}(t_1; t_0; \mathbf{R}_0^{\alpha-1}; \mathbf{Y}_0^{\nu-1}; \mathbf{0}_d; \mathbf{U}_{[t_0,t_1]}^{\mu}) \end{bmatrix} =$$

$$= \mathbf{S}_{IIO1} = \begin{bmatrix} \mathbf{R}_1^{\alpha-1} \\ \mathbf{Y}_1^{\nu-1} \end{bmatrix}. \tag{11.185}$$

11.9.2   Criterion

Equation (11.179) is the same as Equation (11.137) (Subsection 11.7.2 of Section 11.7). The equivalent form of the latter, hence also of the former, is the *EISO* form (11.139), i.e.,

$$\frac{d\widehat{\mathbf{X}}(t)}{dt} = \widehat{A}\widehat{\mathbf{X}}(t) + \widehat{\mathbf{B}}^{(\mu)}\widehat{\mathbf{U}}^{\mu}(t), \forall t \in \mathfrak{T}_0, \ \mu \geq 1, \ \widehat{\mathbf{B}}^{(\mu)} \in \mathfrak{R}^{\widehat{n}\times(\mu+1)r}, \ \widehat{n} = \alpha\rho,$$

(11.186)

where the vector $\widehat{\mathbf{X}}$ is defined as $\mathbf{X}$ in (11.141) that now reads:

$$\left\{ \begin{array}{c} \alpha > 1 \Longrightarrow \widehat{\mathbf{X}}_i = \mathbf{R}^{(i-1)} \in \mathfrak{R}^{\rho}, \ \forall i = 1, 2, ..., \alpha, \ i.e., \\ \widehat{\mathbf{X}} = \left[\widehat{\mathbf{X}}_1^T \vdots \widehat{\mathbf{X}}_2^T \vdots ... \vdots \widehat{\mathbf{X}}_{\alpha}^T\right]^T = \left[\mathbf{R}^T \vdots \mathbf{R}^{(1)T} \vdots ... \vdots \mathbf{R}^{(\alpha-1)^T}\right]^T = \\ = \mathbf{R}^{\alpha-1} \in \mathfrak{R}^{\widehat{n}}, \end{array} \right\}$$

$$\alpha = 1 \Longrightarrow \widehat{\mathbf{X}}_1 = \widehat{\mathbf{X}} = \mathbf{R} \in \mathfrak{R}^{\rho}, \qquad (11.187)$$

and the matrices are determined in Equations (11.142)-(11.145), (Section 11.7), i.e.,

$$\alpha > 1 \Longrightarrow \widehat{A} =$$

$$\begin{bmatrix} O_{\rho} & I_{\rho} & ... & O_{\rho} & O_{\rho} \\ O_{\rho} & O_{\rho} & ... & O_{\rho} & O_{\rho} \\ O_{\rho} & O_{\rho} & ... & I_{\rho} & O_{\rho} \\ ... & ... & ... & ... & ... \\ O_{\rho} & O_{\rho} & ... & O_{\rho} & I_{\rho} \\ -A_{\alpha}^{-1}A_0 & -A_{\alpha}^{-1}A_1 & ... & -A_{\alpha}^{-1}A_{\alpha-2} & -A_{\alpha}^{-1}A_{\alpha-1} \end{bmatrix} \in \mathfrak{R}^{\alpha\rho\times\alpha\rho} \ ,$$

$$\alpha = 1 \Longrightarrow \widehat{A} = -A_1^{-1}A_0 \in \mathfrak{R}^{\rho\times\rho},$$

(11.188)

$$\widehat{B}_{inv} = \left\{ \begin{array}{l} \left[\begin{array}{c} O_{(\alpha-1)\rho,\rho} \\ I_{\rho} \end{array}\right] \in \mathfrak{R}^{\widehat{n}\times\rho}, \ \alpha > 1, \\ I_{\rho} \in \mathfrak{R}^{\rho\times\rho}, \ \alpha = 1, \end{array} \right\} \in \mathfrak{R}^{\widehat{n}\times\rho}, \qquad (11.189)$$

$\widehat{B}^{(\mu)}$ and $\widehat{\mathbf{B}}^{(\mu)}$,

$$\widehat{B}^{(\mu)} = \left[B_0 \vdots B_1 \vdots ... \vdots B_{\mu}\right] \in \mathfrak{R}^{\rho\times(\mu+1)r}, \ \widehat{B}^{(0)} = B_0 \qquad (11.190)$$

$$\widehat{\mathbf{B}}^{(\mu)} = \widehat{B}_{inv}A_{\alpha}^{-1}\widehat{B}^{(\mu)} \in \mathfrak{R}^{\widehat{n}\times(\mu+1)r}. \qquad (11.191)$$

Theorem 212 is valid unchanged for the *IIO* system (11.179), (11.180) internal state controllability:

Theorem 220 *Internal state controllability criteria for the IIO system (11.179), (11.180)*

For the IIO system (11.179), (11.180) to be internal state controllable it is necessary and sufficient that:

a) Any of the following equivalent conditions 1. through 6.a) holds if $\mu = 0$,

b) Any of the following equivalent conditions 1. through 5, 6.b) or 6.c) holds if $\mu > 0$.

In a) and b) Equations (11.187) - (11.191) induced by (11.179) determine the matrices:

1. All rows of both matrices $\widehat{\Phi}(t_1, t)\widehat{B}^{(\mu)}$ and $\widehat{\Phi}(t_0, t)\widehat{B}^{(\mu)}$ are linearly independent on $[t_0, t_1]$ for any $(t_0, t_1 > t_0) \in In\mathfrak{T}_0 \times In\mathfrak{T}$.

2. All rows of $\widehat{\Phi}(s)\widehat{B}^{(\mu)} = \left(sI - \widehat{A}\right)^{-1}\widehat{B}^{(\mu)}$ and $(sI - A)^{-1}B_{inv}$ are linearly independent on $\mathbb{C}$.

3. The Gram matrix $G_{\widehat{\Phi}\widehat{B}}(t_1, t_0)$ (11.192) of $\widehat{\Phi}(t_1, t)\widehat{B}^{(\mu)}$,

$$G_{\widehat{\Phi}\widehat{B}}(t_1, t_0) = \int_{t_0}^{t_1} \widehat{\Phi}(t_1, \tau)\widehat{B}^{(\mu)}\left(\widehat{B}^{(\mu)}\right)^T \widehat{\Phi}^T(t_1, \tau)\, d\tau \qquad (11.192)$$

is nonsingular for any $(t_0, t_1 > t_0) \in In\mathfrak{T}_0 \times In\mathfrak{T}_0$; i.e.,

$$rankG_{\widehat{\Phi}\widehat{B}}(t_1, t_0) = n, \ for \ any \ (t_0, t_1 > t_0) \in In\mathfrak{T}_0 \times In\mathfrak{T}_0. \qquad (11.193)$$

4. The $n \times n(\mu + 1)r$ controllability matrix $\widehat{C}$,

$$\widehat{C} = \left[\widehat{B}^{(\mu)} \vdots \ \widehat{A}\widehat{B}^{(\mu)} \vdots \ \widehat{A}^2\widehat{B}^{(\mu)} \vdots \ ... \ \vdots \ \widehat{A}^{n-1}\widehat{B}^{(\mu)}\right] \in \mathfrak{R}^{\widehat{n} \times \widehat{n}(\mu+1)r}, \qquad (11.194)$$

has the full rank $\widehat{n}$,

$$rank\widehat{C} = \widehat{n}. \qquad (11.195)$$

5. For every eigenvalue $s_i\left(\widehat{A}\right)$ of the matrix $\widehat{A}$, equivalently for every complex number $s \in \mathbb{C}$, the $\widehat{n} \times (\widehat{n} + (\mu + 1)r)$ matrix $\left[sI_{\widehat{n}} - \widehat{A} \vdots \widehat{B}^{(\mu)}\right]$ has the full rank $\widehat{n}$,

$$rank\left[sI_{\widehat{n}} - \widehat{A} \vdots \widehat{B}^{(\mu)}\right] = \widehat{n},$$

$$\forall s = s_i\left(\widehat{A}\right) \in \mathbb{C}, \ \forall i = 1, 2, ...n, \ i.e., \ \forall s \in \mathbb{C}. \qquad (11.196)$$

6. a) If $\mu = 0$ the control vector function $\widehat{\mathbf{U}}(.)$ obeys

$$\widehat{\mathbf{U}}(t) = \left(\widehat{\Phi}(t_1, t)\,\widehat{\mathbf{B}}_0\right)^T G_{\widehat{\Phi}\widehat{B}}^{-1}(t_1, t_0)\left[\widehat{\mathbf{X}}_1 - \widehat{\Phi}(t_1, t_0)\,\widehat{\mathbf{X}}_0\right],$$

$\forall t \in [t_0, t_1]$, for any $(t_0, t_1 > t_0) \in In\mathfrak{T}_0 \times In\mathfrak{T}_0, \mu = 0$.   (11.197)

b) If $\mu > 0$ the control vector function $\widehat{\mathbf{U}}(.)$ obeys either

$$T^{-1}\widehat{\mathbf{U}}^{\mu}(t) = T^T\left(\widehat{\mathbf{B}}^{(\mu)}\right)^T \widehat{\Phi}^T(t_1, t)\,G_{\widehat{\Phi}\widehat{B}T}^{-1}(t_1, t_0)\left[\widehat{\mathbf{X}}_1 - \widehat{\Phi}(t_1, t_0)\,\widehat{\mathbf{X}}_0\right], \mu > 0,$$
        (11.198)

where $T \in \mathfrak{R}^{(\mu+1)r \times (\mu+1)r}$ is any nonsingular matrix and $G_{\widehat{\Phi}\widehat{B}T}(t_1, t_0)$ is the Gram matrix of $\widehat{\Phi}(t_1, t)\,\widehat{\mathbf{B}}^{(\mu)}T$,

$$G_{\widehat{\Phi}\widehat{B}T}(t_1, t_0) = \int_{t_0}^{t_1} \widehat{\Phi}(t_1, \tau)\,\widehat{\mathbf{B}}^{(\mu)}TT^T\left(\widehat{\mathbf{B}}^{(\mu)}\right)^T \widehat{\Phi}^T(t_1, \tau)\,d\tau,$$ (11.199)

c) or the control vector function $\widehat{\mathbf{U}}(.)$ obeys

$$\widehat{\mathbf{B}}^{(\mu)}\widehat{\mathbf{U}}^{\mu}(t) = \widehat{\Phi}^T(t_1, t)\,G_{\widehat{\Phi}}^{-1}(t_1, t_0)\left[\widehat{\mathbf{X}}_1 - \widehat{\Phi}(t_1, t_0)\,\widehat{\mathbf{X}}_0\right], \ \mu > 0,$$ (11.200)

where $G_{\widehat{\Phi}}(t_1, t_0)$ is the Gram matrix of $\widehat{\Phi}(t_1, t)$,

$$G_{\widehat{\Phi}}(t_1, t_0) = \int_{t_0}^{t_1} \widehat{\Phi}(t_1, \tau)\,\widehat{\Phi}^T(t_1, \tau)\,d\tau.$$ (11.201)

Let also

$$\left\{ \begin{array}{c} \nu > 1 \Longrightarrow \widetilde{\mathbf{X}}_i = \mathbf{Y}^{(i-1)} \in \mathfrak{R}^{\rho}, \ \forall i = 1, 2, ..., \nu, \ i.e., \\ \widetilde{\mathbf{X}} = \left[\widetilde{\mathbf{X}}_1^T \vdots \widetilde{\mathbf{X}}_2^T \vdots ... \vdots \widetilde{\mathbf{X}}_{\nu}^T\right]^T = \left[\mathbf{Y}^T \vdots \mathbf{Y}^{(1)T} \vdots ... \vdots \mathbf{Y}^{(\nu-1)^T}\right]^T = \\ = \mathbf{Y}^{\nu-1} \in \mathfrak{R}^{\widetilde{n}}, \ \widetilde{n} = \nu N, \end{array} \right\}$$

$$\nu = 1 \Longrightarrow \widetilde{\mathbf{X}}_1 = \widetilde{\mathbf{X}} = \mathbf{Y} \in \mathfrak{R}^N,$$ (11.202)

$$\nu > 1 \Longrightarrow \widetilde{A} =$$

$$\begin{bmatrix} O_N & I_N & ... & O_N & O_N \\ O_N & O_N & ... & O_N & O_N \\ O_N & O_N & ... & I_N & O_N \\ ... & ... & ... & ... & ... \\ O_N & O_N & ... & O_N & I_N \\ -E_{\nu}^{-1}E_0 & -E_{\nu}^{-1}E_1 & ... & -E_{\nu}^{-1}E_{\nu-2} & -E_{\nu}^{-1}E_{\nu-1} \end{bmatrix} \in \mathfrak{R}^{\nu N \times \nu N},$$

$$\nu = 1 \Longrightarrow \widetilde{A} = -E_1^{-1}E_0 \in \mathfrak{R}^{N \times N},$$

        (11.203)

$$\widetilde{B}_{inv} = \left\{ \begin{array}{l} \left[\begin{array}{c} O_{(\nu-1)N,N} \\ I_N \end{array} \right] \in \mathfrak{R}^{n \times N}, \ \nu > 1, \\ I_N \in \mathfrak{R}^{N \times N}, \ \nu = 1, \end{array} \right\} \in \mathfrak{R}^{\nu N \times N}, \qquad (11.204)$$

$$\widetilde{B}^{(\mu)} = Q^{(\mu)} = \left[Q_0 \vdots Q_1 \vdots ... \vdots Q_\mu \right] \in \mathfrak{R}^{N \times (\mu+1)r}, \ \widetilde{B}^{(0)} = Q_0 \qquad (11.205)$$

$$\widetilde{\mathbf{B}}^{(\mu)} = \widetilde{B}_{inv} E_\nu^{-1} \widetilde{B}^{(\mu)} \in \mathfrak{R}^{\nu N \times (\mu+1)r}, \qquad (11.206)$$

$$\widetilde{C} = \left[I_N \vdots O_N \vdots ... \vdots O_N \right] \in \mathfrak{R}^{N \times \nu N}. \qquad (11.207)$$

$$\widetilde{D}_{inv} = \left\{ \begin{array}{l} \left[\begin{array}{c} O_{(\nu-1)N,N} \\ I_N \end{array} \right] \in \mathfrak{R}^{\nu N \times N}, \ \nu > 1, \\ I_N \in \mathfrak{R}^{N \times N}, \ \nu = 1, \end{array} \right\} \in \mathfrak{R}^{\nu N \times N}, \qquad (11.208)$$

$$\widetilde{R}^{(\alpha-1)} = \left[R_{y0} \vdots R_{y1} \vdots ... \vdots R_{y,\alpha-1} \right] \in \mathfrak{R}^{N \times \alpha \rho}, \ \widetilde{R}^{(0)} = R_{y0} \qquad (11.209)$$

$$\widetilde{D} = \widetilde{D}_{inv} E_\nu^{-1} \widetilde{R}^{(\alpha-1)} \in \mathfrak{R}^{\nu N \times \alpha \rho}, \qquad (11.210)$$

These equations permit us to set Equation (11.180) in the following equivalent forms:

$$\frac{d\widetilde{\mathbf{X}}(t)}{dt} = \widetilde{A}\widetilde{\mathbf{X}}(t) + \widetilde{\mathbf{B}}^{(\mu)} \widetilde{\mathbf{U}}^\mu(t) + \widetilde{D}\widehat{\mathbf{X}}(t), \forall t \in \mathfrak{T}_0, \qquad (11.211)$$

$$\mathbf{Y}(t) = \widetilde{C}\widetilde{\mathbf{X}}(t), \ \forall t \in \mathfrak{T}_0. \qquad (11.212)$$

Equations (11.186), (11.211), and (11.212) lead to the following equivalent description of the IIO system (11.179), (11.180):

$$\frac{d\mathbf{X}(t)}{dt} = A\mathbf{X}(t) + B\mathbf{U}^\mu(t), \forall t \in \mathfrak{T}_0, \qquad (11.213)$$

$$\mathbf{Y}(t) = C\mathbf{X}(t), \ \forall t \in \mathfrak{T}_0, \qquad (11.214)$$

where

$$\mathbf{X} = \left[\begin{array}{c} \widehat{\mathbf{X}} \\ \widetilde{\mathbf{X}} \end{array} \right] \in \mathfrak{R}^{\widehat{n}+\widetilde{n}}, \ n = \widehat{n} + \widetilde{n} = \alpha \rho + \nu N, \qquad (11.215)$$

$$\mathbf{U} = \widehat{\mathbf{U}} = \widetilde{\mathbf{U}}, \qquad (11.216)$$

$$A = \left[\begin{array}{cc} \widehat{A} & O_{\widehat{n},\widetilde{n}} \\ \widetilde{D} & \widetilde{A} \end{array} \right] \in \mathfrak{R}^{n \times n} \Longrightarrow \Phi(t_1, t) = e^{A(t_1-t)}, \qquad (11.217)$$

$$B^{(\mu)} = \left[\begin{array}{c} \widehat{B}^{(\mu)} \\ \widetilde{B}^{(\mu)} \end{array} \right] \in \mathfrak{R}^{n \times (\mu+1)r}, \qquad (11.218)$$

$$C = \left[O_{N,\widehat{n}} \vdots \widetilde{C} \right] = \left[O_{N,\widehat{n}} \vdots I_N \vdots O_N \vdots ... \vdots O_N \right] \in \mathfrak{R}^{N \times n}. \qquad (11.219)$$

The matrices defined in Equations (11.188)-(11.191), (11.203)-(11.210) determine the matrices defined by Equations (11.217)-(11.219).

The system (11.213), (11.214) is the equivalent *EISO* system (11.8), (11.9) to the *IIO* system (11.179), (11.180) .

Theorem 180 for $\mu = 0$, (Section 11.1), and Theorem 205 for $\mu > 0$, (Section 11.5), with the matrices determined by Equations (11.188)-(11.191), (11.203)-(11.210), (11.217)-(11.219) hold for the system (11.213), (11.214), i.e., for the *IIO* system (11.179), (11.180) due to their equivalency.

Theorem 221 *State controllability criteria for the IIO system defined by (11.179), (11.180))*

For the IIO system (11.179), (11.180) to be state controllable it is necessary and sufficient that:

a) Any of the following equivalent conditions 1. through 6.a) holds if $\mu = 0$,

b) Any of the following equivalent conditions 1. through 5, 6.b) or 6.c) holds if $\mu > 0$.

In a) and b) Equations (11.187)-(11.191) induced by (11.179), (11.202)-(11.218) determine the matrices:

1) All rows of the matrices $\Phi(t_1, t) B^{(\mu)}$ and $\Phi(t_0, t) B^{(\mu)}$ are linearly independent on $[t_0, t_1]$ for any $(t_0, t_1 > t_0) \in In\mathfrak{T}_0 \times In\mathfrak{T}_0$.

2) All rows of $\Phi(s) B^{(\mu)} = (sI - A)^{-1} B^{(\mu)}$ are linearly independent on $\mathbb{C}$.

3) The Gram matrix $G_{ctB}(t_1, t_0)$ of $\Phi(t_1, t) B^{(\mu)}$,

$$G_{ctB}(t_1, t_0) = \int_{t_0}^{t_1} \Phi(t_1, \tau) B^{(\mu)} \left(B^{(\mu)} \right)^T \Phi^T(t_1, \tau) \, d\tau, \; \forall t \in \mathfrak{T}_0, \quad (11.220)$$

is nonsingular for any $(t_0, t_1 > t_0) \in In\mathfrak{T}_0 \times In\mathfrak{T}_0$; i.e.,

$$rank G_{ctB}(t_1, t_0) = n,$$
$$for \; any \, (t_0, t_1 > t_0) \in In\mathfrak{T}_0 \times In\mathfrak{T}_0. \qquad (11.221)$$

4) The $n \times n(\mu+1)r$ controllability matrix $\mathcal{C}_B$,

$$\mathcal{C}_B = \left[B^{(\mu)} \vdots AB^{(\mu)} \vdots A^2 B^{(\mu)} \vdots ... \vdots A^{n-1} B^{(\mu)} \right], \qquad (11.222)$$

has the full rank n,

$$rank\ \mathcal{C}_B = n.\qquad(11.223)$$

5) For every eigenvalue $s_i(A)$ of the matrix A, equivalently for every complex number $s \in \mathbb{C}$, the $n \times (n + (\mu + 1) r)$ matrix $\left[sI - A \vdots B^{(\mu)}\right]$ has the full rank n,

$$rank\ \left[sI - A \vdots B^{(\mu)}\right] = n,$$
$$\forall s = s_i(A) \in \mathbb{C},\ \forall i = 1, 2, ..., n,\ \forall s \in \mathbb{C}.\qquad(11.224)$$

6. a) If $\mu = 0$ the control vector function $U(.)$ obeys

$$U(t) = (\Phi(t_1, t) B_0)^T\ G_{\Phi B}^{-1}(t_1, t_0)\ [X_1 - \Phi(t_1, t_0) X_0],$$
$$\forall t \in [t_0, t_1],\ t_1 \in In\mathfrak{T}_0.\qquad(11.225)$$

b) If $\mu > 0$ the control vector function $U(.)$ obeys either

$$T^{-1}U^\mu(t) = T^T\left(B^{(\mu)}\right)^T \Phi^T(t_1, t)\, G_{\Phi BT}^{-1}(t_1, t_0)\, [X_1 - \Phi(t_1, t_0) X_0],\ \mu > 0,$$
$$(11.226)$$

where $T \in \mathfrak{R}^{(\mu+1)r \times (\mu+1)r}$ is any nonsingular matrix and $G_{\Phi BT}(t_1, t_0)$ is the Gram matrix of $\Phi(t_1, t) B^{(\mu)} T$,

$$G_{\Phi BT}(t_1, t_0) = \int_{t_0}^{t_1} \Phi(t_1, \tau) B^{(\mu)} T T^T \left(B^{(\mu)}\right)^T \Phi^T(t_1, \tau)\, d\tau.\qquad(11.227)$$

c) or the control vector function $U(.)$ obeys

$$B^{(\mu)} U^\mu(t) = \Phi^T(t_1, t)\, G_\Phi^{-1}(t_1, t_0)\, [X_1 - \Phi(t_1, t_0) X_0],\ \mu > 0,\quad(11.228)$$

where $G_\Phi(t_1, t_0)$ is the Gram matrix of $\Phi(t_1, t)$,

$$G_\Phi(t_1, t_0) = \int_{t_0}^{t_1} \Phi(t_1, \tau) \Phi^T(t_1, \tau)\, d\tau.\qquad(11.229)$$

Exercise 222 *Test the state controllability of the selected IIO physical plant in Exercise 78, Section 6.3.*

11.10   *IIO* system output controllability

11.10.1   Definition

The general output controllability Definition 165, (Section 10.3), holds unchanged for the *IIO* system (6.1), (6.2):

Definition 223 *Output controllability of the IIO system (6.1), (6.2)*

The IIO system (6.1), (6.2) is the output controllable if and only if for every initial output vector $\mathbf{Y}_0 \in \mathfrak{R}^N$ *at* $t_0 \in \mathfrak{T}$ *and for any final output vector* $\mathbf{Y}_1 \in \mathfrak{R}^N$ *there exist a moment* $t_1 \in In\mathfrak{T}_0$ *and an extended control* $\mathbf{U}^{\mu}_{[t_0,t_1]}$ *on the time interval* $[t_0, t_1]$ *such that*

$$\mathbf{Y}(t_1; t_0; \mathbf{Y}_0; \mathbf{U}^{\mu}_{[t_0,t_1]}) = \mathbf{Y}_1.$$

11.10.2   Criteria

State space-output space product approach

The system (11.213), (11.214) is the *EISO* system (11.8), (11.9), which is equivalent to the unperturbed both the *IO* system (2.15), (Section 11.2), and *IIO* system (6.1), (6.2) with the matrices determined by Equations (11.217)-(11.219). With this in mind, the system matrix $H_{IIO}(t, t_0)$ (11.230),

$$H_{IIO}(t, t_0) = C\Phi(t, t_0)B^{(\mu)} \in \mathfrak{R}^{N \times (\mu+1)r}, \tag{11.230}$$

and its Laplace transform $H_{IIO}(s)$ (11.231),

$$H_{IIO}(s) = C\Phi(s)B^{(\mu)} \in \mathfrak{C}^{N \times (\mu+1)r}, \tag{11.231}$$

correspond to $H_{IO}(t, t_0)$ (11.32) and $H_{IO}(s)$ (11.33), (Section 11.2).

For the *IIO* system (6.1), (6.2) Theorem 190, (Section 11.2), reads:

Theorem 224 *Equivalent conditions for the output controllability of the IIO system (6.1), (6.2)*

The IIO system (6.1), (6.2) obeys invariantly the output controllability necessary rank condition (10.29), i.e.,

$$rankC = N, \ \forall s \in \mathbb{C}. \tag{11.232}$$

For the IIO system (6.1), (6.2) to be output controllable:

A) It is necessary that any of the following conditions 1.-6.a) holds and sufficient that any of the following conditions 1.-4.,6.a) holds if $\mu = 0$*, where the conditions 1.-4.,6.a) are equivalent,*

B) It is necessary that any of the following conditions 1.-5., 6.b) holds and sufficient that any of the following conditions 1.-4., 6.b) holds if $\mu > 0$, where the conditions 1.-4.,6.b) are equivalent.

In A) and B) Equations (11.188)-(11.191), (11.203)-(11.210) induced by Equations (6.1), (6.2) determine the matrices defined by Equations (11.217)-(11.219):

1. All rows of the system matrix $H_{IIO}(t,t_0)$ (11.230) are linearly independent on $[t_0,t_1]$ for any $(t_0,t_1 > t_0) \in In\mathfrak{T}_0 \times In\mathfrak{T}_0$.

2. All rows of the system matrix $H_{IIO}(s)$ (11.231) are linearly independent on $\mathbb{C}$.

3. The Gram matrix $G_{HIIO}(t,t_0)$ of $H_{IIO}(t,t_0)$ (11.230),

$$G_{HIIO}(t,t_0) = \int_{t_0}^{t} H_{IIO}(\tau,t_0) H_{IIO}^T(\tau,t_0)\, d\tau,$$

$$\text{for any } (t_0,t_1 > t_0) \in In\mathfrak{T}_0 \times In\mathfrak{T}_0, \qquad (11.233)$$

is nonsingular; i.e.,

$$rankG_{HIIO}(t,t_0) = N, \text{ any } (t_0,t_1 > t_0) \in In\mathfrak{T}_0 \times In\mathfrak{T}_0. \qquad (11.234)$$

4. The output $N \times n\,(\mu+1)\,r$ controllability matrix $\mathcal{C}_{IIOout}$,

$$\mathcal{C}_{IIOout} = \left[C\mathbf{B}^{(\mu)} \vdots CA\mathbf{B}^{(\mu)} \vdots CA^2\mathbf{B}^{(\mu)} \vdots ... \vdots CA^{n-1}\mathbf{B}^{(\mu)} \right], \ \mu \geq 0,$$

$$(11.235)$$

has the full rank N,

$$rank\mathcal{C}_{IOout} = N. \qquad (11.236)$$

5. For every eigenvalue $s_i(A)$ of the matrix A, equivalently for every complex number $s \in \mathbb{C}$, the matrix $\mathcal{C}_{outIO}(s)$,

$$\mathcal{C}_{outIO}(s) = \left[C(sI - A) \vdots C\mathbf{B}^{(\mu)} \right] \in \mathfrak{R}^{N \times (n+(\mu+1)r)}, \ \mu \geq 0,$$

has the full rank N,

$$rank\mathcal{C}_{outIO}(s) = N,$$
$$\forall s = s_i(A) \in \mathbb{C}, \ \forall i = 1,2,...n, \ i.e., \ \forall s \in \mathbb{C}. \qquad (11.237)$$

6. The control vector function $\mathbf{U}(.)$ obeys:
  a) The following equation if $\mu = 0$:

$$\mathbf{U}(t) = (H_{IIO}(t_1,t))^T G_{HIIO}^{-1}(t_1,t_0) [\mathbf{Y}_1 - C\Phi(t_1,t_0)\mathbf{X}_0], \ \mu = 0,$$

$$(11.238)$$

b) The following equations for any $(\mu + 1)\,r \times (\mu + 1)\,r$ nonsingular matrix R:

$$R^{-1}\mathbf{U}^{\mu}\left(t\right) = \left\{ \begin{array}{c} \left(H_{IIO}\left(t_1, t\right) R\right)^{T} \bullet G_{HIIOR}^{-1}\left(t_1, t_0\right) \bullet \\ \bullet \left[\mathbf{Y_1} - C\Phi\left(t_1, t_0\right) \mathbf{X_0}\right] \end{array} \right\}, \quad \mu > 0, \quad (11.239)$$

where $G_{HIIOR}\left(t_1, t_0\right)$ is the Gram matrix of $H_{IIO}\left(t_1, t\right) R$,

$$G_{HIIOR}\left(t_1, t_0\right) = \int_{t_0}^{t_1} H_{IIO}\left(t_1, \tau\right) R R^{T} H_{IIO}^{T}\left(t_1, \tau\right) d\tau. \qquad (11.240)$$

Output space approach

Equations (6.31)-(6.35), (Section 6.1), determine the system response in terms of its input-output data:

$$\mathbf{Y}(t; \mathbf{R}_{0-}^{\alpha-1}; \mathbf{Y}_0^{\nu-1}; \mathbf{U}^{\mu}) = \int_{0-}^{t} \Gamma_{IIO}(t, \tau)\mathbf{U}(\tau)d\tau +$$

$$+\Gamma_{IIOI_0}(t)\mathbf{U}_{0-}^{\mu-1} + \Gamma_{IIOR_0}(t)\mathbf{R}_{0-}^{\alpha-1} + \Gamma_{IIOY_0}(t)\mathbf{Y}_{0-}^{\nu-1}, \; \forall t \in \mathfrak{T}_0. \quad (11.241)$$

Theorem 225 *Output controllability criterion for the IIO system (6.1), (6.2)*

In order for the IIO system (6.1), (6.2) to be output controllable it is necessary and sufficient that any of the following equivalent conditions holds:

1) For any $(t_0, t_1 > t_0) \in In\mathfrak{T}_0 \times In\mathfrak{T}_0$ the rows of the system output fundamental matrix $\Gamma_{IIO}(t, t_0)$ (6.32), (Section 6.1), are linearly independent on the time interval $[t_0, t_1]$.

2) The rows of the system transfer function matrix $G_{IIOU}(s)$ (6.23), (Section 6.1), are linearly independent on $\mathbb{C}$.

3) For any $(t_0, t_1 > t_0) \in In\mathfrak{T}_0 \times In\mathfrak{T}_0$ the Gram matrix $G_{\Gamma IIO}(t_0, t_1)$ of the system matrix $\Gamma_{IIO}(t, t_0) \in \mathfrak{R}^{N \times r}$,

$$G_{\Gamma IIO}\left(t_0, t_1\right) = \int_{t_0}^{t_1} \Gamma_{IIO}(t_1, \tau)\Gamma_{IIO}^{T}(t_1, \tau)d\tau \in \mathfrak{R}^{N \times N}. \qquad (11.242)$$

is nonsingular, i.e., the condition (11.243),

$$det G_{\Gamma IIO}\left(t_0, t_1\right) \neq 0, \;\; any \;\; (t_0, t_1 > t_0) \in In\mathfrak{T}_0 \times In\mathfrak{T}_0, \qquad (11.243)$$

holds.

4) The control vector function obeys the following linear algebraic vector equation:

$$\mathbf{U}(t) =$$
$$= \Gamma_{IIO}^{T}(t_1, t) G_{\Gamma IIO}^{-1}(t_1) \left[\begin{array}{c} \mathbf{Y}_1 - \Gamma_{IIOI_0}(t_1)\mathbf{U}_{0^-}^{\mu-1} - \\ -\Gamma_{IIOR_0}(t_1)\mathbf{R}_{0^-}^{\alpha-1} - \Gamma_{IIOY_0}(t_1)\mathbf{Y}_{0^-}^{\nu-1} \end{array} \right].$$

$$(11.244)$$

Appendix D.15 contains the proof of Theorem 195.

Exercise 226 *Test the output controllability of the selected IIO physical plant in Exercise 78, Section 6.3.*

Part IV

APPENDIX

Appendix A

Notation

The meaning of the notation is explained in the text at its first use.

A.1 Abbreviations

$\mathcal{C}$ *controller*

$\mathcal{CS}$ *control system*

Cl   *closure*

HISO system   *Higher order Input-State-Output system defined by* (5.1), (5.2)

iff   *if and only if*

I   *Input*

II   *Input-Internal (dynamics)*

IIO   *Input-Internal and Output state*

IIO system   *Input-Internal and Output state system defined by* (6.1), (6.2)

In   *the interior*

IO   *Input-Output*

IO system   *Input-Output system defined by* (2.1)

IS   *Input-State*

ISO   *Input-State-Output*

ISO system   *Input-State-Output system defined by* (3.1), (3.2)

MIMO   *Multiple-Input-Multiple-Output*

$\mathcal{O}$ *object*

$\mathcal{P}$ *plant*

EISO system   *Extended Input-State-Output system* (4.1), (4.2)

SISO   *Single-Input Single-Output*

System	*Continuous-time time-invariant linear dynamical system*

A.2   Indexes

A.2.1   SUBSCRIPTS

d the subscript d denotes "desired"

  e *equilibrium*

  i the subscript i denotes "the i-th"

  j the subscript j denotes "the j-th"

  P for *plant*

  zero   the subscript *zero* denotes "the zero value"

  0   the subscript 0 (zero) associated with a variable $(.)$ denotes its initial value $(.)_0$; however, if $(.) \subset \mathfrak{T}$ then the subscript 0 (zero) associated with $(.)$ denotes the *time* set $\mathfrak{T}_0$, $(.)_0 = \mathfrak{T}_0$

A.2.2   SUPERSCRIPT

$i \in \{0, 1, ..., \eta, \mu\}$ is the highest derivative of the disturbance vector acting on the plant in general

  $k \in \{0, 1, ..., \mu\}$ is the highest derivative of the control vector acting on the plant in general

  $l \in \{0, 1, ..., m\}$ is the highest order of the tracking in general

  $m \in \{1, \alpha, \nu, \alpha + \nu\}$ is the plant order in general

  0   is the highest derivative of both the disturbance vector and control vector acting on the *ISO* plant

  1   is the order of the *ISO* and *EISO* plant

  α is the order of the *HISO* and *IIO* plant

  η is the highest derivative of the disturbance vector acting on the *IO* plant

  μ is the highest derivative of the control vector acting on the *HISO*, *IO*, *IIO* and *EISO* plant, as well as the highest derivative of the disturbance vector acting on the *IIO* and on the *EISO* plant

  ν is the order of the *IO* plant

A.3   Letters

Lower case block or italic letters are used for scalars. Lower case bold block letters denote vectors. Upper case block letters denote matrices, or points. Upper case Fraktur letters designate sets or spaces.

The notation "$; t_{(.)0}$" will be omitted as an argument of a variable if, and only if, a choice of the initial moment $t_{(.)0}$ does not have any influence on the value of the variable

BLACKBOARD BOLD LETTERS

$\mathbb{C}$ *set of complex numbers s*

$\mathbb{C}^k$ *k-dimensional complex vector space*

A.3.1   CALLIGRAPHIC LETTERS

$\mathcal{C}$ *controller*, or *controllability matrix* $\mathcal{C}$,

$$\mathcal{C} = \left[B \vdots AB \vdots A^2 B \cdots \vdots A^{n-1} B \right]$$

$\mathcal{CS}$ *control system*

$\mathcal{I}$ *integral output space*, $\mathcal{I} = \mathfrak{T} \times \mathfrak{R}^N$

$\mathcal{L}^\mp \{\mathbf{i}(.)\}$ *Left (-), right (+), respectively, the Laplace transform of a function* $\mathbf{i}(.)$,

$$\mathcal{L}^\mp \{\mathbf{i}(t)\} = \mathbf{I}^\mp(s) = \int_{0^\mp}^\infty \mathbf{i}(t)e^{-st}dt = \lim \left[\int_{\mp\zeta}^\infty \mathbf{i}(t)e^{-st}dt : \zeta \longrightarrow 0^+ \right]$$

$\mathcal{P}$ *plant*

$\mathcal{S}$ state

$\mathcal{S}(.)$ *motion*

A.3.2   FRAKTUR LETTERS

Capital Fraktur letters are used for spaces or for sets.

$\mathfrak{A} \subseteq \mathfrak{R}^n$ *a nonempty subset of* $\mathfrak{R}^n$

$\mathfrak{B} \subseteq \mathfrak{R}^n$ *a nonempty subset of* $\mathfrak{R}^n$

$\mathfrak{C}$ *the family of all defined and continuous functions on* $\mathfrak{T}_0$

$\mathfrak{C}^k(\mathfrak{S})$ *the family of all functions defined, continuous and k-times continuously differentiable on the set* $\mathfrak{S} \subseteq \mathfrak{T} \cup \mathfrak{R}^i$, $\mathfrak{C}^k(\mathfrak{R}^i) = \mathfrak{C}^{ki}$

$\mathfrak{C}^k(\mathfrak{T}_0)$ *the family of all functions defined, continuous and k-times continuously differentiable on* $\mathfrak{T}_0$

$\mathfrak{C}^0(\mathfrak{S})$ *the family of all functions defined and continuous on the set* $\mathfrak{S}$, $\mathfrak{C}^0(\mathfrak{R}^i) = \mathfrak{C}^{0,i} = \mathfrak{C}(\mathfrak{R}^i)$

$\mathfrak{C}^{k-}(\mathfrak{R}^i)$ *the family of all functions defined everywhere and k-times continuously differentiable on* $\mathfrak{R}^i \setminus \{\mathbf{0}_i\}$, *which have defined and continuous derivatives at the origin* $\mathbf{0}_i$ *of* $\mathfrak{R}^i$ *up to the order* $(k-1)$, *which are defined*

and continuous at *at the origin* $\mathbf{0}_i$ and have defined the left and the right
$k - th$ order derivative *at the origin* $\mathbf{0}_i$

$\mathfrak{D}^k$ is a given, or to be determined, *family of all bounded k-times continu-*
ously differentiable on $\mathfrak{T}_0$ *permitted disturbance vector total functions* $\mathbf{D}(.)$,
or *deviation functions* $\mathbf{d}(.)$, $\mathfrak{D}^k \subset \mathfrak{C}^k$, the Laplace transforms of which are
strictly proper real rational complex functions,

$$\mathfrak{D}^k = \left\{ \mathbf{D}(.) : \mathbf{D}^{(k)}(t) \in \mathfrak{C},\ \exists \zeta \in \mathfrak{R}^+ \Longrightarrow \left\| \mathbf{D}^k(t) \right\| < \zeta,\ \forall t \in \mathfrak{T}_0 \right\},$$

or

$$\mathfrak{D}^k = \left\{ \mathbf{d}(.) : \mathbf{d}^{(k)}(t) \in \mathfrak{C},\ \exists \xi \in \mathfrak{R}^+ \Longrightarrow \left\| \mathbf{d}^k(t) \right\| < \xi,\ \forall t \in \mathfrak{T}_0 \right\}$$

$\mathfrak{D}^k_-$ is a subfamily of $\mathfrak{D}^k$, $\mathfrak{D}^k_- \subset \mathfrak{D}^k$, such that the real part of every pole
of the Laplace transform $\mathbf{D}(s)$ of every $\mathbf{D}(.) \in \mathfrak{D}^k_-$ is negative, $\mathfrak{D}_- = \mathfrak{D}^0_-$

$\mathfrak{D}^0 = \mathfrak{D}$ is the *family of all bounded continuous permitted disturbance*
vector total functions $\mathbf{D}(.)$ *or deviation functions* $\mathbf{d}(.)$, $\mathfrak{D} \subset \mathfrak{C}$, the Laplace
transforms of which are strictly proper real rational complex functions

$\mathfrak{I}^k$ is a given, or to be determined, family of all bounded and k-times
continuously differentiable permitted input vector functions $\mathbf{I}(.)$,

$$\mathfrak{I}^k \subset \mathfrak{C}^k \cap \mathfrak{L}$$

$\mathfrak{I}^0 = \mathfrak{I}$ is the family of all bounded continuous permitted input vector functions $\mathbf{I}(.)$

$$\mathfrak{I} \subset \mathfrak{C} \cap \mathfrak{L}$$

$\mathfrak{I}^k_-$ is a subfamily of $\mathfrak{D}^k$, $\mathfrak{I}^k_- \subset \mathfrak{I}^k$, such that the real part of every pole
of the Laplace transform $\mathbf{I}(s)$ of every $\mathbf{I}(.) \in \mathfrak{I}^k_-$ is negative, $\mathfrak{I}_- = \mathfrak{I}^0_-$

$\mathfrak{L}$ *the family of all strictly proper real rational complex functions, the*
original of which are bounded time-dependent functions,

$$\mathfrak{L} = \left\{ \mathbf{I}(.) : \left(\begin{array}{c} \exists \gamma(\mathbf{I}) \in \mathfrak{R}^+ \Longrightarrow \|\mathbf{I}(t)\| < \gamma(\mathbf{I}),\ \forall t \in \mathfrak{T}_0, \\ \mathcal{L}^{\mp}\{\mathbf{I}(t)\} = \mathbf{I}^{\mp}(s) = \left[I_1^{\mp}(s)\ \ I_2^{\mp}(s)\ \ \ldots\ I_M^{\mp}(s) \right]^T, \\ I_k^{\mp}(s) = \dfrac{\displaystyle\sum_{j=0}^{j=\zeta_k} a_{kj} s^j}{\displaystyle\sum_{j=0}^{j=\psi_k} b_{kj} s^j}, 0 \le \zeta_k < \psi_k,\ \forall k = 1, 2, ..., M, \end{array} \right) \right\}$$

$\mathfrak{R}$ *the set of all real numbers*
$\mathfrak{R}^+$ *the set of all positive real numbers*
$\mathfrak{R}_+$ *the set of all nonnegative real numbers*

$\mathfrak{R}^{\nu N}$ *the extended output space of the IO system*, which is simultaneously the space of its internal dynamics - its *internal dynamics space*

$\mathfrak{R}^n$ an *n-dimensional real vector space, the state space of the ISO system*

$\mathfrak{T}$ *the accepted reference time set*, the arbitrary element of which is an arbitrary moment t and the *time* unit of which is second s, $1_t = s$, $t \langle s \rangle$,

$$\mathfrak{T} = \{t : t[T] \langle s \rangle, \; numt \in \mathfrak{R}, \; dt > 0\}, \; inf \; \mathfrak{T} = -\infty, \; sup \; \mathfrak{T} = \infty$$

$\mathfrak{T}_0$ *the subset of* $\mathfrak{T}$, which has the minimal element $min\mathfrak{T}_0$ that is the initial instant $t_{0\mp}$, $numt_{0\mp} = 0^{\mp}$,

$$\mathfrak{T}_0 = \{t : t \in \mathfrak{T}, \; t \geq t_{0\mp}, \; numt_0 = 0\}, \mathfrak{T}_0 \subset \mathfrak{T},$$
$$min\mathfrak{T}_0 = t_0 \in \mathfrak{T}, \; sup \; \mathfrak{T}_0 = \infty$$

$\mathfrak{Y}_d^k$ a given, or to be determined, *family of all bounded k-times continuously differentiable realizable desired total output vector functions* $\mathbf{Y}_d(.)$, $\mathfrak{Y}_d^k \subset \mathfrak{C}^{kN}$, the Laplace transforms of which are strictly proper real rational complex functions,

$$\mathfrak{Y}_d^k = \left\{ \mathbf{Y}_d(.) : \mathbf{Y}_d(t) \in \mathfrak{C}^{kN}, \; \exists \kappa \in \mathfrak{R}^+ \Longrightarrow \left\| \mathbf{Y}_d^k(t) \right\| < \kappa, \; \forall t \in \mathfrak{T}_0 \right\}$$

$\mathfrak{Y}_{d-}^k$ is a subfamily of $\mathfrak{Y}_d^k$, $\mathfrak{Y}_{d-}^k \subset \mathfrak{Y}_d^k$, such that the real part of every pole of the Laplace transform $\mathbf{Y}_d(s)$ of every $\mathbf{Y}_d(.) \in \mathfrak{Y}_{d-}^k$ is negative, $\mathfrak{Y}_{d-} = \mathfrak{Y}_{d-}^0$

$\mathfrak{Y}_{d0}^k$ *the set of the desired output initial conditions* $\mathbf{Y}_{d0}^k = \mathbf{Y}_d^k(t_0)$ *of* $\mathbf{Y}_d^k(t)$ *of every* $\mathbf{Y}_d(.) \in \mathfrak{Y}_d^k$,

$$\mathfrak{Y}_{d0}^k = \left\{ \mathbf{Y}_{d0}^k : \mathbf{Y}_{d0}^k = \mathbf{Y}_d^k(t_0), \; \mathbf{Y}_d(.) \in \mathfrak{Y}_d^k \right\} \tag{A.1}$$

$\mathfrak{Y}_d = \mathfrak{Y}_d^0$ is the *family of all bounded continuous realizable desired total output vector functions* $\mathbf{Y}_d(.)$, $\mathfrak{Y}_d = \mathfrak{Y}_d^0 \subset \mathfrak{C}^{0d}$, the Laplace transforms of which are strictly proper real rational complex functions

A.3.3 GREEK LETTERS

α *a nonnegative integer,*
β *a nonnegative integer,*
η *a natural number,,*
$\lambda_m(H)$
$\lambda_M(H)$

μ *a nonnegative integer,*

ν *a nonnegative integer,*

τ a subsidiary notation for *time t,*

$\Xi = diag\,\{\epsilon_1 \ \ \epsilon_2 \ \ ... \ \epsilon_N\}$

$\Phi_{IO}(t, t_0) \in \mathfrak{R}^{N \times n}$ the *IO* system *output fundamental matrix* (Equation (11.52), Section 11.2),

$$\Phi_{IO}(t, t_0) = \mathcal{L}^{-1}\left\{\left(A^{(\nu)} S_N^{(\nu)}(s)\right)^{-1}\right\},$$

$$\Phi_{IO}(s) = \left(A^{(\nu)} S_N^{(\nu)}(s)\right)^{-1},$$

$\Phi_{ISO}(t, t_0) \equiv \Phi(t, t_0) \in \mathfrak{R}^{n \times n}$ *the fundamental matrix (3.4) of the ISO* system (3.1), (3.2),

$$\Phi(t, t_0) = e^{At}\left(e^{At_0}\right)^{-1} = e^{A(t-t_0)} \in \mathfrak{R}^{n \times n},$$

$$\Phi(s) = \mathcal{L}\{\Phi(t, t_0)\} = \mathcal{L}\left\{e^{A(t-t_0)}\right\} = (sI - A)^{-1},$$

has the following well known properties, Equations (3.5)-(3.7) (all in Section 3.1):

$$det\Phi(t, t_0) \neq 0, \ \forall t \in \mathfrak{T}_0,$$

$$\Phi(t, t_0)\,\Phi(t_0, t) = e^{A(t-t_0)} e^{A(t_0-t)} \equiv e^{A0} = I_n,$$

$$\Phi^{(1)}(t, t_0) = A\Phi(t, t_0) = \Phi(t, t_0)\,A.$$

ρ *a natural number.*

A.3.4   ROMAN LETTERS

$A \in \mathfrak{R}^{n \times n}$ *the matrix describing the internal dynamics of the ISO system*

$A_k \in \mathfrak{R}^{N \times N}$ *the matrix associated with the $k-th$ derivative $\mathbf{Y}^{(k)}$ of the output vector $\mathbf{Y}$ of the IO system*

$A^{(\nu)} \in \mathfrak{R}^{N \times (\nu+1)N}$ *the extended matrix describing the IO system internal dynamics,* $A^{(\nu)} = \left[A_0 \vdots A_1 \vdots ... \vdots A_\nu\right]$

$B_k \in \mathfrak{R}^{N \times M}$ *the matrix associated with the $k-th$ derivative $\mathbf{I}^{(k)}$ of the input vector $\mathbf{I}$ of the IO system*

$B^{(\mu)} \in \mathfrak{R}^{N \times (\mu+1)r}$ *the extended matrix describing the transmission of the influence of the control vector $\mathbf{U}(t)$ on the system dynamics,* $B^{(\mu)} = \left[B_0 \vdots B_1 \vdots ... \vdots B_\mu\right]$

$B^{(\mu)} \in \mathfrak{R}^{n \times (\mu+1)r}$ *the extended matrix describing the transmission of the influence of the control vector* $\mathbf{U}(t)$ *on the system dynamics,* $B^{(\mu)} =$

$$\left[B_0 \vdots B_1 \vdots ... \vdots B_\mu \right]$$

$C \in \mathfrak{R}^{N \times n}$ *the matrix of the ISO system, which describes the transmission of the state vector action on the system output vector* $\mathbf{Y}$

$\mathbf{C}_0$ is *the vector of all initial conditions* acting on the system,

$$\mathbf{C}_0 = \begin{bmatrix} I_0 \\ I_0^{(1)} \\ ... \\ I_0^{(\mu-1)} \\ X_0 \\ Y_0 \\ Y_0^{(1)} \\ ... \\ Y_0^{(\nu-1)} \end{bmatrix} \in \mathfrak{R}^{\mu M + n + \nu N},$$

d *a natural number*

$\mathbf{d} \in \mathfrak{R}^d$ *the disturbance deviation vector,* (2.53), (Section 2.2),

$$\mathbf{d} = \mathbf{D} - \mathbf{D}_N$$

$\mathbf{D} \in \mathfrak{R}^d$ *the total disturbance vector*

$\mathbf{D}_N \in \mathfrak{R}^d$ *the nominal disturbance vector*

$D \in R^{N \times d}$ *the ISO system matrix describing the transmission of the influence of* $\mathbf{I}(t)$ *on the system output*

$\mathbf{e}$ *the output error vector* $\mathbf{e} \in R^N$,

$$\mathbf{e} = \mathbf{Y_d} - \mathbf{Y} = -\mathbf{y}, \mathbf{e} = \begin{bmatrix} e_1 & e_2 & ... & e_N \end{bmatrix}^T$$

$F(.) : \mathfrak{T}_0 \longrightarrow \mathfrak{R}^{N \times N}$ *a matrix function associated with* $\mathbf{f}(.)$,

$$\mathbf{f} = \begin{bmatrix} f_1 & f_2 & ... & f_N \end{bmatrix}^T \Longrightarrow F = diag \{ f_1 \quad f_2 \quad ... \quad f_N \}$$

$G = G^T \in R^{p \times p}$ *the symmetric matrix of the quadratic form* $v(\mathbf{w}) = \mathbf{w}^T G \mathbf{w}$

$G(s)$ is *the transfer function matrix of a time-invariant continuous-time linear dynamical system*

$h(.)$ *the Heaviside function,* i.e. *the unite step function,* $h(.) : \mathfrak{T} \rightarrow [0,1], h(t) = 0$ *for* $t < 0, h(t) \in [0,1]$ *for* $t = 0, h(t) = 1$ *for* $t > 0$, Figure A.1,

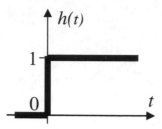

Figure A.1: Heaviside function $h\left(.\right)$.

$H \in R^{N \times r}$ a matrix

$H = H^T \in R^{p \times p}$ the symmetric matrix of the quadratic form $v(\mathbf{w}) = \mathbf{w}^T H \mathbf{w}$

i an arbitrary natural number, or the imaginary unit $\sqrt{-1}$, or the input deviation variable

$\mathbf{i} \in \Re^M$ the input deviation vector, $\mathbf{i} = \begin{bmatrix} i_1 & i_2 & \dots & i_M \end{bmatrix}^T$, (2.54), (Section 2.2),

$$\mathbf{i} = \mathbf{I} - \mathbf{I}_N$$

$\mathbf{i}^\mu(t) \in \Re^{(\mu+1)M}$ the extended input vector at a moment t, $\mathbf{i}^\mu(t) = \begin{bmatrix} \mathbf{i}(t) & \vdots & \mathbf{i}^{(1)}(t) & \vdots & \dots & \vdots & \mathbf{i}^{(\mu)}(t) \end{bmatrix}^T$

$\mathbf{i}_{0\mp}^{\mu-1} \in \Re^{\mu M}$ the initial extended input vector at the initial moment $t_0 = 0$, $\mathbf{i}_{0\mp}^{\mu-1} = \mathbf{i}^{\mu-1}(0^\mp) = \begin{bmatrix} \mathbf{i}_{0(\mp)} & \vdots & \mathbf{i}_{0(\mp)}^{(1)} & \vdots & \dots & \vdots & \mathbf{i}_{0(\mp)}^{(\mu-1)} \end{bmatrix}^T \in \Re^{\mu M}$

I the identity matrix of the n-th order, $I = diag\{1 \ 1 \ \dots \ 1\} \in \Re^{n \times n}$, or the total input variable, $I_n = I$

I_N the identity matrix of the N-th order, $I_N = diag\{1 \ 1 \ \dots \ 1\} \in \Re^{N \times N}$

$\mathbf{I} \in \Re^M$ the total input vector, $\mathbf{I} = \begin{bmatrix} I_1 & I_2 & \dots & I_M \end{bmatrix}^T$

$I_i \in \Re^i$ is the i-th order identity matrix,

$\mathbf{I}_N \in \Re^M$ the nominal input vector, $\mathbf{I}_N = \begin{bmatrix} I_{N1} & I_{N2} & \dots & I_{NM} \end{bmatrix}^T$

$Int\mathfrak{T}_0$ the interior of the set $\mathfrak{T}_0$,

$$Int\mathfrak{T}_0 = \{t : \ t \in \mathfrak{T}_0, \ t > 0\}$$

$J \in \Re^{n \times M}$ a matrix

k an arbitrary natural number

$M(.)$ a complex valued matrix function of any type

$M(s)$ a complex valued matrix of any type

m *a nonnegative integer*

n *a natural number*

N *a natural number*, if N is the dimension of the output vector and if n is the dimension of the state vector then $N \leq n$

O *the zero matrix of the appropriate order*

p *a natural number*

$P \in R^{n \times N}$ *a matrix*

$P_k \in \mathfrak{R}^{\rho \times M}$ *a matrix*

$P^{(\beta)} \in \mathfrak{R}^{\rho \times M(\beta+1)}$ *an extended matrix describing the transmission of the influence of* $\mathbf{i}^{\beta}(t)$ *on the internal dynamics of the IIO system*, $P^{(\alpha)} = [P_0 \vdots P_1 \vdots ... P_{\beta}]$

q *a natural number*

$Q \in R^{N \times N}$ *a matrix*

$Q_k \in \mathfrak{R}^{\rho \times \rho}$ *a matrix*

$Q^{(\alpha)} \in \mathfrak{R}^{\rho \times \rho(\alpha+1)}$ *the extended matrix describing the internal dynamics of the IIO system*, $Q^{(\alpha)} = \left[Q_0 \vdots Q_1 \vdots ... Q_{\alpha} \right]$

$\mathbf{r} \in \mathfrak{R}^{\rho}$ *a subsidiary deviation vector*, which is *the internal dynamics deviation vector*, (5.43), (Section 5.2),

$$\mathbf{r} = \mathbf{R} - \mathbf{R}_N$$

$\mathbf{R} \in \mathfrak{R}^{\rho}$ *a subsidiary total vector*, which is *the internal dynamics total vector of the HISO system and of the IIO system*

$\mathbf{R}_N \in \mathfrak{R}^{\rho}$ *a subsidiary nominal vector*, which is *the internal dynamics nominal vector of the HISO system and of the IIO system*

$R_k \in \mathfrak{R}^{N \times \rho}$ *a matrix*

$R_y^{(\alpha-1)} \in \mathfrak{R}^{N \times \alpha \rho}$ *the extended matrix describing the action of the extended internal dynamics vector* $\mathbf{R}^{\alpha-1}$ *on the output dynamics of the IIO system*, $R_y^{(\alpha-1)} = \left[R_{y0} \vdots R_{y1} \vdots ... R_{y,(\alpha-1)} \right]$

S *a state variable*

$\mathbf{S}$ *a state vector*,

$$\mathbf{S} = \left[S_1 \vdots S_2 \vdots S_3 \vdots ... \vdots S_K \right]^T \in \mathfrak{R}^K$$

$\mathrm{Re}\, s$ *the real part of* $s = \sigma + j\omega$, $\mathrm{Re}\, s = \sigma$

s *the basic time unit: second, or a complex variable or a complex number* $s = \sigma + j\omega$

$$sign(.) : \mathfrak{R} \rightarrow \{-1, 0, 1\} \qquad \textit{the scalar signum function,}$$

$$sign(x) = |x|^{-1} x \ \textit{if} \ x \neq 0, \ \textit{and} \ sign(0) = 0$$

$S_i^{(k)}(.) : \mathfrak{C} \longrightarrow \mathfrak{C}^{i(k+1) \times i}$ *the matrix function of s defined by (2.17)*
in Subsection 1.2:

$$S_i^{(k)}(s) = \left[s^0 I_i \vdots s^1 I_i \vdots s^2 I_i \vdots ... \vdots s^k I_i \right]^T \in \mathfrak{C}^{i(k+1) \times i},$$

$$(k, i) \in \{(\mu, M), \ (\nu, N)\}$$

t *time (temporal variable)*, or *an arbitrary time value (an arbitrary moment, an arbitrary instant)*; and formally mathematically t denotes for short also the numerical *time* value *numt* if it does not create a confusion,

$$t[\mathrm{T}] \langle s \rangle, \ numt \in \mathfrak{R}, \ dt > 0 \ , \ \textit{or equivalently:} \ t \in \mathfrak{T}.$$

It has been the common attitude to use the notation t of *time* and of its arbitrary temporal value also for its numerical value *numt*, e.g. $t = 0$ is used in the sense *numt* $= 0$. We do the same throughout the book if there is not any confusion because we can replace t everywhere by $t1_t^{-1}$, $\left(t1_t^{-1}\right) \in \mathfrak{R}$, that we denote again by t, *numt* $= num\left(t1_t^{-1}\right)$

t_0 a conventionally accepted *initial value of time (initial instant, initial moment)*, $t_0 \in \mathfrak{T}$, *numt*$_0 = 0$, i.e., simply $t_0 = 0$ in the sense *numt*$_0 = 0$

$t_{\inf}$ *the first instant*, which has not happened, $t_{\inf} = -\infty$

$t_{\sup}$ *the last instant*, which will not occur, $t_{\sup} = \infty$

$t_{ZeroTotal}$ *the total zero value of time*, which has not existed and will not happen

t_{zero} *a conventionally accepted relative zero value of time*

T   *the temporal dimension*, "the *time* dimension", which is the physical dimension of *time*

$T \in \mathfrak{R}^+$ *the period of a periodic behavior*

$T_k \in \mathfrak{R}^{N \times M}$ *a matrix*

$T^{(\mu)} \in \mathfrak{R}^{N \times M(\mu+1)}$ *the extended matrix describing the action of the extended input vector* $\mathbf{i}^\mu$ *on the output dynamics of the IIO system*, $T^{(\mu)} = \left[T_0 \vdots T_1 \vdots ... T_\mu \right]$

$\mathbf{U}_{[t_0, t_1]}^\mu$ extended control on the time interval $[t_0, t_1]$

$v(.) : \mathfrak{R}^p \rightarrow \mathfrak{R}$ *a quadratic form*, $v(\mathbf{w}) = \mathbf{w}^T \mathbf{W} \mathbf{w}$

$\mathbf{V}(s)$ is *the Laplace transform of all actions on the system*; it is composed of the Laplace transform $\mathbf{I}(s)$ of the input vector $\mathbf{I}(t)$ and of all (input and output) initial conditions,

$$\mathbf{V}(s) = \begin{bmatrix} \mathbf{I}(s) \\ \mathbf{C}_0 \end{bmatrix}$$

$\mathbf{w} \in \mathfrak{R}^p$ *a subsidiary real valued vector,*

$$\mathbf{w} = [w_1 \;\; w_2 \;\; ... \;\; wp]^T \in \left\{ [\mathbf{r}^{\alpha-1} \;\; \mathbf{y}^{\nu-1}]^T, \; \mathbf{x}, \; \mathbf{y}^{\nu-1} \right\},$$

$$p \in \{\rho, \; n, \; N\}$$

$W = W^T \in R^{p \times p}$ *the symmetric matrix of the quadratic form* $v(\mathbf{w})$,
$v(\mathbf{w}) = \mathbf{w}^T W \mathbf{w}, \; W \in \{G = G^T, \; H = H^T\}$

$x \in \mathfrak{R}$ *a real valued scalar state deviation variable*

$\mathbf{x} \in \mathfrak{R}^n$ *the state vector deviation of the ISO system,* (3.38), (Section 3.2),

$$\mathbf{x} = [x_1 \;\; x_2 \;\; ... \;\; x_n]^T, \; \mathbf{x} = \mathbf{X} - \mathbf{X}_N = \mathbf{X} - \mathbf{X}_d$$

$\mathbf{X} \in \mathfrak{R}^n$ *the total state vector of the ISO system,* $\mathbf{X} = [X_1 \; X_2 \; .. \; X_n]^T$

$\mathbf{X}_N \in \mathfrak{R}^n$ *the total nominal state vector of the ISO system,* $\mathbf{X}_N = [X_{N1} \; X_{N2} \; ... \; X_{Nn}]^T$

$y \in \mathfrak{R}$ *a real valued scalar output deviation variable*

$\mathbf{y} \in \mathfrak{R}^N$ *a real valued vector output deviation variable - the output deviation vector of both the plant and of its control system,* $\mathbf{y} = [y_1 \;\; y_2 \;\; ... \;\; y_N]^T$, (9.3), (Section 2.2),

$$\mathbf{y} = \mathbf{Y} - \mathbf{Y}_d = -\varepsilon$$

$\mathbf{Y} \in \mathfrak{R}^N$ *a real total valued vector output - the total output vector of both the plant and of its control system,* $\mathbf{Y} = [Y_1 \;\; Y_2 \;\; ... \;\; Y_N]^T$

$\mathbf{Y}_d \in \mathfrak{R}^N$ *a desired (a nominal) total valued vector output - the desired total output vector of both the plant and of its control system,* $\mathbf{Y}_d = [Y_{d1} \; Y_{d2} \; ... \; Y_{dN}]^T$

$\mathbf{y}_{0\mp}^{\nu-1} \in \mathfrak{R}^{\nu N}$ *the initial extended output vector at the initial moment*

$t_0 = 0, \; \mathbf{y}_{0\mp}^{\nu-1} = \mathbf{y}^{\nu-1}(0^\mp) = \left[\mathbf{y}_{0(\mp)}^T \; \vdots \; \mathbf{y}_{0(\mp)}^{(1)T} \; \vdots \; ... \; \vdots \; \mathbf{y}_{0(\mp)}^{(\nu-1)T} \right]^T, \; \mathbf{y}_{0\mp}^0 = \mathbf{y}^0(0^\mp) = \mathbf{y}_{0\mp} = \mathbf{y}(0^\mp)$

A.4   Name

Stable (stability) matrix: a square matrix is *stable (stability) matrix* if, and only if, the real parts of all its eigenvalues are negative.

A.5   Symbols, vectors, sets and matrices

(.)   *an arbitrary variable*, or *an index*

$|(.)| : \mathfrak{R} \to \mathfrak{R}_+$ *the absolute value (module) of a (complex valued) scalar variable* (.)

$\|.\| : \mathfrak{R}^n \to \mathfrak{R}_+$ *an accepted norm on* $\mathfrak{R}^n$, which is the *Euclidean norm on* $\mathfrak{R}^n$ iff not stated otherwise:

$$\|\mathbf{x}\| = \|\mathbf{x}\|_2 = \sqrt{\mathbf{x}^T\mathbf{x}} = \sqrt{\sum_{i=1}^{i=n} x_i^2}$$

$\|.\|_1 : \mathfrak{R}^n \to \mathfrak{R}_+$ is the *taxicab norm* or *Manhattan norm:*

$$\|\mathbf{x}\|_1 = \sum_{i=1}^{i=n} |x_i|$$

$\sim$ *equivalent*

$\langle 1.. \rangle$ shows *the units 1... of a physical variable*

$[\,\alpha,\,\beta\,] \subset \mathfrak{R}$ *a compact interval* $[\alpha,\beta] = \{x : x \in \mathfrak{R},\ \alpha \le x \le \beta\}$

$[\,\alpha,\,\beta\,[\,\subseteq \mathfrak{R}$ *a left closed, right open interval,* $[\alpha,\beta[= \{x : x \in \mathfrak{R},$
$\alpha \le x < \beta\}$

$]\,\alpha,\,\beta\,]\subseteq \mathfrak{R}$ *a left open, right closed interval,* $[\alpha,\beta[= \{x : x \in \mathfrak{R},\ \alpha < x \le \beta\}$

$]\,\alpha,\,\beta\,[\,\subseteq \mathfrak{R}$ *an open interval,* $]\alpha,\beta[= \{x : x \in \mathfrak{R},\ \alpha < x < \beta\}$

$(\sigma,\infty[\in \{]\sigma,\infty[,\ [\sigma,\infty[\}$,

$(\,\alpha,\,\beta\,) \subseteq \mathfrak{R}$ *a general interval,* $(\,\alpha,\beta\,) \in \{[\alpha,\beta],\ [\alpha,\beta[,\ ]\alpha,\beta],\ ]\alpha,\beta[\}$

$\lambda_i(A)$ *the eigenvalue* $\lambda_i(A)$ *of the matrix* A

$[\,A..\,]$ shows *the physical dimension* A... *of a physical variable*

$\left[A_1 \vdots A_2 \vdots ... \vdots A_\nu \right]$ *a structured matrix* composed of the submatrices $A_1,\ A_2,\ ...,\ A_\nu$

$\mathbf{0}_k = [0\ \ 0\ ...0]^T \in \mathfrak{R}^k,$ *the elementwise zero vector,* $\mathbf{0}_n = \mathbf{0}$

$\mathbf{1}_k = [1\ \ 1...1]^T \in \mathfrak{R}^k,$ *the elementwise unity vector,* $\mathbf{1}_n = \mathbf{1}$

$\mathbf{w} = \varepsilon$ *the elementwise vector equality,*

$$\varepsilon = [\varepsilon_1\ \ \varepsilon_2\ \ ...\ \varepsilon_N]^T \in \mathfrak{R}^N,\ \mathbf{w} = [w_1\ \ w_2\ \ ...\ w_N]^T,$$

$$\mathbf{w} = \varepsilon \Longleftrightarrow w_i = \varepsilon_i, \forall i = 1, 2, ..., N$$

$\mathbf{w} \ne \varepsilon$ *the elementwise vector inequality,*

$$\mathbf{w} \ne \varepsilon \Longleftrightarrow w_i \ne \varepsilon_i,\ \forall i = 1, 2, ..., N$$

∀   *for every*
adj A   *the adjoint matrix of the nonsingular square matrix* A, $det A \neq$
$0 \Longrightarrow A adj A = (det A) I$
det A   *the determinant of the matrix* A, $det A = |A|$
A^{-1} *the inverse matrix of the nonsingular square matrix* A, $det A \neq$
$0 \Longrightarrow A^{-1} = adj A / det A$
blockdiag $\{ \cdot \quad \cdot \quad \cdot \quad .... \cdot \}$ *block diagonal matrix*, the entries of which
are matrices, e.g.,

$$blockdiag \{ E_k \quad B_k \} = \begin{bmatrix} E_k & O_{N,r} \\ O_{N,d} & B_k \end{bmatrix}, \ k = 0, 1, .., \nu$$

In $\mathfrak{T}_{0F}$ *the interior of* $\mathfrak{T}_{0F}$,

$$In \ \mathfrak{T}_{0F} = \{ t : \ t \in Cl\mathfrak{T}, \ 0 < t < t_F \} \tag{A.2}$$

$\min(\delta, \Delta)$ denotes *the smaller between* δ *and* Δ,

$$\min(\delta, \Delta) = \left\{ \begin{matrix} \delta, \ \delta \leq \Delta, \\ \Delta, \ \Delta \leq \delta \end{matrix} \right\}$$

$Re \lambda_i(A)$ *the real part* of the eigenvalue $\lambda_i(A)$ of the matrix A
∃   *there exist(s)*
∈   *belong(s) to, are (is) members (a member) of*, respectively
$\sqrt{-1}$ *the imaginary unit* denoted by i, $i = \sqrt{-1}$
inf   *infimum*
max   *maximum*
min   *minimum*
numx   *the numerical value of* x, if $x = 50V$ then $numx = 50$,
phdim $x(.)$ *the physical dimension of a variable* $x(.)$,

$$x(.) = t \Longrightarrow phdim \ x(.) = phdim \ t = \text{T}, \ but \ dim \ t = 1$$

sup   *supremum*
$\|.\|$ can be any *norm* if not otherwise specified

A.6   Units

$1_{(.)}$ *the unit of a physical variable* (.)
    1_t *the time unit of the reference time axis* T, $1_t = s$

Appendix B

Example

B.1 *IO* system example

Example 227 *Let the IO system (2.1) be defined by*

$$
\begin{bmatrix} 1 & 2 \\ 3 & -4 \end{bmatrix} \begin{bmatrix} Y_1^{(3)}(t) \\ Y_2^{(3)}(t) \end{bmatrix} + \begin{bmatrix} 2 & 3 \\ 1 & 0 \end{bmatrix} \begin{bmatrix} Y_1^{(1)}(t) \\ Y_2^{(1)}(t) \end{bmatrix} + \begin{bmatrix} 4 & 0 \\ 1 & 1 \end{bmatrix} \begin{bmatrix} Y_1(t) \\ Y_2(t) \end{bmatrix} =
$$

$$
= \begin{bmatrix} 5 & 3 \\ 4 & 1 \end{bmatrix} \begin{bmatrix} U_1^{(2)}(t) \\ U_2^{(2)}(t) \end{bmatrix} + \begin{bmatrix} 1 & 0 \\ 0 & 3 \end{bmatrix} \begin{bmatrix} U_1^{(1)}(t) \\ U_2^{(1)}(t) \end{bmatrix} + \begin{bmatrix} 4 & 2 \\ 3 & 6 \end{bmatrix} \begin{bmatrix} U_1(t) \\ U_2(t) \end{bmatrix}.
$$

$$\tag{B.1}$$

In this case:

$$
N = 2, \ \nu = 3, \ r = 2, \ \mu = 2 < 3 = \nu, \tag{B.2}
$$

$$
A_0 = \begin{bmatrix} 4 & 0 \\ 1 & 1 \end{bmatrix}, \ A_1 = \begin{bmatrix} 2 & 3 \\ 1 & 0 \end{bmatrix}, \ A_2 = \begin{bmatrix} 0 & 0 \\ 0 & 0 \end{bmatrix} = O_2, \ A_3 = \begin{bmatrix} 1 & 2 \\ 3 & -4 \end{bmatrix},
$$

$$
A^{(3)} = \begin{bmatrix} 4 & 0 & 2 & 3 & 0 & 0 & 1 & 2 \\ 1 & 1 & 1 & 0 & 0 & 0 & 3 & -4 \end{bmatrix}, \tag{B.3}
$$

$$
B_0 = \begin{bmatrix} 4 & 2 \\ 3 & 6 \end{bmatrix}, \ B_1 = \begin{bmatrix} 1 & 0 \\ 0 & 3 \end{bmatrix}, \ B_2 = \begin{bmatrix} 5 & 3 \\ 4 & 1 \end{bmatrix},
$$

$$
B^{(2)} = \begin{bmatrix} 4 & 2 & 1 & 0 & 5 & 3 \\ 3 & 6 & 0 & 3 & 4 & 1 \end{bmatrix}, \tag{B.4}
$$

The complex functions $S_i^{(k)}(s)$ (2.17) and $Z_k^{(s-1)}(s)$ (2.18) read:

$$S_2^{(3)}(s) = \begin{bmatrix} s^0 & 0 & s^1 & 0 & s^2 & 0 & s^3 & 0 \\ 0 & s^0 & 0 & s^1 & 0 & s^2 & 0 & s^3 \end{bmatrix}^T =$$

$$= \begin{bmatrix} 1 & 0 & s & 0 & s^2 & 0 & s^3 & 0 \\ 0 & 1 & 0 & s & 0 & s^2 & 0 & s^3 \end{bmatrix}^T \implies \tag{B.5}$$

$$A^{(3)}S_2^{(3)}(s) = \begin{bmatrix} 4+2s+s^3 & 3s+2s^3 \\ 1+s+3s^3 & 1-4s^3 \end{bmatrix} \tag{B.6}$$

$$S_2^{(2)}(s) = \begin{bmatrix} s^0 & 0 & s^1 & 0 & s^2 & 0 \\ 0 & s^0 & 0 & s^1 & 0 & s^2 \end{bmatrix}^T = \begin{bmatrix} 1 & 0 & s & 0 & s^2 & 0 \\ 0 & 1 & 0 & s & 0 & s^2 \end{bmatrix}^T, \tag{B.7}$$

$$Z_2^{(3-1)}(s) = \begin{bmatrix} O_2 & O_2 & O_2 \\ s^0 U_2 & O_2 & O_2 \\ s^1 U_2 & s^0 U_2 & O_2 \\ s^{3-1}U_2 & s^{3-2}U_2 & s^{3-3}U_2 \end{bmatrix} = \tag{B.8}$$

$$= \begin{bmatrix} 0 & 0 & 0 & 0 & 0 & 0 \\ 0 & 0 & 0 & 0 & 0 & 0 \\ 1 & 0 & 0 & 0 & 0 & 0 \\ 0 & 1 & 0 & 0 & 0 & 0 \\ s & 0 & 1 & 0 & 0 & 0 \\ 0 & s & 0 & 1 & 0 & 0 \\ s^2 & 0 & s & 0 & 1 & 0 \\ 0 & s^2 & 0 & s & 0 & 1 \end{bmatrix} = Z_2^{(2)}(s),$$

$$A^{(3)}Z_2^{(2)}(s) = \begin{bmatrix} 2+s^2 & 3+2s^2 & s & 2s & 1 & 2 \\ 1+3s^2 & 3-4s^2 & 3s & -4s & 3 & -4 \end{bmatrix},$$

and

$$Z_2^{(2-1)}(s) = \begin{bmatrix} O_2 & O_2 \\ s^0 I_2 & O_2 \\ s^{2-1}I_2 & s^{1-1}I_2 \end{bmatrix} = \begin{bmatrix} 0 & 0 & 0 & 0 \\ 0 & 0 & 0 & 0 \\ 1 & 0 & 0 & 0 \\ 0 & 1 & 0 & 0 \\ s & 0 & 1 & 0 \\ 0 & s & 0 & 1 \end{bmatrix} = Z_2^{(1)}(s). \tag{B.9}$$

$$B^{(2)}Z_2^{(1)}(s) = \begin{bmatrix} 1+5s & 3s & 5 & 3 \\ 4s & 3+s & 4 & 1 \end{bmatrix} \tag{B.10}$$

Equations (B.2)-(B.9) give the following specific form to the equations (2.20) and (2.22) for the specific system (B.1):

$$
\mathcal{L}^{\mp}\left\{\sum_{k=0}^{k=3} A_k \mathbf{Y}^{(k)}(t)\right\} = A^{(3)} S_2^{(3)}(s) \mathbf{Y}^{\mp}(s) - A^{(3)} Z_3^{(2)}(s) \mathbf{Y}_{0\mp}^2 =
$$

$$
= \underbrace{\begin{bmatrix} 4 + 2s + s^3 & 3s + 2s^3 \\ 1 + s + 3s^3 & 1 - 4s^3 \end{bmatrix}}_{A^{(3)} S_3^{(3)}(s)} \underbrace{\begin{bmatrix} Y_1^{\mp}(s) \\ Y_2^{\mp}(s) \end{bmatrix}}_{\mathbf{Y}^{\mp}(s)} -
$$

$$
- \underbrace{\begin{bmatrix} 2 + s^2 & 3 + 2s^2 & s & 2s & 1 & 2 \\ 1 + 3s^2 & 3 - 4s^2 & 3s & -4s & 3 & -4 \end{bmatrix}}_{A^{(3)} Z_2^{(2)}(s)} \underbrace{\begin{bmatrix} Y_{10\mp} \\ Y_{20\mp} \\ Y_{10\mp}^{(1)} \\ Y_{20\mp}^{(1)} \\ Y_{10\mp}^{(2)} \\ Y_{20\mp}^{(2)} \end{bmatrix}}_{\mathbf{Y}_{0\mp}^2}, \qquad (B.11)
$$

i.e.,

$$
\mathcal{L}^{\mp}\left\{\sum_{k=0}^{k=3} A_k \mathbf{Y}^{(k)}(t)\right\} = \begin{bmatrix} \left(4 + 2s + s^3\right) Y_1^{\mp}(s) + \left(3s + 2s^3\right) Y_2^{\mp}(s) \\ \left(1 + s + 3s^3\right) Y_1^{\mp}(s) + \left(1 - 4s^3\right) Y_2^{\mp}(s) \end{bmatrix} -
$$

$$
- \begin{bmatrix} \left(2 + s^2\right) Y_{10\mp} + \left(3 + 2s^2\right) Y_{20\mp} + sY_{10\mp}^{(1)} + 2sY_{20\mp}^{(1)} + Y_{10\mp}^{(2)} + 2Y_{20\mp}^{(2)} \\ \left(1 + 3s^2\right) Y_{10\mp} + \left(3 - 4s^2\right) Y_{20\mp} + 3sY_{10\mp}^{(1)} - 4sY_{20\mp}^{(1)} + 3Y_{10\mp}^{(2)} - 4Y_{20\mp}^{(2)} \end{bmatrix},
$$

$$
\mathcal{L}^{\mp}\left\{\sum_{k=0}^{k=2} B_k \mathbf{U}^{(k)}(t)\right\} = B^{(2)} S_2^{(2)}(s) \mathbf{U}^{\mp}(s) - B^{(2)} Z_2^{(2-1)}(s) \mathbf{U}^{2-1}(0^{\mp}) =
$$

$$
= \underbrace{\begin{bmatrix} 4 + s + 5s^2 & 2 + 3s^2 \\ 3 + 4s^2 & 6 + 3s + s^2 \end{bmatrix}}_{B^{(2)} S_2^{(2)}(s)} \underbrace{\begin{bmatrix} U_1^{\mp}(s) \\ U_2^{\mp}(s) \end{bmatrix}}_{\mathbf{U}^{\mp}(s)} -
$$

$$
- \underbrace{\begin{bmatrix} 1 + 5s & 3s & 5 & 3 \\ 4s & 3 + s & 4 & 1 \end{bmatrix}}_{B^{(2)} Z_2^{(1)}(s)} \underbrace{\begin{bmatrix} U_{10\mp} \\ U_{20\mp} \\ U_{10\mp}^{(1)} \\ U_{20\mp}^{(1)} \end{bmatrix}}_{\mathbf{B}_{0\mp}^1} \qquad (B.12)
$$

i.e.,

$$\mathcal{L}^{\mp}\left\{\sum_{k=0}^{k=2} B_k \mathbf{U}^{(k)}(t)\right\} = \left[\begin{array}{c} \left(4 + s + 5s^2\right) U_1^{\mp}(s) + \left(2 + 3s^2\right) U_2^{\mp}(s) \\ \left(3 + 4s^2\right) U_1^{\mp}(s) + \left(6 + 3s + s^2\right) U_2^{\mp}(s) \end{array}\right] -$$

$$- \left[\begin{array}{c} (1 + 5s) U_{10\mp} + 3sU_{20\mp} + 5U_{10\mp}^{(1)} + 3U_{20\mp}^{(1)} \\ 4sU_{10\mp} + (3 + s) U_{20\mp} + 4U_{10\mp}^{(1)} + U_{20\mp}^{(1)} \end{array}\right].$$

Besides,

$$\left(A^{(3)} S_2^{(3)}(s)\right)^{-1} = \left[\begin{array}{cc} 4 + 2s + s^3 & 3s + 2s^3 \\ 1 + s + 3s^3 & 1 - 4s^3 \end{array}\right]^{-1} =$$

$$= \left[\begin{array}{cc} \frac{4s^3 - 1}{10s^6 + 19s^4 + 17s^3 + 3s^2 + s - 4} & \frac{2s^3 + 3s}{10s^6 + 19s^4 + 17s^3 + 3s^2 + s - 4} \\ \frac{3s^3 + s + 1}{10s^6 + 19s^4 + 17s^3 + 3s^2 + s - 4} & -\frac{s^3 + 2s + 4}{10s^6 + 19s^4 + 17s^3 + 3s^2 + s - 4} \end{array}\right]. \qquad \text{(B.13)}$$

Equations (B.11), (B.13) determine (2.24) for the system (B.1) as follows:

$$\mathbf{Y}^{\mp}(s) = \left[\begin{array}{cc} \frac{4s^3 - 1}{10s^6 + 19s^4 + 17s^3 + 3s^2 + s - 4} & \frac{2s^3 + 3s}{10s^6 + 19s^4 + 17s^3 + 3s^2 + s - 4} \\ \frac{3s^3 + s + 1}{10s^6 + 19s^4 + 17s^3 + 3s^2 + s - 4} & -\frac{s^3 + 2s + 4}{10s^6 + 19s^4 + 17s^3 + 3s^2 + s - 4} \end{array}\right] \bullet$$

$$\bullet \left[\begin{array}{c} \left[\begin{array}{cc} 4 + s + 5s^2 & 2 + 3s^2 \\ 3 + 4s^2 & 6 + 3s + s^2 \end{array}\right] \\ \left[\begin{array}{cccc} -1 - 5s & -3s & -5 & -3 \\ -4s & -3 - s & -4 & -1 \end{array}\right] \\ \left[\begin{array}{cccccc} 2 + s^2 & 3 + 2s^2 & s & 2s & 1 & 2 \\ 1 + 3s^2 & 3 - 4s^2 & 3s & -4s & 3 & -4 \end{array}\right] \end{array}\right]^T \left[\begin{array}{c} \mathbf{B}^{\mp}(s) \\ \mathbf{B}_{0\mp}^1 \\ \mathbf{Y}_{0\mp}^2 \end{array}\right] =$$

$$= F_{IO}(s)\, \mathbf{V}_{IO}^{\mp}(s), \qquad \text{(B.14)}$$

where
- $F_{IO}(s)$,

$$F_{IO}(s) = \left[G_{IO}(s) \;\vdots\; G_{IOI_0}(s) \;\vdots\; G_{IOY_0}(s)\right] =$$

$$= \left[\begin{array}{cc} \frac{4s^3 - 1}{10s^6 + 19s^4 + 17s^3 + 3s^2 + s - 4} & \frac{2s^3 + 3s}{10s^6 + 19s^4 + 17s^3 + 3s^2 + s - 4} \\ \frac{3s^3 + s + 1}{10s^6 + 19s^4 + 17s^3 + 3s^2 + s - 4} & -\frac{s^3 + 2s + 4}{10s^6 + 19s^4 + 17s^3 + 3s^2 + s - 4} \end{array}\right] \bullet$$

$$\bullet \left[\begin{array}{c} \left[\begin{array}{cc} 4 + s + 5s^2 & 2 + 3s^2 \\ 3 + 4s^2 & 6 + 3s + s^2 \end{array}\right] \\ \left[\begin{array}{cccc} -1 - 5s & -3s & -5 & -3 \\ -4s & -3 - s & -4 & -1 \end{array}\right] \\ \left[\begin{array}{cccccc} 2 + s^2 & 3 + 2s^2 & s & 2s & 1 & 2 \\ 1 + 3s^2 & 3 - 4s^2 & 3s & -4s & 3 & -4 \end{array}\right] \end{array}\right]^T, \qquad \text{(B.15)}$$

is the full transfer function matrix of the *IO* system (B.1), which determines:

- The system transfer function $G_{IO}(s)$ relative to the input vector U:

$$G_{IO}(s) = \begin{bmatrix} \frac{4s^3-1}{10s^6+19s^4+17s^3+3s^2+s-4} & \frac{2s^3+3s}{10s^6+19s^4+17s^3+3s^2+s-4} \\ \frac{3s^3+s+1}{10s^6+19s^4+17s^3+3s^2+s-4} & -\frac{s^3+2s+4}{10s^6+19s^4+17s^3+3s^2+s-4} \end{bmatrix} \bullet$$

$$\bullet \begin{bmatrix} 4+s+5s^2 & 2+3s^2 \\ 3+4s^2 & 6+3s+s^2 \end{bmatrix} \Longrightarrow$$

$$G_{IO}(s) =$$

$$= \begin{bmatrix} \frac{-4+8s-5s^2+34s^3+4s^4+28s^5}{10s^6+19s^4+17s^3+3s^2+s-4} & \frac{-2+18s+6s^2+23s^3+6s^4+14s^5}{10s^6+19s^4+17s^3+3s^2+s-4} \\ \frac{-8-s+6s^2+6s^3+3s^4+11s^5}{10s^6+19s^4+17s^3+3s^2+s-4} & -\frac{26+26s+13s^2+17s^3+3s^4+10s^5}{10s^6+19s^4+17s^3+3s^2+s-4} \end{bmatrix}, \quad \text{(B.16)}$$

- The system transfer function $G_{IOU_0}(s)$ relative to the extended initial control vector $U_{0\mp}^{\mu-1}$:

$$G_{IOU_0}(s) = \begin{bmatrix} \frac{4s^3-1}{10s^6+19s^4+17s^3+3s^2+s-4} & \frac{2s^3+3s}{10s^6+19s^4+17s^3+3s^2+s-4} \\ \frac{3s^3+s+1}{10s^6+19s^4+17s^3+3s^2+s-4} & -\frac{s^3+2s+4}{10s^6+19s^4+17s^3+3s^2+s-4} \end{bmatrix} \bullet$$

$$\bullet \begin{bmatrix} -1-5s & -3s & -5 & -3 \\ -4s & -3-s & -4 & -1 \end{bmatrix} \Longrightarrow$$

$$G_{IOU_0}(s) =$$

$$= \begin{bmatrix} \frac{1+5s-12s^2-4s^3-28s^4}{10s^6+19s^4+17s^3+3s^2+s-4} & \frac{-1+10s+3s^2-3s^3-11s^4}{10s^6+19s^4+17s^3+3s^2+s-4} \\ \frac{-6s-3s^2-6s^3-14s^4}{10s^6+19s^4+17s^3+3s^2+s-4} & \frac{12+7s-s^2+3s^3-8s^4}{10s^6+19s^4+17s^3+3s^2+s-4} \\ \frac{5-12s-28s^3}{10s^6+19s^4+17s^3+3s^2+s-4} & \frac{11+3s-11s^3}{10s^6+19s^4+17s^3+3s^2+s-4} \\ \frac{3-3s+10s^3}{10s^6+19s^4+17s^3+3s^2+s-4} & \frac{1-s-8s^3}{10s^6+19s^4+17s^3+3s^2+s-4} \end{bmatrix}^T, \quad \text{(B.17)}$$

- The system transfer function $G_{IOY_0}(s)$ relative to the extended initial output vector $Y_{0\mp}^{\nu-1}$:

$$G_{IOY_0}(s) = \begin{bmatrix} \frac{4s^3-1}{10s^6+19s^4+17s^3+3s^2+s-4} & \frac{2s^3+3s}{10s^6+19s^4+17s^3+3s^2+s-4} \\ \frac{3s^3+s+1}{10s^6+19s^4+17s^3+3s^2+s-4} & -\frac{s^3+2s+4}{10s^6+19s^4+17s^3+3s^2+s-4} \end{bmatrix} \bullet$$

$$\bullet \begin{bmatrix} 2+s^2 & 3+2s^2 & s & 2s & 1 & 2 \\ 1+3s^2 & 3-4s^2 & 3s & -4s & 3 & -4 \end{bmatrix} \Longrightarrow$$

$$G_{IOY_0}(s) =$$

$$= \begin{bmatrix} \dfrac{-2+3s-s^2+19s^3+10s^5}{10s^6+19s^4+17s^3+3s^2+s-4} & \dfrac{-2-11s^2}{10s^6+19s^4+17s^3+3s^2+s-4} \\[2mm] \dfrac{-3+9s-2s^2+6s^3}{10s^6+19s^4+17s^3+3s^2+s-4} & \dfrac{-9-3s+18s^2+16s^3+20s^5}{10s^6+19s^4+17s^3+3s^2+s-4} \\[2mm] \dfrac{-s+9s^2+10s^4}{10s^6+19s^4+17s^3+3s^2+s-4} & \dfrac{-11s\ -5s^2}{10s^6+19s^4+17s^3+3s^2+s-4} \\[2mm] \dfrac{-2s-12s^2}{10s^6+19s^4+17s^3+3s^2+s-4} & \dfrac{-14s-6s^2+2s^4}{10s^6+19s^4+17s^3+3s^2+s-4} \\[2mm] \dfrac{-1+9s+10s^3}{10s^6+19s^4+17s^3+3s^2+s-4} & \dfrac{-11-5s}{10s^6+19s^4+17s^3+3s^2+s-4} \\[2mm] \dfrac{-2-12s}{10s^6+19s^4+17s^3+3s^2+s-4} & \dfrac{-14-6s+2s^3}{10s^6+19s^4+17s^3+3s^2+s-4} \end{bmatrix}^T, \qquad (B.18)$$

- The Laplace transform $\mathbf{V}_{IO}^{\mp}(s)$ of the system action vector $\mathbf{V}_{IO}(t)$ reads

$$\mathbf{V}_{IO}^{\mp}(s) = \begin{bmatrix} \mathbf{U}^{\mp}(s) \\ \mathbf{U}_{0\mp}^{1} \\ \mathbf{Y}_{0\mp}^{2} \end{bmatrix} = \begin{bmatrix} \mathbf{U}^{\mp}(s) \\ \mathbf{C}_{IO0\mp} \end{bmatrix}, \qquad (B.19)$$

- And the vector $\mathbf{C}_{IO0\mp}$ of all the initial conditions is found to be

$$\mathbf{C}_{IO0\mp} = \begin{bmatrix} \mathbf{U}_{0\mp}^{1} \\ \mathbf{Y}_{0\mp}^{2} \end{bmatrix}. \qquad (B.20)$$

Appendix C

Transformations

C.1 Transformation of *IO* into *ISO* system

The state space theory of the linear dynamical and control systems has been mainly established and effective for the *ISO* systems (3.1), (3.2) (Section 3.1). In order to transform the *IO* system (2.1), i.e., (2.15) (Section 2.1.1) into the *ISO* systems (3.1), (3.2) the well-known formal mathematical transformation has been used. It has to satisfy the condition that the transformed system should not contain any derivative of the input vector despite the influence of derivatives of the input vector on the original system and the condition that the only accepted derivative is the first derivative of the state vector and only in the state equation. We will illustrate it for the *IO* system (2.1) subjected to the external action of the input vector $\mathbf{I}$ and its derivatives, i.e., subjected to the action of the extended input vector $\mathbf{I}^\mu$.

The *IO* system (2.1):

$$\sum_{k=0}^{k=\nu} A_k \mathbf{Y}^{(k)}(t) = \sum_{k=0}^{k=\mu} H_k \mathbf{I}^{(k)}(t), \ detA_\nu \neq 0, \forall t \in \mathfrak{T}_0, \ \nu \geq 1, \ 0 \leq \mu \leq \nu,$$

(C.1)

can be formally mathematical transformed into the mathematically equivalent *ISO* system (3.1), (3.2),

$$\frac{d\mathbf{X}(t)}{dt} = A\mathbf{X}(t) + H\mathbf{I}(t), \ \forall t \in \mathfrak{T}_0, \ A \in \mathfrak{R}^{n \times n}, \ \mathbf{U} \in \mathfrak{R}^M, \ P \in \mathfrak{R}^{n \times M}, \ (C.2)$$

$$\mathbf{Y}(t) = C\mathbf{X}(t) + Q\mathbf{I}(t), \ \forall t \in \mathfrak{T}_0, \ C \in \mathfrak{R}^{N \times n}, \ C \neq O_{N,n}, \ Q \in \mathfrak{R}^{N \times M}.$$

(C.3)

by applying the following formal mathematical transformations:

$$\mathbf{X}_1 = \mathbf{Y} - H_\nu \mathbf{I}, \tag{C.4}$$

$$\mathbf{X}_2 = \dot{\mathbf{X}}_1 + A_{\nu-1}\,\mathbf{Y} - H_{\nu-1}\mathbf{I}, \tag{C.5}$$

$$\mathbf{X}_3 = \dot{\mathbf{X}}_2 + A_{\nu-2}\,\mathbf{Y} - H_{\nu-2}\mathbf{I}, \tag{C.6}$$

$$.... \tag{C.7}$$

$$\mathbf{X}_{\nu-1} = \dot{\mathbf{X}}_{\nu-2} + A_2\,\mathbf{Y} - H_2\mathbf{I} \tag{C.8}$$

$$\mathbf{X}_\nu = \dot{\mathbf{X}}_{\nu-1} + A_1\mathbf{Y} - H_1\mathbf{I}, \tag{C.9}$$

where $H_k = O_{N,r}$ for $k = \mu + 1, \mu + 2, ..., \nu$ if $\mu < \nu$. The vectors $\mathbf{X}_1, \mathbf{X}_2, ...$ $\mathbf{X}_\nu \in \mathfrak{R}^N$ are the mathematical state subvectors of the vector $\mathbf{X} \in \mathfrak{R}^n$ that is *the mathematical state vector* of the *IO* system (C.1) and of the equivalent *ISO* system (C.2), (C.3),

$$\mathbf{X} = \begin{bmatrix} \mathbf{X}_1^T & \mathbf{X}_2^T & ... & \mathbf{X}_\nu^T \end{bmatrix}^T \in \mathfrak{R}^n, \ n = \nu N. \tag{C.10}$$

Comment 228 *The state subvectors $\mathbf{X}_1, \mathbf{X}_2, ... \mathbf{X}_\nu$ (C.4)-(C.9) and the state vector $\mathbf{X}$ (C.10) do not any physical sense, i.e., they are physically meaningless, if $\mu > 0$, equivalently if $H_k = O_{N,r}$ for $k \in \{1, 2, ..., \nu\}$. This is the consequence of their definitions to be linear combinations of the input vector, the output vector and the derivative of the preceding state subvector if it exists. Their physical nature and properties are most often inherently different.*

The transformations (C.4)-(C.9) are formal mathematical, physically useless in general. They lead to the following matrices of the *ISO* system (C.2), (C.3) mathematically formally equivalent to the *IO* system (C.1):

$$A = \begin{bmatrix} -A_{\nu-1} & I_N & ... & O_N & O_N \\ -A_{\nu-2} & O_N & ... & O_N & O_N \\ ... & ... & ... & ... & ... \\ -A_1 & O_N & ... & O_N & I_N \\ -A_0 & O_N & ... & O_N & O_N \end{bmatrix}, \tag{C.11}$$

$$H = \begin{bmatrix} H_{\nu-1} - A_{\nu-1}H_\nu \\ H_{\nu-2} - A_{\nu-2}H_\nu \\ .... \\ H_1 - A_1 H_\nu \\ H_0 - A_0 H_\nu \end{bmatrix}, \tag{C.12}$$

$$C = \begin{bmatrix} I_N & O_N & O_N & ... & O_N & O_N & O_N \end{bmatrix}, \tag{C.13}$$

$$Q = H_\nu. \tag{C.14}$$

Conclusion 229 *The aim of the book and the transformations (C.4)-(C.9)*
   The aim of the book to further develop and generalize the control theory with the simultaneous physical and mathematical, i.e., the full engineering, sense, excludes the use of the pure formal mathematical transformations (C.4)-(C.9) if $\mu > 0$.

C.2 *ISO* and *EISO* forms of *IIO* system

The *ISO* and *EISO* forms of the *IIO* system
   The compact form of the overall mathematical model of the *IIO* system (6.1), (6.2), (Section 6.1), reads in terms of the total coordinates:

$$\begin{bmatrix} A_\alpha & O_{\rho,N} \\ O_{N,\alpha} & E_\nu \end{bmatrix} \begin{bmatrix} \mathbf{R}^{(\alpha)}(t) \\ \mathbf{Z}^{(\nu)}(t) \end{bmatrix} + \begin{bmatrix} A^{(\alpha-1)} & O_{\rho,\nu+1} \\ -R^{(\alpha-1)} & E^{(\nu-1)} \end{bmatrix} \begin{bmatrix} \mathbf{R}^{\alpha-1}(t) \\ \mathbf{Z}^{\nu-1}(t) \end{bmatrix} =$$
$$= \begin{bmatrix} H^{(\mu)} \\ Q^{(\mu)} \end{bmatrix} \mathbf{I}^\mu(t), \ \ \mathbf{Y}(t) = \mathbf{Z}(t), \tag{C.15}$$

where we use a subsidiary vector $\mathbf{Z}$,

$$\mathbf{Z}(t) = \mathbf{Y}(t) = \mathbf{S}_{\alpha+1}(t), \ \mathbf{Z}^{(k)}(t) = \mathbf{Y}^{(k)}(t) = \mathbf{S}_{\alpha+k+1}(t), \ k = 0, 1, .., \nu - 1,$$
$$\mathbf{Z}^{\nu-1}(t) = \mathbf{Y}^{\nu-1}(t) = \mathbf{S}_O(t). \tag{C.16}$$

In terms of the deviations the system model is given by (6.65), (6.66) (Section 6.2), which can be set in the form of (C.15), (C.16):

$$\begin{bmatrix} A_\alpha & O_{\rho,N} \\ O_{N,\alpha} & E_\nu \end{bmatrix} \begin{bmatrix} \mathbf{r}^{(\alpha)}(t) \\ \mathbf{z}^{(\nu)}(t) \end{bmatrix} + \begin{bmatrix} A^{(\alpha-1)} & O_{\rho,\nu+1} \\ -R^{(\alpha-1)} & E^{(\nu-1)} \end{bmatrix} \begin{bmatrix} \mathbf{r}^{\alpha-1}(t) \\ \mathbf{z}^{\nu-1}(t) \end{bmatrix} =$$
$$= \begin{bmatrix} H^{(\mu)} \\ Q^{(\mu)} \end{bmatrix} \mathbf{i}^\mu(t), \ \ \mathbf{y}(t) = \mathbf{z}(t), \tag{C.17}$$

$$\mathbf{z}(t) = \mathbf{y}(t) = \mathbf{s}_{\alpha+1}(t), \ \mathbf{z}^{(k)}(t) = \mathbf{y}^{(k)}(t) = \mathbf{s}_{\alpha+k+1}(t), \ k = 0, 1, .., \nu - 1, \tag{C.18}$$
$$\mathbf{z}^{\nu-1}(t) = \mathbf{y}^{\nu-1}(t) = \mathbf{s}_O(t).$$

We continue to use the system model (C.17), (C.18) in terms of the deviations by recalling the fact that the system models (C.15), (C.16), and (C.17), (C.18) have the same properties.

Condition 70 (Section 6.1) and (9.38) (Section 9.5) permit us to transform Equation (C.17) into

$$
\begin{bmatrix} s_\alpha^{(1)}(t) \\ s_{\alpha+\nu}^{(1)}(t) \end{bmatrix} + \begin{bmatrix} A_\alpha^{-1}A^{(\alpha-1)} & O_{\rho,\nu+1} \\ -E_\nu^{-1}R^{(\alpha-1)} & E_\nu^{-1}E^{(\nu-1)} \end{bmatrix} \begin{bmatrix} r^{\alpha-1}(t) \\ z^{\nu-1}(t) \end{bmatrix} =
$$

$$
= \begin{bmatrix} A_\alpha^{-1}H^{(\mu)} \\ E_\nu^{-1}Q^{(\mu)} \end{bmatrix} i^\mu(t),
$$

$$
y(t) = \left[\overbrace{O_{N,\rho} \ \vdots \ ... \ \vdots \ O_{N,\rho}}^{O_{N,\rho} \ repeats\ \alpha-times} \underbrace{\phantom{O_{N,\rho} ... O_{N,\rho}}}_{O_{N,\alpha\rho}=C_I} \ \vdots \ I_N \ \vdots \ \overbrace{O_N \ \vdots \ ... \ \vdots \ O_N}^{O_N\ repeats\ (\nu-1)-times} \underbrace{}_{O_{(\nu-1)N}} \right] \begin{bmatrix} r^{\alpha-1}(t) \\ z^{\nu-1}(t) \end{bmatrix}.
$$

$$(C.19)$$

In view of Equation (9.38), the extended form of Equation (9.39) (Section 9.5) reads:

$$s_1 = r_1$$

$$s_2 = s_1^{(1)} = r^{(1)} \Longrightarrow s_1^{(1)} = s_2$$

$$s_3 = s_2^{(1)} = r^{(2)} \Longrightarrow s_2^{(1)} = s_3$$

$$...$$

$$s_\alpha = s_{\alpha-1}^{(1)} = r^{(\alpha-1)} \Longrightarrow s_{\alpha-1}^{(1)} = s_\alpha$$

$$s_{\alpha+1} = y$$

$$s_{\alpha+2} = s_{\alpha+1}^{(1)} = y^{(1)} \Longrightarrow s_{\alpha+1}^{(1)} = s_{\alpha+2},$$

$$s_{\alpha+3} = s_{\alpha+2}^{(1)} = y^{(2)} \Longrightarrow s_{\alpha+2}^{(1)} = s_{\alpha+3},$$

$$...$$

$$s_{\alpha+\nu} = s_{\alpha+\nu-1}^{(1)} = y^{(\nu-1)} \Longrightarrow s_{\alpha+\nu-1}^{(1)} = s_{\alpha+\nu}, \qquad (C.20)$$

Equation (C.19) determines:

$$\mathbf{s}_\alpha^{(1)}(t) = -A_\alpha^{-1}A^{(\alpha-1)}\underbrace{\left[\mathbf{s}_1^T(t)\ \mathbf{s}_2^T(t)\ ...\ \mathbf{s}_\alpha^T(t)\right]^T}_{\mathbf{s}_I(t)} + A_\alpha^{-1}H^{(\mu)}\mathbf{i}^\mu(t),$$

$$\mathbf{s}_{\alpha+\nu}^{(1)}(t) = \begin{pmatrix} E_\nu^{-1}R^{(\alpha-1)}\underbrace{\left[\mathbf{s}_1^T(t)\ \mathbf{s}_2^T(t)\ ...\ \mathbf{s}_\alpha^T(t)\right]^T}_{\mathbf{s}_I(t)} - \\ -E_\nu^{-1}E^{(\nu-1)}\underbrace{\left[\mathbf{s}_{\alpha+1}^T(t)\ \mathbf{s}_{\alpha+2}^T(t)\ ...\ \mathbf{s}_{\alpha+\nu}^T(t)\right]^T}_{\mathbf{s}_O(t)} + \\ +E_\nu^{-1}Q^{(\mu)}\mathbf{i}^\mu(t) \end{pmatrix}, \qquad \text{(C.21)}$$

$$\mathbf{s}_I(t) = \left[\mathbf{s}_1^T(t)\ \mathbf{s}_2^T(t)\ ...\ \mathbf{s}_\alpha^T(t)\right]^T \in \mathfrak{R}^{\alpha\rho}, \qquad \text{(C.22)}$$

$$\mathbf{s}_O(t) = \left[\mathbf{s}_{\alpha+1}^T(t)\ \mathbf{s}_{\alpha+2}^T(t)\ ...\ \mathbf{s}_{\alpha+\nu}^T(t)\right]^T \in \mathfrak{R}^{\nu N}. \qquad \text{(C.23)}$$

Equations (C.19) - (C.21) imply:

$$A = \begin{bmatrix} A_{11} & O_{\alpha\rho,\nu N} \\ A_{21} & A_{22} \end{bmatrix} \in \mathfrak{R}^{n\times n},\ n = \alpha\rho + \nu N, \qquad \text{(C.24)}$$

$$A_{11} = \begin{bmatrix} O_\rho & I_\rho & ... & O_\rho & O_\rho \\ O_\rho & O_\rho & ... & O_\rho & O_\rho \\ \vdots & \vdots & \vdots & \vdots & \vdots \\ O_\rho & O_\rho & ... & O_\rho & I_\rho \\ -A_\alpha^{-1}A_0 & -A_\alpha^{-1}A_1 & ... & -A_\alpha^{-1}A_{\alpha-2} & -A_\alpha^{-1}A_{\alpha-1} \end{bmatrix},$$

$$A_{11} \in \mathfrak{R}^{\alpha\rho\times\alpha\rho}, \qquad \text{(C.25)}$$

$$A_{21} = \begin{bmatrix} O_{N,\rho} & O_{N,\rho} & ... & O_{N,\rho} & O_{N,\rho} \\ O_{N,\rho} & O_{N,\rho} & ... & O_{N,\rho} & O_{N,\rho} \\ \vdots & \vdots & \vdots & \vdots & \vdots \\ E_\nu^{-1}R_{y0} & E_\nu^{-1}R_{y1} & ... & E_\nu^{-1}R_{y,\alpha-2} & E_\nu^{-1}R_{y,\alpha-1} \end{bmatrix},$$

$$A_{21} \in \mathfrak{R}^{\nu N\times\alpha\rho}, \qquad \text{(C.26)}$$

$$A_{22} = \begin{bmatrix} O_N & I_N & ... & O_N & O_N \\ O_N & O_N & ... & O_N & O_N \\ \vdots & \vdots & \vdots & \vdots & \vdots \\ O_N & O_N & ... & O_N & I_N \\ -E_\nu^{-1}E_0 & -E_\nu^{-1}E_1 & ... & -E_\nu^{-1}E_{\nu-2} & -E_\nu^{-1}E_{\nu-1} \end{bmatrix},$$

$$A_{22} \in \mathfrak{R}^{\nu N\times\nu N}, \qquad \text{(C.27)}$$

$$C = \left[\underbrace{O_{N,\alpha\rho}}_{C_I} \vdots \underbrace{I_N \vdots O_{N,(\nu-1)N}}_{C_O} \right] = \left[C_I \vdots C_O \right] \in \mathfrak{R}^{N \times n}, \qquad (C.28)$$

$$P^{(\mu)} = W = \left[\begin{array}{c} W_1 \\ W_2 \end{array} \right] \in \mathfrak{R}^{n \times (\mu+1)M}, \qquad (C.29)$$

$$W_1 = \left[\begin{array}{ccccc} O_{\rho,M} & O_{\rho,M} & \ldots & O_{\rho,M} & O_{\rho,M} \\ O_{\rho,M} & O_{\rho,M} & \ldots & O_{\rho,M} & O_{\rho,M} \\ \vdots & \vdots & \vdots & \vdots & \vdots \\ O_{\rho,M} & O_{\rho,M} & \ldots & O_{\rho,M} & O_{\rho,M} \\ A_\alpha^{-1} H_0 & A_\alpha^{-1} H_1 & \ldots & A_\alpha^{-1} H_{\mu-1} & A_\alpha^{-1} H_\mu \end{array} \right],$$

$$W_1 \in \mathfrak{R}^{\alpha\rho \times (\mu+1)M}, \qquad (C.30)$$

$$W_2 = \left[\begin{array}{ccccc} O_{N,M} & O_{N,M} & \ldots & O_{N,M} & O_{N,M} \\ O_{N,M} & O_{N,M} & \ldots & O_{N,M} & O_{N,M} \\ \vdots & \vdots & \vdots & \vdots & \vdots \\ O_{N,M} & O_{N,M} & \ldots & O_{N,M} & O_{N,M} \\ E_\nu^{-1} Q_0 & E_\nu^{-1} Q_1 & \ldots & E_\nu^{-1} Q_{\mu-1} & E_\nu^{-1} Q_\mu \end{array} \right]$$

$$W_2 \in \mathfrak{R}^{\nu N \times (\mu+1)M}, \qquad (C.31)$$

$$\mathbf{W}(t) = \mathbf{I}^\mu(t) \in \mathfrak{R}^{(\mu+1)M}, \quad \mathbf{w}(t) = \mathbf{i}^\mu(t) \in \mathfrak{R}^{(\mu+1)M}. \qquad (C.32)$$

Altogether,

$$\frac{d\mathbf{S}(t)}{dt} = A\mathbf{S}(t) + W\mathbf{W}(t) = A\mathbf{S}(t) + P^{(\mu)}\mathbf{I}^\mu(t), \qquad (C.33)$$

$$\mathbf{Y}(t) = C\mathbf{S}(t), \qquad (C.34)$$

These equations represent the *ISO* form for $\mathbf{I}^\mu(t)$ replaced by $\mathbf{W}(t)$, and *EISO* form of the *IIO* system (6.65), (6.66). In terms of the deviations of all variables which in the free regime, i.e., for $\mathbf{w}(t) \equiv \mathbf{0}_m$, Equations (C.33), (C.34) take the following form:

$$\frac{d\mathbf{s}(t)}{dt} = A\mathbf{s}(t), \qquad (C.35)$$

$$\mathbf{y}(t) = C\mathbf{s}(t). \qquad (C.36)$$

It is the system (8.11), (8.12) (Section 8.2).

Appendix D

Proofs

D.1 Proof of Lemma 97

Proof. This is the proof of Lemma 101, (Section 7.4).

For any nonzero row $1 \times \nu$ vector $\mathbf{g} = [\gamma_1\ \gamma_2 ... \gamma_\nu]$, $\mathbf{g} \in \mathfrak{R}^{1 \times \nu}$, and for the matrix function $R(.)$ (7.45) the following holds:

$$
\mathbf{g}R(t) = [\gamma_1\ \gamma_2 ... \gamma_\nu]
\begin{bmatrix}
\rho_1(t) \\
\rho_2(t) \\
\vdots \\
\rho_\nu(t)
\end{bmatrix}
= \gamma_1 \rho_1(t) + ... + \gamma_\nu \rho_\nu(t) =
$$

$$
= \gamma_1 \left[\rho_{11}(t) \quad ... \quad \rho_{1\mu}(t) \right] + .. + \gamma_\nu \left[\rho_{\nu 1}(t) \quad ... \quad \rho_{\nu\mu}(t) \right] =
$$

$$
= \left[\sum_{i=1}^{i=\nu} \gamma_i \rho_{i1}(t) \quad \sum_{i=1}^{i=\nu} \gamma_i \rho_{i2}(t) \quad ... \quad \sum_{i=1}^{i=\nu} \gamma_i \rho_{i,\mu-1}(t) \quad \sum_{i=1}^{i=\nu} \gamma_i \rho_{i\mu}(t) \right]. \quad (\text{D.1})
$$

Linear independence of the rows $\rho_i(.)$ of the matrix function $R(.)$ (7.45) on $\mathfrak{T}_0$ guarantees (Definition 98) that there exists $\tau \in \mathfrak{T}_0$ such that their linear combination (D.1) vanishes at $\tau \in \mathfrak{T}_0$:

$$
\mathbf{g}R(\tau) = \gamma_1 \rho_1(\tau) + ... + \gamma_\nu \rho_\nu(\tau) = \left[0 \quad ... \quad 0 \right] \Longleftrightarrow
$$

$$
\Longleftrightarrow \left[\sum_{i=1}^{i=\nu} \gamma_i \rho_{i1}(\tau) = 0 \quad ... \quad \sum_{i=1}^{i=\nu} \gamma_i \rho_{i\mu}(\tau) = 0 \right],
$$

if and only if all coefficients γ_i are zero:

$$\text{All } \rho_i\left(t\right) \text{ are linearly independent on } \mathfrak{T}_0 \Longleftrightarrow$$

$$\Longleftrightarrow \exists \tau \in \mathfrak{T}_0 \Longrightarrow \left(\begin{array}{c} \gamma_1 \rho_1\left(\tau\right) + ... + \gamma_\nu \rho_\nu\left(\tau\right) = \mathbf{0}_\nu^T \Longleftrightarrow \\ \Longleftrightarrow \gamma_1 = ... = \gamma_\nu = 0 \end{array} \right) \Longleftrightarrow$$

$$\Longleftrightarrow \exists \tau \in \mathfrak{T}_0 \Longrightarrow \left(\begin{array}{c} \displaystyle\sum_{i=1}^{i=\nu} \gamma_i \rho_{ik}\left(\tau\right) = 0, \ \forall k = 1, 2, ..., \mu \Longleftrightarrow \\ \Longleftrightarrow \gamma_1 = ... = \gamma_\nu = 0. \end{array} \right). \qquad (D.2)$$

This implies the following:

$$\text{All } \rho_i\left(t\right) \text{ are linearly independent on } \mathfrak{T}_0 \Longrightarrow$$

$$\forall \left(\mathbf{g} \neq \mathbf{0}_\nu^T\right) \in \mathfrak{R}^{1\times\nu}, \ \exists \sigma \in \mathfrak{T}_0 \Longrightarrow$$

$$\mathbf{g} R\left(\sigma\right) = \gamma_1 \rho_1\left(\sigma\right) + ... + \gamma_\nu \rho_\nu\left(\sigma\right) =$$

$$= \left[\sum_{i=1}^{i=\nu} \gamma_i \rho_{i1}\left(\sigma\right) \quad ... \quad \sum_{i=1}^{i=\nu} \gamma_i \rho_{i\mu}\left(\sigma\right) \right] \neq \mathbf{0}_\nu^T \Longrightarrow$$

$$\Longrightarrow \exists k \in \{1, 2, ..., \mu\} \Longrightarrow \sum_{i=1}^{i=\nu} \gamma_i \rho_{ik}\left(\sigma\right) \neq \mathbf{0}. \qquad (D.3)$$

This proves Lemma 101. ∎

D.2   Proof of Theorem 110

Lemma 230 *Linear independence and the sign of the Gram matrix quadratic form*

1) If the rows $\mathbf{r}_i\left(.\right)$ of the matrix function $R\left(.\right)$ (7.45) are linearly independent on $\mathfrak{T}_0$ then for any $1 \times \nu$ vector $\mathbf{a}$, $\mathbf{a} \in \mathfrak{R}^{1\times\nu}$, there exists $t \in \mathfrak{T}_0$ for which the integrand $R\left(t\right) R^T\left(t\right)$ of the Gram matrix (7.69) satisfies the following:

$$\forall \left(\mathbf{a} \neq \mathbf{0}_\nu^T\right) \in \mathfrak{R}^{1\times\nu}, \ \exists t \in \mathfrak{T}_0 \Longrightarrow \mathbf{a} R\left(t\right) R^T\left(t\right) \mathbf{a}^T > 0. \qquad (D.4)$$

2) There exists $t \in \mathfrak{T}_0$ such that $R\left(t\right) R^T\left(t\right)$ is positive definite at $t \in \mathfrak{T}_0$.

Proof. 1) Let the rows $\mathbf{r}_i\left(.\right)$ of the matrix function $R\left(.\right)$ (7.45), $R\left(t\right) \in \mathfrak{C}\left(\mathfrak{T}_0\right)$, be linearly independent on $\mathfrak{T}_0$. For any $1\times\nu$ vector $\mathbf{a}$, $\mathbf{a} \in \mathfrak{R}^{1\times\nu}$, there exists $t \in \mathfrak{T}_0$ for which $\mathbf{a} R\left(t\right)$ is nonzero vector, i.e., (7.65) holds (Lemma

101). The product $\mathbf{a}R(t) R^T(t) \mathbf{a}^T$ is the norm $\|R^T(t) \mathbf{a}^T\|$ of the nonzero vector $R^T(t) \mathbf{a}^T = (\mathbf{a}R(t))^T$, which is positive and proves Lemma 230.

2) By definition, a $\nu \times \nu$ matrix $M(t) = R(t) R^T(t)$ is *positive definite* at $t \in \mathfrak{T}_0$ if and only if its quadratic form $\mathbf{a}R(t) R^T(t) \mathbf{a}^T$ is positive for every nonzero vector $\mathbf{a} \in \mathfrak{R}^{1 \times \nu}$, i.e., if and only if

i) $\mathbf{a}R(t) R^T(t) \mathbf{a}^T = 0 \Longleftrightarrow \mathbf{a} = \mathbf{0}_\nu^T$, $t \in \mathfrak{T}_0$,

ii) $\mathbf{a}R(t) R^T(t) \mathbf{a}^T > 0$, $\forall \left(\mathbf{a} \neq \mathbf{0}_\nu^T\right) \in \mathfrak{R}^{1 \times \nu}$, $t \in \mathfrak{T}_0$.

This definition, the fact that $\mathbf{a} = \mathbf{0}_\nu^T$ implies $\mathbf{a}R(t) R^T(t) \mathbf{a}^T = 0$ and (D.4) prove positive definiteness of $\mathbf{a}R(t) R^T(t) \mathbf{a}^T$ at $t \in \mathfrak{T}_0$. ∎

Lemma 230 enables us to prove Theorem 114, (Section 7.4).

Proof. This is the proof of Theorem 114, (Section 7.4).

If the functions $\rho_i(.) : \mathfrak{T} \longrightarrow \mathfrak{R}$, $i = 1, 2$, are integrable on a connected nonempty nonsingleton subset $\mathfrak{T}^* = (t_a, t_b)$ of $\mathfrak{T}$, $\mathfrak{T}^* \subseteq \mathfrak{T}$, then their product $\rho_1(.)\rho_2(.)$ is also integrable on $\mathfrak{T}^*$. Let $R(.)$ (7.45) be integrable on $\mathfrak{T}_0$. Then the product $R(t) R^T(t)$ is also integrable on $\mathfrak{T}_0$, i.e., the Gram matrix (7.69) (Section 7.4) of $R(t)$ exists on $\mathfrak{T}_0$.

Necessity. Let the rows $\mathbf{r}_i(.)$ of the integrable matrix function $R(.)$ (7.45) be linearly independent on $\mathfrak{T}_0$. For any $1 \times \nu$ vector $\mathbf{a}$, $\mathbf{a} \in \mathfrak{R}^{1 \times \nu}$, there exists $t \in \mathfrak{T}_0$ for which the integrand $R(t) R^T(t)$ of the Gram matrix (7.69) satisfies (D.4) (Lemma 230), which is possible if, and only if, $R(t) R^T(t)$ is positive definite at the moment $t \in \mathfrak{T}_0$. This and the fact that $R(\tau) R^T(\tau)$ is at least positive semidefinite or positive definite at any $\tau \in \mathfrak{T}_0$ guarantee that $G(t)$ (7.69) is positive definite, hence nonsingular, on $\mathfrak{T}_0$.

That can be proved also by contradiction. Let be assumed that $G(t)$ (7.69) is singular for every $t \in \mathfrak{T}_0$ in spite of the linear independence of the rows of the matrix function $R(.)$ (7.45). There exists an $1 \times \nu$ nonzero vector $\mathbf{a}$, $\left(\mathbf{a} \neq \mathbf{0}_\nu^T\right) \in \mathfrak{R}^{1 \times \nu}$, such that for every $t \in \mathfrak{T}_0$ the product $\mathbf{a}G(t)$ is zero vector:

$$\mathbf{a}G(t) = \mathbf{0}_\nu^T, \ \forall t \in \mathfrak{T}_0 \Longrightarrow \mathbf{a}G(t) \mathbf{a}^T = 0, \ \forall t \in \mathfrak{T}_0 \Longrightarrow$$

$$\mathbf{a}\left(\int_0^t R(\tau) R^T(\tau) d\tau\right) \mathbf{a}^T = 0, \ \forall t \in \mathfrak{T}_0 \Longrightarrow$$

$$\int_0^t (\mathbf{a}R(\tau))\left(R^T(\tau) \mathbf{a}^T\right) dt = 0, \ \forall t \in \mathfrak{T}_0,$$

which is possible only if $\mathbf{a}R(t) = \mathbf{0}_\nu^T$, $\forall t \in \mathfrak{T}_0$. Since $\mathbf{a} \neq \mathbf{0}_\nu^T$ then $\mathbf{a}R(t) = \mathbf{0}_\nu^T$, $\forall t \in \mathfrak{T}_0$ implies the linear dependence of the rows of the matrix function $R(.)$ (7.45) (Lemma 100), which contradicts the linear independence of the rows of the matrix function $R(.)$ (7.45). The contradiction rejects the assumption that the Gram matrix $G(t) \in \mathfrak{R}^{\nu \times \nu}$ (7.69) is singular on $\mathfrak{T}_0$, and proves its nonsingularity on $\mathfrak{T}_0$.

Sufficiency. Let the Gram matrix $G(t) \in \mathfrak{R}^{\nu \times \nu}$ (7.69) be nonsingular on $\mathfrak{T}_0$. Let, in spite of that, the rows of the matrix function $R(.)$ (7.45) be linearly dependent. Then there exists an $1 \times \nu$ nonzero vector $\mathbf{a}$, $\left(\mathbf{a} \neq \mathbf{0}_\nu^T\right) \in \mathfrak{R}^{1 \times \nu}$, such that for every $t \in \mathfrak{T}_0$ the product $\mathbf{a}R(t)$ is zero vector (Lemma 100). We pre-multiply the Gram matrix $G(t)$ (7.69) by $\mathbf{a}$ and post-multiply it by $\mathbf{a}^T$:

$$\mathbf{a}G(t)\mathbf{a}^T = \int_0^t (\mathbf{a}R(\tau))\left(R^T(\tau)\mathbf{a}^T\right) d\tau = \mathbf{0}_\nu^T, \ \forall t \in \mathfrak{T}_0.$$

Since $G(t)$ is positive definite at some $t \in \mathfrak{T}_0$ then $\mathbf{a}G(t)\mathbf{a}^T = \mathbf{0}_\nu^T$ implies $\mathbf{a} = \mathbf{0}_\nu^T$ that contradicts $\mathbf{a} \neq \mathbf{0}_\nu^T$. The contradiction is the consequence of the assumed linear dependence of the rows of the matrix function $R(.)$ (7.45) on $\mathfrak{T}_0$. The contradiction rejects the assumption and proves that the rows of the matrix function $R(.)$ (7.45) are linearly independent on $\mathfrak{T}_0$.

We can prove the sufficiency also as follows. Let, in spite of the nonsingularity of $G(t)$ on $\mathfrak{T}_0$, the rows of the matrix function $R(.)$ (7.45) be linearly dependent. Then there exists an $1 \times \nu$ nonzero vector $\mathbf{a}$, $\left(\mathbf{a} \neq \mathbf{0}_\nu^T\right) \in \mathfrak{R}^{1 \times \nu}$, such that for every $t \in \mathfrak{T}_0$ the product $\mathbf{a}R(t)$ is zero vector (Lemma 100). We premuliply the Gram matrix $G(t)$ (7.69) by $\mathbf{a}$:

$$\mathbf{a}G(t) = \int_0^t (\mathbf{a}R(\tau)) R^T(\tau) d\tau = \mathbf{0}_\nu^T, \ \forall t \in \mathfrak{T}_0.$$

The nonsingularity of $G(t)$ on $\mathfrak{T}_0$ implies the existence of its inverse $G^{-1}(t)$ on $\mathfrak{T}_0$. The multiplication of the preceding equation on the right-hand side by $G^{-1}(t)$ results in:

$$\mathbf{a}G(t)G^{-1}(t) = \mathbf{a} = \mathbf{0}_\nu^T, \ \forall t \in \mathfrak{T}_0.$$

The result $\mathbf{a} = \mathbf{0}_\nu^T$ contradicts the chosen $\mathbf{a} \neq \mathbf{0}_\nu^T$. The contradiction disproves the assumption on the linear dependence of the rows of the matrix function $R(.)$ (7.45) on $\mathfrak{T}_0$, which proves that the rows of the matrix function $R(.)$ (7.45) are linearly independent on $\mathfrak{T}_0$. $\blacksquare$

D.3   Proof of Theorem 121

Definition 231 *Analytic function* $\mathbf{r}_i(.) : \mathfrak{T} \longrightarrow \mathfrak{R}^{1 \times \mu}$

*A function $\mathbf{r}_i(.) : \mathfrak{T} \longrightarrow \mathfrak{R}^{1 \times \mu}$ is **analytic on** $[t_0, t_1]$, $t_1 \in In\mathfrak{T}$, if, and only if:*

1. $\mathbf{r}_i(.)$ is infinitely times differentiable on $[t_0, t_1]$: $\mathbf{r}_i(t) \in C^\infty([t_0, t_1])$.

2. $\mathbf{r}_i(t)$ can be represented by the Taylor series on the ε-neighborhood of t_0 for some $\varepsilon \in \mathfrak{R}^+$,

$$\mathbf{r}_i(t) = \sum_{k=0}^{k=\infty} \frac{(t - t_0)^k}{k!} \mathbf{r}_i^{(k)}(t_0), \ \forall t \in [t_0 - \varepsilon, t_0 + \varepsilon].$$

The following theorem is helpful to prove observability and controllability criteria.

Proof. This is the proof of Theorem 131.

As explained in Section 8.1 the system observability does not depend on the input vector. The system can be considered in the free regime in terms of the variables deviations. It is described by (8.11), (8.12) repeated as

$$\frac{d\mathbf{s}(t)}{dt} = A\mathbf{s}(t), \tag{D.5}$$

$$\mathbf{y}(t) = C\mathbf{s}(t). \tag{D.6}$$

Its motion $\mathbf{s}(t; \mathbf{s}_0; \mathbf{0}_M)$ and its output response $\mathbf{y}(t; \mathbf{s}_0; \mathbf{0}_M)$ have the following well-known forms (8.13), (8.14):

$$\mathbf{s}(t; \mathbf{s}_0; \mathbf{0}_M) = e^{At}\mathbf{s}_0 = \Phi(t)\mathbf{s}_0, \tag{D.7}$$

$$\mathbf{y}(t; \mathbf{s}_0; \mathbf{0}_M) = Ce^{At}\mathbf{s}_0 = C\Phi(t)\mathbf{s}_0. \tag{D.8}$$

Necessity. Let the system (8.11), (8.12), i.e., (D.5), (D.6), be observable. Definition 130 holds. Let the output response $\mathbf{y}(t; \mathbf{s}_0; \mathbf{0}_M)$ be known for every $t \in \mathfrak{T}_0$.

1. We multiply (D.8) by $\Phi^T(t) C^T$ on the left, integrate the product from 0 to any $t \in In\mathfrak{T}_0$ and use (8.15):

$$\int_0^{t \in In\mathfrak{T}_0} \Phi^T(t) C^T \mathbf{y}(t; \mathbf{s}_0; \mathbf{0}_M) \, dt = \underbrace{\left(\int_0^{t \in In\mathfrak{T}_0} \Phi^T(t) C^T C \Phi(t) \, dt \right)}_{G_{OB}(t)} \mathbf{s}_0 \Longrightarrow$$

$$\int_0^{t \in In\mathfrak{T}_0} \Phi^T(\tau) C^T \mathbf{y}(\tau; \mathbf{s}_0; \mathbf{0}_M) \, d\tau = G_{OB}(t) \mathbf{s}_0.$$

The left-hand side of this equation is known due to the knowledge of the system response $\mathbf{y}\,(t; \mathbf{s}_0; \mathbf{0}_M)$, the matrices C and $\Phi\,(t)$. The system observability means that the left-hand side of that equation uniquely determines the initial state $\mathbf{s}_0$, which implies the nonsingularity of $G_{OB}\,(t)$ on $[0,t]$ for any $t \in In\mathfrak{T}_0$. This proves the necessity of the condition 1.

2. The necessity of the nonsingularity of $G_{OB}\,(t)$ and Theorem 114 imply the linear independence of the rows of $\Phi^T\,(t)\,C^T$ that are the columns of $C\Phi\,(t)$ on $[0,t]$ for some $t \in In\mathfrak{T}_0$. This proves the necessity of the condition 2.

3. The matrix $C(sI_N - A)^{-1}$ is the Laplace transform of $C\Phi\,(t, t_0)$. This, the linearity of the Laplace transform (which does not influence the rank) of the matrix, and the necessity of the condition under 2. imply the necessity of the condition under 3.

4. The fundamental matrix $\Phi\,(t) = e^{At}$ is analytic, i.e., it is infinitely times differentiable and can be represented by the Taylor series (Definition 231), which, together with the Cayley-Hamilton theorem, permits us to express $\Phi\,(t)$ in terms of some continuously differentiable linearly independent functions $\alpha_k\,(.) : \mathfrak{T} \longrightarrow \mathfrak{R}, \forall k = 0, 1, ..., n - 1$, and the powers A^k of the matrix A up to $k = n - 1$ [2, pp. 124, 125], [79, p. 489]:

$$\Phi\,(t) = e^{At} = \sum_{k=0}^{k=n-1} \alpha_k\,(t)\,A^k. \tag{D.9}$$

Let

$$\begin{aligned} H_k &= \left(CA^k\right)^T = A^{k^T} C^T \in \mathfrak{R}^{n \times N}, \ i.e., \\ H_k^T &= CA^k \in \mathfrak{R}^{N \times n}, \forall k = 0, 1, ..., n - 1. \end{aligned} \tag{D.10}$$

Equations (D.6) and (D.9) yield

$$\mathbf{y}(t) = \mathbf{y}\,(t; \mathbf{s}_0; \mathbf{0}_M) = \sum_{k=0}^{k=n-1} \alpha_k\,(t)\,H_k^T \mathbf{s}_0 = \sum_{k=0}^{k=n-1} \alpha_k\,(t)\,I_N H_k^T \mathbf{s}_0 \in \mathfrak{R}^N, \tag{D.11}$$

or, equivalently:

$$\mathbf{y}(t) = \left[\alpha_0\,(t)\,I_N \vdots \alpha_1\,(t)\,I_N \vdots ... \vdots \alpha_{n-1}\,(t)\,I_N\right] \begin{bmatrix} H_0^T \mathbf{s}_0 \\ H_1^T \mathbf{s}_0 \\ ... \\ H_{n-1}^T \mathbf{s}_0 \end{bmatrix} \in \mathfrak{R}^N. \tag{D.12}$$

Let

$$\Lambda(t) = \begin{bmatrix} \alpha_0(t)\, I_N \\ \alpha_1(t)\, I_N \\ ... \\ \alpha_{n-1}(t)\, I_N \end{bmatrix} \in \mathfrak{R}^{nN \times N} \Longrightarrow$$

$$\underbrace{}_{\Lambda(t)}$$

$$\Lambda^T(t) = \begin{bmatrix} \alpha_0(t)\, I_N & \vdots & \alpha_1(t)\, I_N & \vdots & ... & \vdots & \alpha_{n-1}(t)\, I_N \end{bmatrix} \in \mathfrak{R}^{N \times nN} \qquad (D.13)$$

This sets Equation (D.12) into

$$\mathbf{y}(t) = \Lambda^T(t) \begin{bmatrix} H_0^T \mathbf{s}_0 \\ H_1^T \mathbf{s}_0 \\ ... \\ H_{n-1}^T \mathbf{s}_0 \end{bmatrix} \in \mathfrak{R}^N. \qquad (D.14)$$

Let this equation be premultiplied by $\Lambda(t)$ and then be integrated from 0 to t:

$$\int_0^t \Lambda(\tau)\Lambda^T(\tau)\, d\tau \begin{bmatrix} H_0^T \mathbf{s}_0 \\ H_1^T \mathbf{s}_0 \\ ... \\ H_{n-1}^T \mathbf{s}_0 \end{bmatrix} = \int_0^t \Lambda(\tau)\mathbf{y}(\tau)d\tau. \qquad (D.15)$$

The linear independence of the functions $\alpha_k(.) : \mathfrak{T} \longrightarrow \mathfrak{R}$, $\forall k = 0, 1, ..., n - 1$, and Theorem 114, (Section 7.4), imply the nonsingularity of the Gram matrix $G_\alpha(t) \in \mathfrak{R}^{nN \times nN}$ of the functions $\alpha_k(.)\, I_N : \mathfrak{T} \longrightarrow \mathfrak{R}^{N \times N}$, $\forall k = 0, 1, ..., n - 1$:

$$G_\alpha(t) = \int_0^t \Lambda(\tau)\Lambda^T(\tau)\, d\tau, \ detG_\alpha(t) \neq 0. \qquad (D.16)$$

This permits to set (D.15) into

$$G_\alpha(t) \begin{bmatrix} H_0^T \mathbf{s}_0 \\ H_1^T \mathbf{s}_0 \\ ... \\ H_{n-1}^T \mathbf{s}_0 \end{bmatrix} = \int_0^t \Lambda(\tau)\mathbf{y}(\tau)d\tau$$

or, after multiplying this equation on the left by $G_\alpha^{-1}(t, t_0)$,

$$
\begin{bmatrix} H_0^T \mathbf{s}_0 \\ H_1^T \mathbf{s}_0 \\ \cdots \\ H_{n-1}^T \mathbf{s}_0 \end{bmatrix} = \begin{bmatrix} H_0^T \\ H_1^T \\ \cdots \\ H_{n-1}^T \end{bmatrix} \mathbf{s}_0 = \begin{bmatrix} H_0 \vdots H_1 \vdots \cdots \vdots H_{n-1} \end{bmatrix}^T \mathbf{s}_0 =
$$

$$
= \quad G_\alpha^{-1}(t) \int_0^t \Lambda(\tau) \mathbf{y}(\tau) d\tau. \tag{D.17}
$$

In view of the necessity of the condition 1, Equation (8.15), and in view of Equations (D.9)-(D.14), the observability of the system implies the the unique solution $\mathbf{s}_0$ of Equation (D.17) that is possible if and only if:

$$
rank \begin{bmatrix} H_0 \vdots H_1 \vdots \cdots \vdots H_{n-1} \end{bmatrix}^T = n, \tag{D.18}
$$

or, equivalently, due to $H_k = \left(CA^k\right)^T$, $\forall k = 0, 1, ..., n-1$:

$$
rank \begin{bmatrix} C^T \vdots (CA)^T \vdots \cdots \vdots \left(CA^{n-1}\right)^T \end{bmatrix}^T = rank O_{OBS} = n. \tag{D.19}
$$

This proves the necessity of the condition 4.

5. In order to prove the necessity of the condition 5) let us first prove the following: ∎

Lemma 232 *Initial state and zero response of the observable system*

If the system (8.1), (8.12), equivalently, the system (D.5), (D.6), is observable then

$$
\forall t \in In\mathfrak{T}_0, \ \left(\mathbf{y}(\sigma; \mathbf{s}_0; \mathbf{0}_m) = Ce^{A\sigma}\mathbf{s}_0 = \mathbf{0}_N, \ \forall \sigma \in [0, t]\right) \Longleftrightarrow \mathbf{s}_0 = \mathbf{0}_n, \tag{D.20}
$$

$$
\left(\mathbf{y}(s) = C\left(sI_n - A\right)^{-1}\mathbf{s}_0 = \mathbf{0}_N, \ on \ \mathbb{C}\right) \Longleftrightarrow \mathbf{s}_0 = \mathbf{0}_n. \tag{D.21}
$$

Proof. This is the proof of Lemma 232.

Let the system (8.1), (8.12), equivalently, the system (D.5), (D.6), be observable. All columns of $C\Phi(t)$ are linearly independent on [0,t] for any $t \in In\mathfrak{T}_0$, i.e., on $\mathfrak{T}_0$, due to the necessity of the condition 2, and all columns of $C(sI_N - A)^{-1}$ are linearly independent on $\mathbb{C}$ due to the necessity of the condition 3. This means that the rows of $[C\Phi(t)]^T$ are linearly independent on $[0,t]$ for any $t \in In\mathfrak{T}_0$ and rows of $\left[(sI_n - A)^{-1}\right]^T C^T$ are linearly independent on $\mathbb{C}$. These facts and Lemma 101 (Section 7.4) guarantee that for any nonzero initial state vector $s_0 \in R^n$ the following statements hold:

$$\forall (s_0 \neq \mathbf{0}_n) \in \mathfrak{R}^n \Longrightarrow$$

$$\Longrightarrow \left\{ \begin{array}{l} \forall t \in \mathfrak{T}_0,\ \exists \tau \in [0,t] \Longrightarrow \mathbf{y}(\tau; s_0; \mathbf{0}_m) = C\Phi(\tau)\, s_0 \neq \mathbf{0}_N, \\ \forall (s_0 \neq \mathbf{0}_N) \in \mathbb{C},\ \exists s \in \mathbb{C} \Longrightarrow \mathbf{y}(s) = C\, (sI_n - A)^{-1}\, s_0 \neq \mathbf{0}_N, \end{array} \right\}.$$

These statements are equivalent to the statements (D.20) and (D.21) of the Lemma. **This completes the proof of the Lemma.** ∎

Let us continue the proof of Theorem 131.

Proof. We continue with the proof of the necessity of the condition 5) of Theorem 131. The system observability and the explanation in Section 8.1 permit us to apply the Laplace transform to the system (D.5), (D.6). The result is

$$(sI_n - A)\,\mathbf{s}(s) = s_0 \tag{D.22}$$

$$C\mathbf{s}(s) = \mathbf{y}(s). \tag{D.23}$$

i.e.,

$$\left[\begin{array}{c} sI_n - A \\ C \end{array} \right] \mathbf{s}(s) = \left[\begin{array}{c} s_0 \\ \mathbf{y}(s) \end{array} \right]. \tag{D.24}$$

From Equation (8.13) follows

$$\mathbf{s}(t; s_0; \mathbf{0}_M) = e^{At} s_0 = \mathbf{0}_n,\ \forall t \in \mathfrak{T}_0 \Longleftrightarrow s_0 = \mathbf{0}_n, \tag{D.25}$$

which implies

$$\mathbf{s}(s) = \mathbf{s}(s; s_0; \mathbf{0}_M) = \mathbf{0}_n,\ \forall s \in \mathbb{C} \Longleftrightarrow s_0 = \mathbf{0}_n. \tag{D.26}$$

Equation (D.25) and Equation (D.21) (Lemma 232) show that the system response $\mathbf{y}(t; s_0; \mathbf{0}_m)$ and its Laplace transform $\mathbf{y}(s)$ are equal to the zero vector for all $t \in \mathfrak{T}_0$ and for all $s \in \mathbb{C}$, respectively, if and only if the initial state vector is the zero vector, $s_0 = \mathbf{0}_n$, which is, therefore, observable. Consequently we should analyze the observability of any nonzero zero initial state vector $s_0 \neq \mathbf{0}_n$ for which the system response $\mathbf{y}(t; s_0; \mathbf{0}_m)$ and its Laplace transform $\mathbf{y}(s)$ are different from the zero vector for some $t \in \mathfrak{T}_0$ and $s \in \mathbb{C}$, respectively, due to Lemma 232. We accept that there is $t \in \mathfrak{T}_0$

at which $\mathbf{y}(t;\mathbf{s}_0;\mathbf{0}_m) \neq \mathbf{0}_N$ and $\mathbf{y}(s) \neq \mathbf{0}_N$ for some $s \in \mathbb{C}$. The equations
(D.22), (D.23) show, equivalently the equation (D.24) shows, that for the
unique determination of the nonzero initial state vector $\mathbf{s}_0$ from the nonzero
output response it is necessary that the state $\mathbf{s}(t;\mathbf{s}_0;\mathbf{0}_m)$ and its Laplace
transform are also uniquely determined. The system observability guarantees
that the nonzero initial state $\mathbf{s}_0$ is uniquely determined from the nonzero
output response, hence the state $\mathbf{s}(t;\mathbf{s}_0;\mathbf{0}_m)$ and its Laplace transform are
uniquely determined from the nonzero output response. This implies the full
rank of the matrix

$$\left[\begin{array}{c} sI_n - A \\ C \end{array} \right] \in \mathbb{C}^{(n+N)\times n}$$

in the equation (D.24), which is n because the number n of its columns is
less than the number $n + N$ of its rows:

$$rank \left[\begin{array}{c} sI_n - A \\ C \end{array} \right] = n, \ \forall s \in \mathbb{C}^{(n+N)\times n}.$$

This proves the necessity of the condition 5.

Sufficiency. Let all the conditions of the theorem statement be valid.

1. Let $\mathbf{s}_0 \in \mathfrak{R}^n$ be arbitrarily chosen. The multiplication of the equation
(8.13) or the equation (D.7) by $\Phi^T(t) C^T$ on the left transforms them into

$$\Phi^T(t) C^T \mathbf{y}(t;\mathbf{s}_0;\mathbf{0}_m) = \Phi^T(t) C^T C\Phi(t) \mathbf{s}_0.$$

The integral of this equation on the time interval $[0, t]$ on which the observability Gram matrix $G_{OB}(t)$ (8.15) is nonsingular due to the condition 1.,
gives:

$$\int_0^t \Phi^T(\tau) C^T \mathbf{y}(\tau;\mathbf{s}_0;\mathbf{0}_m) d\tau = \int_0^t \Phi^T(\tau) C^T C\Phi(\tau) d\tau \mathbf{s}_0 =$$

$$= G_{OB}(t) \mathbf{s}_0.$$

The nonsingularity of $G_{OB}(t)$ permits to multiply the preceding equation
on the left by its inverse $G_{OB}^{-1}(t)$:

$$\mathbf{s}_0 = G_{OB}^{-1}(t) \int_0^t \Phi^T(\tau) C^T \mathbf{y}(\tau;\mathbf{s}_0;\mathbf{0}_m) d\tau, \ \forall \mathbf{s}_0 \in \mathfrak{R}^n.$$

Every initial state $\mathbf{s}_0 \in \mathfrak{R}^n$ is uniquely determined by the system response
$\mathbf{y}(t;\mathbf{s}_0;\mathbf{0}_m)$. The system (8.1), (8.2) is observable in view of Definition 130.

2.   The linear independence of all columns of $C\Phi\left(t\right)$ on $[0,t]$ for some $t \in In\mathfrak{T}_0$ ensures the linear independence of all rows of $\Phi^T\left(t\right)C^T$ on $[0,t]$ for some $t \in In\mathfrak{T}_0$, which implies the nonsingularity of the observability Gram matrix $G_{OB}\left(t\right)$ due to Theorem 114. This satisfies the condition 1. The system (8.1), (8.2) is observable in view of Definition 130.

3. The application of the inverse Laplace transform to $C(sI_N - A)^{-1}$ preserves the linear independence of its columns to the columns of its original $C\Phi\left(t\right)$. The condition 3 guarantees the validity of the condition 2. The system is observable due to the condition 2.

4. In view of (D.10),

$$H_k = \left(CA^k\right)^T = A^{k^T}C^T \in \mathfrak{R}^{n\times N}, \ i.e.,$$

$$H_k^T = CA^k \in \mathfrak{R}^{N\times n}, \forall k = 0, 1, ..., n-1, \qquad (D.27)$$

and (8.16) we conclude that $H = \left[H_0 \vdots H_1 \vdots ... \vdots H_{n-1}\right]^T \in \mathfrak{R}^{nN\times n}$ is O_{OBS} :

$$H = \begin{bmatrix} H_0^T \\ H_1^T \\ \vdots \\ H_{n-1}^T \end{bmatrix} = \begin{bmatrix} C \\ CA \\ CA^2 \\ ... \\ CA^{n-1} \end{bmatrix} = O_{OBS} \in \mathfrak{R}^{nN\times n}.$$

This permits us to set Equation (D.17) into the following form:

$$O_{OBS}\mathbf{s}_0 = G_\alpha^{-1}\left(t\right)\int_0^t \Lambda\left(\tau\right)\mathbf{y}(\tau)d\tau. \qquad (D.28)$$

Since the number nN of the rows of O_{OBS} is not less than the number n of its columns then the condition (8.17) means that the matrix O_{OBS} has the full rank n. The matrix $O_{OBS}^T O_{OBS} \in \mathfrak{R}^{n\times n}$ is nonsingular. It forms the left inverse

$$O_{OBS}^+ = \left(O_{OBS}^T O_{OBS}\right)^{-1} O_{OBS}^T \in \mathfrak{R}^{n\times nN}$$

of the matrix O_{OBS}. It has the following property:

$$O_{OBS}^+ O_{OBS} = I. \qquad (D.29)$$

We premultiply (D.28) by O_{OBS}^+,

$$O_{OBS}^+ O_{OBS}\mathbf{s}_0 = O_{OBS}^+ G_\alpha^{-1}\left(t\right)\int_0^t \Lambda\left(\tau\right)\mathbf{y}(\tau)d\tau,$$

and apply (D.29) to obtain:

$$\mathbf{s}_0 = O_{OBS}^{+} G_\alpha^{-1}(t) \int_0^t \Lambda(\tau) \mathbf{y}(\tau) d\tau.$$

The initial state vector $\mathbf{s}_0$ is uniquely determined by the system response

$$\mathbf{y}(t) = \mathbf{y}(t; \mathbf{s}_0; \mathbf{0}_M).$$

The initial state vector $\mathbf{s}_0$ was arbitrarily chosen, (it can be the zero initial state vector $\mathbf{s}_0 = \mathbf{0}_n$ that is observable). The system (8.1), (8.2) is observable in view of Definition 130.

5. Let the equations (D.22), (D.23) be set into the following form:

$$(sI_n - A)\,\mathbf{s}(s) - I_n\mathbf{s}_0 = O_n \tag{D.30}$$
$$C\mathbf{s}(s) + O_{N,n}\mathbf{s}_0 = \mathbf{y}(s), \tag{D.31}$$

or

$$\left[\begin{array}{c} sI_n - A \\ C \end{array} \right] \mathbf{s}(s) = \left[\begin{array}{c} \mathbf{s}_0 \\ \mathbf{y}(s) \end{array} \right] \tag{D.32}$$

The first form of the proof of the sufficiency of the condition under 5.

The condition under 5. of the theorem statement, i.e., (8.19), implies that the matrix $O_O(s)$ (8.18),

$$O_O(s) = \left[\begin{array}{c} sI_n - A \\ C \end{array} \right] \in \mathbb{C}^{(n+N) \times n}, \tag{D.33}$$

has the full rank n,

$$rank O_O(s) = n, \ \forall s \in \mathbb{C}, \tag{D.34}$$

due to the general rule:

$$rank O_O(s) \leq min(n, n+N) = n.$$

The same holds for its transpose $O_O^T(s)$,

$$O_O^T(s) = \left[\begin{array}{cc} sI_n - A^T & C^T \end{array} \right] \in \mathbb{C}^{n \times (n+N)},$$
$$rank O_O^T(s) = n, \ \forall s \in \mathbb{C}. \tag{D.35}$$

This and (D.34) imply the nonsingularity of the $n \times n$ matrix $O_O^T(s) O_O(s)$,

$$O_O^T(s) O_O(s) = \begin{bmatrix} sI_n - A^T & C^T \end{bmatrix} \begin{bmatrix} sI_n - A \\ C \end{bmatrix} =$$

$$= \left[(sI_n - A^T)(sI_n - A) + C^T C \right] \in \mathbb{C}^{n \times n},$$

$$\det \left(O_{OO}^T(s) O_O(s) \right) \neq 0, \ \forall s \in \mathbb{C}.$$

The left inverse $O_{OL}^+(s)$ of $O_O(s)$ reads

$$O_{OL}^+(s) = \left(O_{OO}^T(s) O_O(s) \right)^{-1} O_{OO}^T(s), \tag{D.36}$$

so that:

$$O_{OL}^+(s) O_O(s) = I, \ \forall s \in \mathbb{C}, \tag{D.37}$$

and

$$O_{OL}^+(s) =$$
$$\left[(sI_n - A^T)(sI_n - A) + C^T C \right]^{-1} \begin{bmatrix} sI_n - A^T & C^T \end{bmatrix} =$$
$$= \begin{bmatrix} G_{11}(s) & G_{12}(s) \end{bmatrix}, \tag{D.38}$$

where

$$G_{11}(s) = \left[(sI_n - A^T)(sI_n - A) + C^T C \right]^{-1} (sI_n - A^T) \in \mathbb{C}^{n \times n}, \tag{D.39}$$

$$G_{12}(s) = \left[(sI_n - A^T)(sI_n - A) + C^T C \right]^{-1} C^T \in \mathbb{C}^{n \times N}. \tag{D.40}$$

We premultiply (D.32) by $O_{OL}^+(s)$ and use (D.37)-(D.40):

$$\mathbf{s}(s) = O_{OL}^+(s) \begin{bmatrix} \mathbf{s}_0 \\ \mathbf{y}(s) \end{bmatrix} = \begin{bmatrix} G_{11}(s) & G_{12}(s) \end{bmatrix} \begin{bmatrix} \mathbf{s}_0 \\ \mathbf{y}(s) \end{bmatrix}.$$

This and (D.30) furnish

$$(sI - A)\mathbf{s}(s) = \mathbf{s}_0 = (sI_n - A) G_{11}(s) \mathbf{s}_0 + (sI_n - A) G_{12}(s) \mathbf{y}(s),$$

or

$$\left[I - (sI - A) G_{11}(s) \right] \mathbf{s}_0 = (sI_n - A) G_{12}(s) \mathbf{y}(s), \ \forall \mathbf{s}_0 \in \mathbb{C}^n. \tag{D.41}$$

The eigenvalues λ_i of the matrix $(sI_n - A) G_{11}(s)$ depend on s, $\lambda_i = \lambda_i(s)$, and are determined as solutions of

$$\det \left[\lambda I_n - (sI_n - A) G_{11}(s) \right] = 0.$$

Let s^* be such that $\lambda_i = \lambda_i(s^*) \neq 1$, $\forall i = 1, 2, .., n$. Then,

$$\det\left[I - (s^*I - A)G_{11}(s^*)\right] \neq 0$$

so that (D.41) becomes

$$\mathbf{s}_0 = \left[I - (s^*I - A)G_{11}(s^*)\right]^{-1}(s^*I - A)G_{12}(s^*)\mathbf{y}(s^*).$$

Every initial state $\mathbf{s}_0$ is uniquely determined by the Laplace transform $\mathbf{y}(s)$ of the system response at $s = s^*$. The system is observable (Definition 130).

The second form of the proof of the sufficiency of the condition under 5.
In order to determine $\mathbf{s}(s)$ from (D.32), i.e., from

$$\begin{bmatrix} sI_n - A \\ C \end{bmatrix}\mathbf{s}(s) = O_O(s)\mathbf{s}(s) = \begin{bmatrix} \mathbf{s}_0 \\ \mathbf{y}(s) \end{bmatrix} \tag{D.42}$$

we will transform the matrix $O_O(s)$ into the equivalent matrix by introducing two nonsingular matrices $P(s)$ and $Q(s)$ defined as follows:

$$\exists P(s) \in \mathbb{C}^{(n+N)\times(n+N)}, \ detP(s) \neq 0, \forall s \in \mathbb{C},$$

$$P(s) = \begin{bmatrix} P_{11}(s) & P_{12}(s) \\ P_{21}(s) & P_{22}(s) \end{bmatrix},$$

$$P_{11}(s) \in \mathbb{C}^{n\times n}, \ detP_{11}(s) \neq 0, \forall s \in \mathbb{C},$$

$$P_{12}(s) \in \mathbb{C}^{n\times N}, \ P_{21}(s) \in \mathbb{C}^{N\times n},$$

$$P_{22}(s) \in \mathbb{C}^{N\times N}, \ detP_{22}(s) \neq 0, \forall s \in \mathbb{C}, \tag{D.43}$$

and

$$\exists Q(s) \in \mathbb{C}^{n\times n}, \ detQ(s) \neq 0, \forall s \in \mathbb{C} \Longrightarrow \tag{D.44}$$

$$P(s)\begin{bmatrix} sI - A \\ C \end{bmatrix}Q(s) = \begin{bmatrix} D(s) \\ O_{N,n} \end{bmatrix},$$

$$D(s) \in \mathbb{C}^{n\times n}, \ detD(s) \neq 0, \forall s \in \mathbb{C}.$$

This and (D.42) imply

$$P(s)\begin{bmatrix} sI - A \\ C \end{bmatrix}Q(s)Q^{-1}(s)\mathbf{s}(s) = P(s)\begin{bmatrix} \mathbf{s}_0 \\ \mathbf{y}(s) \end{bmatrix} \Longrightarrow$$

$$\begin{bmatrix} D(s) \\ O_{N,n} \end{bmatrix}Q^{-1}(s)\mathbf{s}(s) = P(s)\begin{bmatrix} \mathbf{s}_0 \\ \mathbf{y}(s) \end{bmatrix} \Longrightarrow$$

$$\left[D^T \left(s \right) \vdots O_{n,N} \right] \left[\begin{array}{c} D \left(s \right) \\ O_{N,n} \end{array} \right] Q^{-1} \left(s \right) \mathbf{s}(s) = D^T \left(s \right) D \left(s \right) Q^{-1} \left(s \right) \mathbf{s}(s) =$$

$$= \left[D^T \left(s \right) \vdots O_{n,N} \right] P \left(s \right) \left[\begin{array}{c} \mathbf{s}_0 \\ \mathbf{y} \left(s \right) \end{array} \right] \Longrightarrow$$

$$\mathbf{s}(s) = Q \left(s \right) D^{-1} \left(s \right) \left(D^T \left(s \right) \right)^{-1} \left[D^T \left(s \right) \vdots O_{n,N} \right] P \left(s \right) \left[\begin{array}{c} \mathbf{s}_0 \\ \mathbf{y} \left(s \right) \end{array} \right] \Longrightarrow$$

$$\mathbf{s}_0 = \left(sI - A \right) \mathbf{s}(s) = \left(sI - A \right) Q \left(s \right) D^{-1} \left(s \right) \left[I \vdots O_{n,N} \right] \cdot$$

$$\cdot \left[\begin{array}{cc} P_{11} \left(s \right) & P_{12} \left(s \right) \\ P_{21} \left(s \right) & P_{22} \left(s \right) \end{array} \right] \left[\begin{array}{c} \mathbf{s}_0 \\ \mathbf{y} \left(s \right) \end{array} \right] \Longrightarrow$$

$$\mathbf{s}_0 = \left(sI - A \right) Q \left(s \right) D^{-1} \left(s \right) \left[I \vdots O_{n,N} \right] \cdot$$

$$\cdot \left[\begin{array}{c} P_{11} \left(s \right) \mathbf{s}_0 + P_{12} \left(s \right) \mathbf{y} \left(s \right) \\ P_{21} \left(s \right) \mathbf{s}_0 + P_{22} \left(s \right) \mathbf{y} \left(s \right) \end{array} \right] \Longrightarrow$$

$$\mathbf{s}_0 =$$
$$= \left(sI - A \right) Q \left(s \right) D^{-1} \left(s \right) P_{11} \left(s \right) \mathbf{s}_0 + \left(sI - A \right) Q \left(s \right) D^{-1} \left(s \right) P_{12} \left(s \right) \mathbf{y} \left(s \right)$$

$$\left[I - \left(sI - A \right) Q \left(s \right) D^{-1} \left(s \right) P_{11} \left(s \right) \right] \mathbf{s}_0 =$$
$$= \left(sI - A \right) Q \left(s \right) D^{-1} \left(s \right) P_{12} \left(s \right) \mathbf{y} \left(s \right). \tag{D.45}$$

The eigenvalues λ_k of the matrix $\left(sI - A \right) Q \left(s \right) D^{-1} \left(s \right) P_{11} \left(s \right)$ depend on $s, \lambda_k = \lambda_k \left(s \right)$. They are determined as solutions of

$$\det \left[\lambda I - \left(sI - A \right) Q \left(s \right) D^{-1} \left(s \right) P_{11} \left(s \right) \right] = 0.$$

Let s^* be such that $\lambda_k = \lambda_k \left(s^* \right) \neq 1, \forall i = 1, 2, .., n$. Then,

$$\det \left[I - \left(s^* I - A \right) Q \left(s^* \right) D^{-1} \left(s^* \right) P_{11} \left(s^* \right) \right] \neq 0$$

so that (D.45) becomes

$$\mathbf{s}_0 = \left[I - \left(s^* I - A \right) Q \left(s^* \right) D^{-1} \left(s^* \right) P_{11} \left(s^* \right) \right]^{-1} \cdot$$
$$\cdot \left(s^* I - A \right) Q \left(s^* \right) D^{-1} \left(s^* \right) P_{12} \left(s^* \right) \mathbf{y} \left(s^* \right).$$

Every initial state $\mathbf{s}_0$ is uniquely determined by the Laplace transform $\mathbf{y} \left(s \right)$ of the system response at $s = s^*$. The system is observable (Definition 130). This completes the proof. ■

D.4   Proof of Theorem 128

Proof. This is the proof of Theorem 138, (Section 8.4).

This theorem can be proved by using *the Lyapunov method* or by following the Bellman *application* of *the matrix infinitesimal calculus.*

Proof by using the Lyapunov method

Necessity. Let the matrix A be stable, equivalently, let the zero equilibrium state $\mathbf{s}_e = \mathbf{0}_n$ of the system (8.30), (8.31), (Section 8.1), be (globally) asymptotically stable. Then the zero equilibrium state $\mathbf{s}_e = \mathbf{0}_n$ is unique. Let a function $v\,(.) : \mathfrak{R}^n \longrightarrow \mathfrak{R}^n$ be the quadratic form of the symmetric matrix H, $H = H^T$,

$$v\,(\mathbf{s}) = \mathbf{s}^T H \mathbf{s}, \tag{D.46}$$

where the matrix   is the matrix solution of Equation (8.35). Let $q\,(\mathbf{s}) = -\mathbf{s}^T C^T C \mathbf{s}$.Observability of the system guarantees (the condition 2. of Theorem 131 and Lemma 232 in Appendix D.3) that along the system motions $\mathbf{s}\,(t; \mathbf{s}_0)$,

$$\forall t \in In\mathfrak{T}_0, \ \left(\mathbf{y}(\sigma; \mathbf{s}_0; \mathbf{0}_m) = Ce^{A\sigma}\mathbf{s}_0 = \mathbf{0}_N, \ \forall \sigma \in [0, t]\right) \Longleftrightarrow \mathbf{s}_0 = \mathbf{0}_n.$$

Let us assume that for any $\mathbf{s}_0 \neq \mathbf{0}_n$ there is $\theta \in In\mathfrak{T}_0$ such that $Ce^{A\theta}\mathbf{s}_0 = \mathbf{0}_N$. Let θ be the first moment that obeys $Ce^{A\theta}\mathbf{s}_0 = \mathbf{0}_N$. This means that $Ce^{A\theta}\mathbf{s}_0$ is in the hyperplane $C\mathbf{s} = \mathbf{0}_N$ relative to the system motions. Let us assume that the hyperplane $C\mathbf{s} = \mathbf{0}_N$ is positively invariant relative to the system motions, i.e., that

$$\mathbf{s}\,(t; \mathbf{s}_0) \in \{\mathbf{s} : C\mathbf{s} = \mathbf{0}_N\}, \forall\,(t \geq \theta) \in In\mathfrak{T}_0.$$

Then, for $\mathbf{s}_\theta = \mathbf{s}\,(\theta; \mathbf{s}_0) = e^{A\theta}\mathbf{s}_0$:

$$Ce^{At}\mathbf{s}_0 = \mathbf{0}_N \Longrightarrow Ce^{A(t-\theta)}e^{A\theta}\mathbf{s}_0 = Ce^{A(t-\theta)}\mathbf{s}_t = \mathbf{0}_N, \ \forall\,(t < \theta) \in In\mathfrak{T}_0.$$

This and Equation (D.20), (Appendix D.3), imply

$$\mathbf{s}_\theta = \mathbf{s}\,(\theta; \mathbf{s}_0) = e^{A\theta}\mathbf{s}_0 = \mathbf{0}_N.$$

This holds if and only if $\mathbf{s}_0 = \mathbf{0}_N$ due to the nonsingularity of $e^{A\theta}$. However, the obtained result $\mathbf{s}_0 = \mathbf{0}_N$ contradicts the selected $\mathbf{s}_0 \neq \mathbf{0}_N$. The contradiction is the consequence of the assumption on the positive invariance of the hyperplane $C\mathbf{s} = \mathbf{0}_N$ relative to the system motions. The assumption fails, i.e., the hyperplane $C\mathbf{s} = \mathbf{0}_N$ is not positive invariant relative to the system motions, which guarantees that $q\,[\mathbf{s}\,(t; \mathbf{s}_0)]$ vanishes for all $t \in \mathfrak{T}_0$

only at the zero equilibrium state $\mathbf{s}_e = \mathbf{0}_n$, that it is negative out of the zero equilibrium state $\mathbf{s}_e = \mathbf{0}_n$ for almost all $t \in In\mathfrak{T}_0$ and that it is never positive. Since the total derivative of the function $v\,(.)$ along the motion

$$\mathbf{s}\,(t; \mathbf{s}_0) = \Phi\,(t)\,\mathbf{s}_0 = e^{At}\mathbf{s}_0$$

of the system (8.30), (8.31) together with Equation (8.35) has the following form :

$$\frac{d}{dt}v\,(\mathbf{s}) = \mathbf{s}^T\left(A^T H + H A\right)\mathbf{s} = -\mathbf{s}^T C^T C\mathbf{s} <0, \forall \mathbf{s} = \mathbf{s}\,(t; \mathbf{s}_0) \neq \mathbf{0}_n, \quad \text{(D.47)}$$

then the value of the function $v\,(.)$ is almost strictly monotonously decreasing along system motions. If $\tau \in \mathfrak{T}_0$ is such that $C\mathbf{s}\,(\tau; \mathbf{s}_0) = \mathbf{0}_n$ so that $\mathbf{s}^T\,(\tau; \mathbf{s}_0)\,C^T C\mathbf{s}\,(\tau; \mathbf{s}_0) = \mathbf{0}_n$ then, due to the fact that the hyperplane $C\mathbf{s} = \mathbf{0}_n$ is not positively invariant relative to the system motions, there is $(t_1 > \tau) \in \mathfrak{T}_0$ for which $C\mathbf{s}\,(t_1; \mathbf{s}_0) \neq \mathbf{0}_n$.and $-\mathbf{s}^T\,(t_1; \mathbf{s}_0)\,C^T C\mathbf{s}\,(t_1; \mathbf{s}_0) <0$. If the function $v\,(.)$ were not positive definite then it would escape to $-\infty$ as $t \longrightarrow \infty$. It is well known that asymptotic stability of the system (8.30), (8.31) guarantees that system motions asymptotically approach $\mathbf{s}_e = \mathbf{0}_n$ as $t \longrightarrow \infty$. Therefore, $v\,[\mathbf{s}\,(t; \mathbf{s}_0)] \longrightarrow 0$ as $t \longrightarrow \infty$. Since it vanishes only at the origin, it is continuous and its value is almost strictly decreasing for $(\mathbf{s}_0 \neq \mathbf{0}_n) \in \mathfrak{R}^n$ then it obeys $v\,[\mathbf{s}\,(t; C)] > 0$ for every $t \in \mathfrak{T}_0$, $\forall\,(\mathbf{s}_0 \neq \mathbf{0}_n) \in \mathfrak{R}^n$ so that $v\,(\mathbf{s}_0) > 0$ for every $(\mathbf{s}_0 \neq \mathbf{0}_n) \in \mathfrak{R}^n$.This, continuity of $v\,(\mathbf{s})$ on $\mathfrak{R}^n$ and $v\,(\mathbf{0}_n) = 0$ prove that the function $v\,(.)$ (D.46) is positive definite that implies positive definiteness of the symmetric matrix H due to Equation (D.46). It obeys the relaxed Lyapunov matrix equation (8.35) by its above definition. The proof of the necessity is complete.

Sufficiency. Let the symmetric positive definite matrix H, $H = H^T$, be the solution of (8.35). Its quadratic form (D.46) is positive definite. From (D.47) follows that the $v\,[\mathbf{s}\,(t; \mathbf{s}_0)]$ is almost strictly monotonously decreasing for every $\mathbf{s}\,(t; \mathbf{s}_0) \neq \mathbf{0}_n$, that it is never increasing and that the hyperplane $C\mathbf{s} = \mathbf{0}_N$ is not positive invariant relative to the system motions. For every $\mathbf{s}_0 \neq \mathbf{0}_n$, $v\,[\mathbf{s}\,(t; \mathbf{s}_0)] \longrightarrow 0$ as $t \longrightarrow \infty$ that further implies $\mathbf{s}\,(t; \mathbf{s}_0) \longrightarrow \mathbf{0}_n$ as $t \longrightarrow \infty$ for every $\mathbf{s}_0 \neq \mathbf{0}_n$. This proves the global both attraction and, due to the system linearity and *time* invariance, asymptotic stability of the zero equilibrium state $\mathbf{s}_e = \mathbf{0}_n$ of the system (8.30), (8.31), i.e., that the matrix A is stable. This completes the proof by using the Lyapunov method.

Proof by applying the matrix infinitesimal calculus

Necessity. Let the matrix A be stable. Let, by following Bellman, $X = X^T : \mathfrak{T} \longrightarrow \mathfrak{R}^{n \times n}$ be the symmetric matrix solution of the following linear *time*-invariant matrix differential equation:

$$\frac{dX}{dt} = A^T X + XA. \tag{D.48}$$

It is easy to verify that

$$X(t; X_0) = e^{A^T t} X_0 e^{At} \tag{D.49}$$

is the unique and symmetric matrix solution of Equation (D.48), which obeys the initial condition:

$$X(0; X_0) = e^{A^T 0} X_0 e^{A0} = I X_0 I = X_0, \ \forall X_0 \in \Re^{n \times n}.$$

Let $X_0 = C^T C$ so that the solution of Equation (D.48) becomes

$$X(t; X_0) = e_0^{A^T t} C^T C e^{At} = \left(C e^{At}\right)^T C e^{At}. \tag{D.50}$$

Let

$$H = \int_0^\infty \left(C e^{At}\right)^T C e^{At} dt = H^T \in \Re^{n \times n}. \tag{D.51}$$

The observability of the pair (A, C) guarantees the linear independence of the rows of $C e^{At}$ on $\mathfrak{T}_0$ (condition 2. of Theorem 131) and the existence of $\tau \in \mathfrak{T}_0$ such that $\left(C e^{A\tau}\right)^T C e^{A\tau}$ is positive definite matrix. The Gram matrix H (D.51) of $C e^{At}$ is nonsingular (Theorem 114, Section 7.4) and positive definite. The integration of Equation (D.48), together with $X_0 = C^T C$, Equation D.50 and Equation D.51, leads to

$$X(\infty; X_0) - C^T C = A^T H + HA. \tag{D.52}$$

The stability of the matrix A guarantees $X(\infty; X_0) = O_n$ that simplifies the preceding equation to:

$$-C^T C = A^T H + HA.$$

This is the Lyapunov matrix equation (8.35). The positive definite matrix H (D.51) is the unique solution of the Lyapunov matrix equation (8.35).

Sufficiency. Let the positive definite symmetric matrix H be defined by (D.51). It satisfies Equation D.52 for $X_0 = C^T C$. It is also the unique solution of the Lyapunov matrix equation (8.35) due to the condition of the theorem. Equation D.52 and Equation (8.35) furnish $X(\infty; X_0) = O_n$ that implies stability of the matrix A. ∎

D.5 Proof of Theorem 130

Proof. Since the system observability is independent of the input vector $\mathbf{I}^\mu(t)$ we consider the system (2.15), equivalently, its compact form (9.4), in the free regime characterized by

$$\mathbf{I}^\mu(t) = \mathbf{0}_{(\mu+1)M}, \ \forall t \in \mathfrak{T}_0. \tag{D.53}$$

The *IO* system (2.15) can be transformed (Theorem 49 in Section 4.1) into the *EISO* form (4.1), (4.2) written herein in terms of the total coordinates for the free regime, i.e., by applying (D.53):

$$\frac{d\mathbf{S}(t)}{dt} = A\mathbf{S}(t), \ A \in \mathfrak{R}^{n \times n}, \ \forall t \in \mathfrak{T}_0, \ n = \nu N, \tag{D.54}$$

$$\mathbf{Y}(t) = C\mathbf{S}(t), \ \forall t \in \mathfrak{T}_0. \tag{D.55}$$

or in terms of the deviations of all variables:

$$\frac{d\mathbf{s}(t)}{dt} = A\mathbf{s}(t), \ \forall t \in \mathfrak{T}_0, \ n = \nu N, \tag{D.56}$$

$$\mathbf{y}(t) = C\mathbf{s}(t), \ \forall t \in \mathfrak{T}_0. \tag{D.57}$$

where the two cases should be distinguished:
  Case 1: $\nu > 1$,
  and
  Case 2: $\nu = 1$.
  In both cases, *Case 1* and *Case 2*, we refer to Equations (4.4) through (4.11) (Section 4.1) repeated as

$$\nu \geq 1, \ \mu \geq 1, \ n = \nu N, \tag{D.58}$$

$$\left\{ \begin{array}{c} \nu > 1 \Longrightarrow \mathbf{S}_i = \mathbf{Y}^{(i-1)}, \ \forall i = 1, 2, ..., \nu, \ i.e., \\ \mathbf{S} = \left[\mathbf{S}_1^T \vdots \mathbf{S}_2^T \vdots ... \vdots \mathbf{S}_n^T \right]^T = \left[\mathbf{Y}^T \vdots \mathbf{Y}^{(1)T} \vdots ... \vdots \mathbf{Y}^{(\nu-1)^T} \right]^T \Longrightarrow \\ \mathbf{S} = \mathbf{Y}^{\nu-1} \in \mathfrak{R}^n \end{array} \right\}$$

$$\nu = 1 \Longrightarrow \mathbf{S} = \mathbf{S}_1 = \mathbf{Y}, \tag{D.59}$$

$$\nu > 1 \Longrightarrow A =$$

$$\begin{bmatrix} O_N & -I_N & O_N & O_N & ... & O_N \\ O_N & O_N & -I_N & O_N & ... & O_N \\ O_N & O_N & O_N & -I_N & ... & O_N \\ ... & ... & ... & ... & ... & ... \\ O_N & O_N & O_N & O_N & ... & -I_N \\ -A_\nu^{-1}A_0 & -A_\nu^{-1}A_1 & -A_\nu^{-1}A_2 & -A_\nu^{-1}A_3 & ... & -A_\nu^{-1}A_{\nu-1} \end{bmatrix},$$

$$\nu = 1 \Longrightarrow A = -A_1^{-1}A_0,$$

$$\text{(D.60)}$$

$$C = [I_N \ \ O_N \ \ O_N \ \ O_N \ ... \ O_N] \in \mathfrak{R}^{N \times n}, \ \ Q = O_{N,M} \in \mathfrak{R}^{N \times M}. \quad \text{(D.61)}$$

The preceding equations (D.58) through (D.61) show that the $(n+N) \times n$ complex matrix $O_O(s)$ (8.18) (Section 8.3),

$$O_O(s) = \begin{bmatrix} sI_n - A \\ C \end{bmatrix} \in \mathfrak{R}^{(n+N) \times n}, \quad \text{(D.62)}$$

takes the following form for the (2.15), equivalently, for its compact form (2.57) in terms of deviations:

$$\nu > 1 \Longrightarrow O_O(s) = \begin{bmatrix} sI_n - A \\ C \end{bmatrix} =$$

$$= \begin{bmatrix} sI_N & -I_N & O_N & ... & O_N \\ O_N & sI_N & -I_N & ... & O_N \\ O_N & O_N & sI_N & ... & O_N \\ ... & ... & ... & ... & ... \\ O_N & O_N & O_N & ... & -I_N \\ A_\nu^{-1}A_0 & A_\nu^{-1}A_1 & A_\nu^{-1}A_3 & ... & sI_N + A_\nu^{-1}A_{\nu-1} \\ I_N & O_N & O_N & ... & O_N \end{bmatrix} \in \mathbb{C}^{(n+N) \times n}, \ n = \nu N,$$

$$\nu = 1 \Longrightarrow n = N \Longrightarrow O_O(s) = \begin{bmatrix} sI_N + A_1^{-1}A_0 \\ C \end{bmatrix}.$$

Case 1: $\nu > 1$. The rank of $O_O(s)$ is n due to the following:

$$rank \begin{bmatrix} sI_n - A \\ C \end{bmatrix} = rank \begin{bmatrix} sI_N & -I_N & O_N & O_N & ... & O_N \\ O_N & sI_N & -I_N & O_N & ... & O_N \\ O_N & O_N & sI_N & -I_N & ... & O_N \\ ... & ... & ... & ... & ... & ... \\ O_N & O_N & O_N & O_N & ... & -I_N \\ I_N & O_N & O_N & O_N & ... & O_N \end{bmatrix} =$$

$$\forall (s, A_k) \in \mathbb{C} \times \mathfrak{R}^{N \times N}, \ \forall k = 1, 2, ..., \nu,$$

$$
= N + rank \begin{bmatrix} -I_N & O_N & O_N & ... & O_N \\ sI_N & -I_N & O_N & ... & O_N \\ O_N & sI_N & -I_N & ... & O_N \\ ... & ... & ... & ... & ... \\ O_N & O_N & O_N & ... & -I_N \end{bmatrix} = (\nu - 1)\, N = \nu N = n,
$$

$$
\forall\, (s, A_k) \in \mathbb{C} \times \mathfrak{R}^{N \times N},\ \forall k = 1, 2, ..., \nu.
$$

The rank of $O_O(s)$ is full n and invariant, i.e., independent of the system matrices $A_i,\ \forall i = 1, 2, ..., \nu$, and $B_{ki},\ \forall k = 1, 2, ..., \mu$, i.e., independent of the system matrices $A^{(\nu)}$ and $B^{(\mu)}$. Since the rank of the matrix $O_O(s)$ (D.62) is full and invariant,

$$
rank O_O(s) = rank \begin{bmatrix} sI_n - A \\ C \end{bmatrix} = n,\ \forall\, (s, A_k) \in \mathbb{C} \times \mathfrak{R}^{N \times N},\ \forall k = 1, 2, ..., \nu,
$$

then it follows from the condition 5. of Theorem 131 (Section 8.3) that the system (2.15), equivalently, its compact forms (9.2), (9.4), is observable invariably relative to its matrices $A_i,\ \forall i = 0, 1, 2, ..., \nu$, and $B_{ki},\ \forall k = 1, 2, ..., \mu$, i.e., independent of the system matrices $A^{(\nu)}$ and $B^{(\mu)}$ in *Case 1:* $\nu > 1$.

Case 2: $\nu = 1$. The matrix $O_O(s)$ has the following form:

$$
\nu = 1 \Longrightarrow n = N \Longrightarrow O_O(s) = \begin{bmatrix} sI_N + A_1^{-1} A_0 \\ I_N \end{bmatrix},\ \forall\, (s, A_k) \in \mathbb{C} \times \mathfrak{R}^{N \times N},
$$

$$
rank O_O(s) = rank \begin{bmatrix} sI_N + A_1^{-1} A_0 \\ I_N \end{bmatrix} = rank I_N = N = n,
$$

$$
\forall\, (s, A_k) \in \mathbb{C} \times \mathfrak{R}^{N \times N},\ \forall k = 0, 1.
$$

It follows from the condition 5. of Theorem 131 that the system (2.15), equivalently (2.57), is observable invariably relative to its matrices $A_i,\ \forall i = 0, 1 = \nu$, and $B_{ki},\ \forall k = 1, 2, ..., \mu$, i.e., independent of the system matrices $A^{(1)}$ and $B^{(\mu)}$ in *Case 2:* $\nu = 1$. ∎

D.6 Proof of Theorem 141

Proof. The extended form of Equation (5.3), (Section 5.1), reads:

$$\mathbf{S}_1 = \mathbf{R}, \; \mathbf{s}_1 = \mathbf{r},$$

$$\mathbf{S}_2 = \mathbf{R}^{(1)} = \mathbf{S}_1^{(1)}, \; \mathbf{s}_2 = \mathbf{r}^{(1)} = \mathbf{s}_1^{(1)},$$

$$\ldots\ldots$$

$$\mathbf{S}_k = \mathbf{R}^{(k-1)} = \mathbf{S}_{k-1}^{(1)}, \; \mathbf{s}_k = \mathbf{r}^{(k-1)} = \mathbf{s}_{k-1}^{(1)},$$

$$\ldots\ldots$$

$$\mathbf{S}_\alpha = \mathbf{R}^{(\alpha-1)} = \mathbf{S}_{\alpha-1}^{(1)}, \; \mathbf{s}_\alpha = \mathbf{r}^{(\alpha-1)} = \mathbf{s}_{\alpha-1}^{(1)},$$

$$\text{and}$$

$$\mathbf{S}_\alpha^{(1)} = \mathbf{R}^{(\alpha)}, \; \mathbf{s}_\alpha^{(1)} = \mathbf{r}^{(\alpha)}.$$

The first α equations read also

$$\mathbf{S}_1 = \mathbf{R}, \; \mathbf{s}_1 = \mathbf{r},$$

and

$$\mathbf{S}_1^{(1)} = \mathbf{S}_2, \; \mathbf{s}_1^{(1)} = \mathbf{s}_2,$$

$$\ldots\ldots$$

$$\mathbf{S}_{k-1}^{(1)} = \mathbf{S}_k, \; \mathbf{s}_{k-1}^{(1)} = \mathbf{s}_k,$$

$$\ldots\ldots$$

$$\mathbf{S}_{\alpha-1}^{(1)} = \mathbf{S}_\alpha, \; \mathbf{s}_{\alpha-1}^{(1)} = \mathbf{s}_\alpha. \qquad (D.63)$$

We link these equations with the *HISO* system equations (5.1), (5.2) in order to determine $\mathbf{S}_\alpha^{(1)} = \mathbf{R}^{(\alpha)}$ and $\mathbf{s}_\alpha^{(1)} = \mathbf{r}^{(\alpha)}$:

$$A_\alpha \mathbf{S}_\alpha^{(1)}(t) + \sum_{k=0}^{k=\alpha-1} A_k \mathbf{S}_{k+1}(t) = \sum_{k=0}^{k=\mu} H_k \mathbf{I}^{(k)}(t), \; \forall t \in \mathfrak{T}_0. \qquad (D.64)$$

The compact vector-matrix form of the united equations (D.63) and (D.64) multiplied on the left by A_α^{-1} (which is nonsingular due to (5.1)) reads:

$$\frac{d\mathbf{S}(t)}{dt} = A\mathbf{S}(t) + P^{(\mu)}\mathbf{I}^\mu(t), \; \mathbf{S} = \left[\mathbf{S}_1^T \vdots \mathbf{S}_2^T \vdots \ldots \vdots \mathbf{S}_\alpha^T\right]^T \in \mathfrak{R}^n, \qquad (D.65)$$

where the matrices A, $P^{(\mu)}$, $H^{(\mu)}$ and C are defined by Equations (9.28)-(11.146), (Section 9.4), respectively. The output vector equation (5.2) gets the following form by recalling $R_{y\alpha} = O_{N,\rho}$:

$$\mathbf{Y}(t) = C\mathbf{S} + Q\mathbf{I}(t), \ \forall t \in \mathfrak{T}_0. \tag{D.66}$$

Analogously, the *HISO* system (5.44), (5.45) becomes:

$$\frac{d\mathbf{s}(t)}{dt} = A\mathbf{s}(t) + P^{(\mu)}\mathbf{i}^\mu(t), \ \forall t \in \mathfrak{T}_0, \ \mathbf{s} = \left[\mathbf{s}_1^T \ \vdots \ \mathbf{s}_2^T \ \vdots \ ... \ \vdots \mathbf{s}_\alpha^T\right]^T \in \mathfrak{R}^n,$$
$$\tag{D.67}$$

$$\mathbf{y}(t) = C\mathbf{s} + Q\mathbf{i}(t), \ \forall t \in \mathfrak{T}_0. \tag{D.68}$$

Equations (D.65), (D.66) determine the (8.1), (8.2) with the matrices defined by Equations (9.28)-(9.30), (11.146). This implies the validity of Theorem 131 for the *HISO* system (5.1), (5.2), and for the *HISO* system (5.44), (5.45). Theorem 151 is the *HISO* form of Theorem 131. ∎

D.7 Proof of Theorem 147

Proof. Appendix C.2 shows how the *IIO* system (6.1), (6.2) in the total coordinates, or in terms of the deviations of all variables (6.65), (6.66), can be transformed into the following *ISO* forms, respectively:

$$\frac{d\mathbf{S}(t)}{dt} = A\mathbf{S}(t) + P^{(\mu)}\mathbf{I}^\mu(t), \tag{D.69}$$

$$\mathbf{Y}(t) = C\mathbf{S}(t), \tag{D.70}$$

$$\frac{d\mathbf{s}(t)}{dt} = A\mathbf{s}(t), \tag{D.71}$$

$$\mathbf{y}(t) = C\mathbf{s}(t), \tag{D.72}$$

with the matrices A, C, and V determined by (C.24)-(C.31)(Appendix C.2). The system (D.74), (D.75) is the system (8.1) and (8.2) (Section 8.1), and the system (D.76), (D.77) is the system the *IIO* system (6.65), (6.66) in the form of the system (8.11) and (8.12) (Section 8.2). Since Theorem 131 holds for the system (8.1) and (8.2), and for the system (8.11) and (8.12), i.e., for the *IIO* system (6.65), (6.66), then it is valid also for both the *IIO* system (D.74), (D.75) and the *IIO* system (D.76), (D.77). Furthermore, Condition (9.20) (Theorem 131), i.e.,

$$O_O(s) = \begin{bmatrix} sI_n - A \\ C \end{bmatrix} \in \mathfrak{R}^{(n+N) \times n}, \ n = \alpha\rho + \nu N. \tag{D.73}$$

reads

$$O_O\left(s\right) = \begin{bmatrix} sI_{\alpha\rho} & O_{\alpha\rho,\nu N} & & A_{11} & O_{\alpha\rho,\nu N} \\ O_{\nu N,\alpha\rho} & sI_{\nu N} & - & A_{21} & A_{22} \\ & O_{N,\alpha\rho} & & I_N \vdots O_{N,(\nu-1)N} \end{bmatrix}$$

due to Equations (C.24)-(C.28) (Appendix C.2), i.e.,

$$O_O\left(s\right) = \begin{bmatrix} sI_{\alpha\rho} - A_{11} & O_{\alpha\rho,\nu N} \\ -A_{21} & sI_{\nu N} - A_{22} \\ O_{N,\alpha\rho} & I_N \vdots O_{N,(\nu-1)N} \end{bmatrix}.$$

This, Equations (C.26), (C.27) (Appendix C.2) determine that the rank of the matrix $rankO_O\left(s\right)$ is the rank of the following matrix:

$$\begin{bmatrix} sI_{\alpha\rho} - A_{11} & & & & O_{\alpha\rho,\nu N} & & \\ O_{N,\rho} & . & O_{N,\rho} & sI_N & . & O_N \\ \vdots & \vdots & \vdots & \vdots & \vdots & \vdots \\ O_{N,\rho} & . & O_{N,\rho} & O_N & . & -I_N \\ -E_\nu^{-1}R_{y0} & . & -E_\nu^{-1}R_{y,\alpha-1} & sI_N + E_\nu^{-1}E_0 & . & sI_N + E_\nu^{-1}E_{\nu-1} \\ & O_{N,\alpha\rho} & & I_N & \vdots & O_{N,(\nu-1)N} \end{bmatrix}.$$

The application of the elementary matrix transformation shows that the rank of the matrix $rankO_O\left(s\right)$ is equal to the rank of the following matrix:

$$\begin{bmatrix} sI_{\alpha\rho} - A_{11} & & & O_{\alpha\rho,\nu N} & & \\ & & sI_N & -I_N & .. & O_N & O_N \\ & & \vdots & \vdots & \vdots & \vdots & \vdots \\ O_{(\nu-1)N,\alpha\rho} & & O_N & \vdots & .. & -I_N & O_N \\ & & O_N & \vdots & .. & sI_N & -I_N \\ & O_{N,\alpha\rho} & I_N & \vdots & & O_{N,(\nu-1)N} & \\ -E_\nu^{-1}R_{y0} & . & -E_\nu^{-1}R_{y,\alpha-1} & sI_N + E_\nu^{-1}E_0 & . & sI_N + E_\nu^{-1}E_{\nu-1} \end{bmatrix},$$

which yields

$$rankO_O\left(s\right) =$$

$$rank \begin{bmatrix} sI_{\alpha\rho} - A_{11} & & & O_{\alpha\rho,\nu N} & & \\ & & sI_N & -I_N & ... & O_N & O_N \\ & & \vdots & \vdots & \vdots & \vdots & \vdots \\ O_{(\nu-1)N,\alpha\rho} & & O_N & \vdots & ... & -I_N & O_N \\ & & O_N & \vdots & ... & sI_N & -I_N \\ & O_{N,\alpha\rho} & I_N & \vdots & O_{N,(\nu-2)N} & \vdots & O_N \end{bmatrix} \implies$$

$$rankO_O(s) =$$

$$= rank \begin{bmatrix} sI_{\alpha\rho} - A_{11} & & & O_{\alpha\rho,\nu N} & & \\ & sI_N & -I_N & \cdots & O_N & O_N \\ & O_N & sI_N & \cdots & O_N & O_N \\ O_{\nu N,\alpha\rho} & \cdots & \cdots & \cdots & \cdots & \cdots \\ & O_N & O_N & \cdots & sI_N & -I_N \\ & I_N & O_N & \cdots & O_N & O_N \end{bmatrix} =$$

$$rankO_O(s) =$$

$$= N + rank \begin{bmatrix} sI_{\alpha\rho} - A_{11} & & O_{\alpha\rho,(\nu-1)N} & & \\ & -I_N & \cdots & O_N & O_N \\ & sI_N & \cdots & O_N & O_N \\ O_{(\nu-1)N,\alpha\rho} & \cdots & \cdots & \cdots & \cdots \\ & O_N & \cdots & sI_N & -I_N \end{bmatrix} \implies$$

$$rankO_O(s) = \nu N + rank(sI_{\alpha\rho} - A_{11}).$$

From this result follows that for the matrix $O_O(s)$ (D.73), i.e., (9.20), to have the full rank $n = \alpha\rho + \nu N$ it is necessary and sufficient that the following holds:

$$rank(sI_{\alpha\rho} - A_{11}) = \alpha\rho, \ \forall s \in C,$$

which is violated for every eigenvalue of the matrix A_{11}. The system violates Condition (9.20) of Theorem 131. The system is not full state observable. ∎

D.8   Proof of Theorem 148

Proof. This is the proof of Theorem 158, (Section 9.5).

Appendix C.2 shows how the internal dynamics parts of the *IIO* system (6.1), (6.2) in total coordinates, or in terms of the deviations of all variables (6.65), (6.66), together with the ouput equation, can be transformed into the following *ISO* forms, respectively:

$$\frac{d\mathbf{S}_{IIOI}(t)}{dt} = A_{11}\mathbf{S}_{IIOI}(t) + W_1\mathbf{W}(t), \tag{D.74}$$

$$\mathbf{Y}(t) = O_{N,\alpha\rho}\mathbf{S}_{IIOI}(t), \tag{D.75}$$

and in the free regime in terms of deviations:

$$\frac{d\mathbf{s}_{IIOI}(t)}{dt} = A_{11}\mathbf{s}_{IIOI}(t), \tag{D.76}$$

$$\mathbf{y}(t) = O_{N,\alpha\rho}\mathbf{s}_{IIOI}(t), \tag{D.77}$$

with the matrices A_{11}, C and W_1 determined by (C.25), (C.28), (C.29), and with $\mathbf{W}(t)$ determined in Equation (C.32) (Appendix C.2). Condition (9.20) (Theorem 131) for the internal state observability demands that the matrix $O_{OSIIOI}(s)$,

$$O_{OSIIOI}(s) = \begin{bmatrix} sI_{\alpha\rho} - A_{11} \\ O_{N,\alpha\rho} \end{bmatrix}$$

has the full rank $\alpha\rho$ for every $s \in \mathbb{C}$. However, its rank is defective for every eigenvalue $s_i(A_{11})$ of the matrix A_{11},

$$rankO_{OSIIOI}(s) = \begin{bmatrix} s_i(A_{11})I_{\alpha\rho} - A_{11} \\ O_{N,\alpha\rho} \end{bmatrix} < \alpha\rho,$$

$$\forall s_i(A_{11}) \in C, \ \forall i = 1, 2, ..., \alpha\rho.$$

The system violates Condition (9.20) of Theorem 131 applied to the system internal state observability. The system is not internal state observable. ∎

D.9   Proof of Theorem 149

Proof. Let the initial input state deviation $\mathbf{s}_I(0)$ be arbitrarily chosen, fixed and known. Equations (C.22)-(C.34) (Appendix C.2) in the free regime (i.e., for $\mathbf{i}^\mu(t) \equiv \mathbf{0}_{(\mu+1)M}$) simplify to:

$$\begin{bmatrix} \mathbf{s}_I^{(1)}(t) \\ \mathbf{s}_O^{(1)}(t) \end{bmatrix} = \begin{bmatrix} A_{11} & O_{\alpha\rho,\nu N} \\ A_{21} & A_{22} \end{bmatrix} \begin{bmatrix} \mathbf{s}_I(t) \\ \mathbf{s}_O(t) \end{bmatrix}, \tag{D.78}$$

$$\mathbf{y}(t) = C \begin{bmatrix} \mathbf{s}_I(t) \\ \mathbf{s}_O(t) \end{bmatrix}. \tag{D.79}$$

The Laplace transforms of these equations have the following forms in view of (C.28) (Appendix C.2):

$$\begin{bmatrix} s\mathbf{s}_I(s) - \mathbf{s}_I(0) \\ s\mathbf{s}_O(s) - \mathbf{s}_O(0) \end{bmatrix} = \begin{bmatrix} A_{11} & O_{\alpha\rho,\nu N} \\ A_{21} & A_{22} \end{bmatrix} \begin{bmatrix} \mathbf{s}_I(s) \\ \mathbf{s}_O(s) \end{bmatrix} \Longrightarrow$$

$$\begin{bmatrix} (sI_{\alpha\rho} - A_{11})\mathbf{s}_I(s) \\ A_{21}\mathbf{s}_I(s) + (sI_{\nu N} - A_{22})\mathbf{s}_O(s) \end{bmatrix} = \begin{bmatrix} \mathbf{s}_I(0) \\ \mathbf{s}_O(0) \end{bmatrix},$$

$$\mathbf{y}(s) = \begin{bmatrix} C_I \vdots C_O \end{bmatrix} \begin{bmatrix} \mathbf{s}_I(s) \\ \mathbf{s}_O(s) \end{bmatrix}, \ C_I = O_{N,\alpha\rho}, \ C_O = \begin{bmatrix} I_N \vdots O_{N,(\nu-1)N} \end{bmatrix}.$$

These equations imply

$$
\begin{bmatrix} sI_{\nu N} - A_{22} & \vdots & -I_{\nu N} \\ I_N & \vdots O_{N,(\nu-1)N} & \vdots O_{N,\nu N} \end{bmatrix} \begin{bmatrix} \mathbf{s}_O(s) \\ \mathbf{s}_O(0) \end{bmatrix} = \begin{bmatrix} -A_{21}\mathbf{s}_I(s) \\ \mathbf{y}(s) \end{bmatrix}.
$$

The matrix

$$
\begin{bmatrix} sI_{\nu N} - A_{22} & \vdots & -I_{\nu N} \\ I_N & \vdots O_{N,(\nu-1)N} & \vdots O_{N,\nu N} \end{bmatrix} \in C^{(\nu+1)N \times (\nu+1)N}.
$$

In order to test its rank we proceed as follows:

$$
rank \begin{bmatrix} sI_N & I_N & .. & O_N & O_N & & \\ O_N & sI_N & .. & O_N & O_N & & \\ \vdots & \vdots & \vdots & \vdots & \vdots & \vdots & -I_{\nu N} \\ O_N & O_N & .. & sI_N & I_N & & \\ E_\nu^{-1}E_0 & E_\nu^{-1}E_1 & .. & E_\nu^{-1}E_{\nu-2} & sI_N+ & & \\ & & & & +E_\nu^{-1}E_{\nu-1} & & \\ I_N & \vdots & & O_{N,(\nu-1)N} & & \vdots O_{N,\nu N} \end{bmatrix} =
$$

$$
rank \begin{bmatrix} sI_N & .. & O_N & & -I_N & O_N & .. & O_N & O_N \\ O_N & .. & O_N & & O_N & -I_N & .. & O_N & O_N \\ \vdots & \vdots & \vdots & \vdots & \vdots & \vdots & \vdots & \vdots & \vdots \\ O_N & .. & I_N & & O_N & O_N & .. & -I_N & I_N \\ E_\nu^{-1}E_0 & .. & \begin{smallmatrix} sI_N+ \\ +E_\nu^{-1}E_{\nu-1} \end{smallmatrix} & & O_N & O_N & .. & O_N & O_N \\ I_N & \vdots & O_{N,(\nu-1)N} & \vdots & & & O_{N,\nu N} & \end{bmatrix} =
$$

$$
= rank \begin{bmatrix} sI_N & -I_N & ... & O_N & O_N \\ O_N & O_N & ... & O_N & O_N \\ \vdots & \vdots & \vdots & \vdots & \vdots \\ E_\nu^{-1}E_0 & O_N & ... & O_N & -I_N \\ I_N & O_N & ... & O_N & O_N \end{bmatrix} =
$$

$$
= N + rank \begin{bmatrix} -I_N & O_N & ... & O_N \\ O_N & -I_N & ... & O_N \\ \vdots & \vdots & \vdots & \vdots \\ O_N & O_N & ... & -I_N \end{bmatrix} = N + \nu N = (\nu + 1) N..
$$

The matrix

$$
\begin{bmatrix} sI_{\nu N} - A_{22} & \vdots & -I_{\nu N} \\ I_N & \vdots O_{N,(\nu-1)N} & \vdots O_{N,\nu N} \end{bmatrix}
$$

has the full rank $(\nu + 1) N$ for every $s \in \mathbb{C}$ independently of the system matrices. Equation

$$
\begin{bmatrix} sI_{\nu N} - A_{22} & \vdots & -I_{\nu N} \\ I_N & \vdots O_{N,(\nu-1)N} & \vdots O_{N,\nu N} \end{bmatrix} \begin{bmatrix} \mathbf{s}_O(s) \\ \mathbf{s}_O(0) \end{bmatrix} = \begin{bmatrix} -A_{21}\mathbf{s}_I(s) \\ \mathbf{y}(s) \end{bmatrix}.
$$

has the unique solution $\mathbf{s}_O(0)$ in terms of the vector

$$
\begin{bmatrix} -A_{21}\mathbf{s}_I(s) \\ \mathbf{y}(s) \end{bmatrix} = \begin{bmatrix} -A_{21}\left(sI_{\alpha\rho} - A_{11}\right)^{-1}\mathbf{s}_I(0), \\ \mathbf{y}(s) \end{bmatrix}
$$

for any chosen, fixed and known $\mathbf{s}_I(0)$ so that $\mathbf{s}_O(0)$ is well determined by the system response $\mathbf{y}(t; \mathbf{s}_I(0); \mathbf{s}_O(0))$. The initial output state $\mathbf{s}_O(0)$ is observable. The system is output state observable in view of Definition 156. The IIO system (6.1), (6.2), equivalently the IIO system (6.65), (6.66), is invariably output state observable, i.e., it is output state observable independently of the system matrices. ∎

D.10   Proof of Theorem 161

Theorem 233 *[18, Theorem 5-3, p. 174]*
  Assume that for each i, $\mathbf{r}_i$ is analytic function (Definition 231, Appendix D.3) on $[t_1, t_2]$. Let $R(.)$ (7.45) be the $\nu \times \mu$ matrix with $\mathbf{r}_i$ as its i-th row, and let $R^{(k)}(t)$ be the k-th derivative of $R(t)$. Let t_0 be any fixed point in $[t_1, t_2]$. Then the $\mathbf{r}_i$'s are linearly independent on $[t_1, t_2]$ if and only if

$$
rank \left[R(t_0) \;\vdots\; R^{(1)}(t_0) \;\vdots\; R^{(2)}(t_0) \;\vdots\; ... \;\vdots\; R^{(\nu-1)}(t_0) \;\vdots\; ... \right] = \nu \quad (D.80)
$$

Proof. This is the proof of Theorem 171.
  The proof is a slight modification and extension of the proof by Chen [18, Theorem 5-4, pp. 177, 178] generalized by adjusting it to the system (10.3), (10.4).
  Proof of the necessity of the conditions 1.-6. and the proof of the sufficiency of the conditions 1.-5.
  Let the system (10.3), (10.4) be state controllable. Definition 161 holds.
  We treat simultaneously the case a) $\mu = 0$ and the case b) $\mu > 0$ in the proof of the necessity.
     1. The fundamental matrix $\Phi(t_1, \tau) = e^{A(t_1 - \tau)}$ and the matrix $\Phi(t_0, \tau) = e^{A(t_0 - \tau)} = e^{-A(\tau - t_0)}$ are integrable, which ensures the existence

of the Gram matrix $G_{\Phi V}(t_1, t_0)$ of $\Phi(t_1, t)B^{(\mu)}$ and of the Gram matrix $G_{\Phi V}(t_0, t_1)$ of $\Phi(t_0, t)B^{(\mu)}$. Let be assumed that the rows of $\Phi(t_1, t)B^{(\mu)}$ are linearly dependent on $[t_0, t_1]$, any $(t_0, t_1 > t_0) \in In\mathfrak{T}_0 \times In\mathfrak{T}_0$. This guarantees the existence of a nonzero vector $\mathbf{a} \in \mathfrak{R}^{1 \times n}$ such that (statement under 1) of Lemma 100, Section 7.4):

$$\mathbf{a}\Phi(t_1, t)B^{(\mu)} = \mathbf{0}_r^T \text{ on } [t_0, t_1], \text{ any } (t_0, t_1 > t_0) \in In\mathfrak{T}_0 \times In\mathfrak{T}_0. \quad \text{(D.81)}$$

Let, for an accepted $(t_0, t_1 > t_0) \in In\mathfrak{T}_0 \times In\mathfrak{T}_0$, $\mathbf{S}(t_0)$ be chosen so that

$$\mathbf{S}(t_0) = \mathbf{S}_0 = \Phi(t_0, t_1)\mathbf{a}^T = \Phi^{-!}(t_1, t_0)\mathbf{a}^T \neq \mathbf{0}_n.$$

The solution of the state equation (10.3) is given by Equations (10.5). We apply now the first Equation (10.5) for $t = t_1$:

$$\mathbf{S}\left(t_1; t_0; \mathbf{S}_0; B^{(\mu)\mu}\right) = \Phi(t_1, t_0)\mathbf{S}_0 + \int_{t_0}^{t_1} \Phi(t_1, \tau)B^{(\mu)}\mathbf{U}^{\mu}(\tau)\,d\tau \quad \text{(D.82)}$$

For the chosen $\mathbf{S}_0 = \Phi(t_0, t_1)\mathbf{a}^T = \Phi^{-!}(t_1, t_0)\mathbf{a}^T$ the system motion becomes

$$\mathbf{S}\left(t_1; t_0; \Phi(t_0, t_1)\mathbf{a}^T; B^{(\mu)\mu}\right) =$$

$$+\Phi(t_1, t_0)\Phi^{-!}(t_1, t_0)\mathbf{a}^T + \int_{t_0}^{t_1} \Phi(t_1, \tau)B^{(\mu)}\mathbf{U}^{\mu}(\tau)\,d\tau.$$

Definition 161 permits an arbitrary choice of the final state vector $\mathbf{S}(t_1)$,

$$\mathbf{S}\left(t_1; t_0; \Phi(t_0, t_1)\mathbf{a}^T; \mathbf{U}^{\mu}\right) = \mathbf{S}(t_1) = \mathbf{S}_1,$$

so that it can be at the origin: $\mathbf{S}(t_1) = \mathbf{0}_n$, which we accept. For this choice of $\mathbf{S}(t_1)$ follows

$$\mathbf{S}\left(t_1; t_0; \Phi(t_0, t_1)\mathbf{a}^T; \mathbf{U}^{\mu}\right) = \mathbf{0}_n = \mathbf{a}^T + \int_{t_0}^{t_1} \Phi(t_1, \tau)B^{(\mu)}\mathbf{U}^{\mu}(\tau)\,d\tau,$$

or, by multiplying this equation on the left by $\mathbf{a}$:

$$\mathbf{a0}_n = 0 = \mathbf{aa}^T + \int_{t_0}^{t_1} \mathbf{a}\Phi(t_1, \tau)B^{(\mu)}\mathbf{U}^{\mu}(\tau)\,d\tau. \quad \text{(D.83)}$$

Equation (D.81) reduces Equation (D.83) to:

$$0 = \mathbf{a}\mathbf{a}^T = \|\mathbf{a}\|.$$

This implies $\mathbf{a} = \mathbf{0}_n^T$ and contradicts the choice of $\mathbf{a} \neq \mathbf{0}_n^T$. The contradiction is the consequence of the assumption that the rows of $\Phi(t_1, t)\mathrm{B}^{(\mu)}$ are linearly dependent on $[t_0, t_1]$ for any accepted $(t_0, t_1 > t_0) \in In\mathfrak{T}_0 \times In\mathfrak{T}_0$. This proves the linear independence of the rows of the matrix function $\Phi(t_1, .)$ on $]t_0, t_1[$ for any $(t_0, t_1 > t_0) \in In\mathfrak{T}_0 \times In\mathfrak{T}_0$. Let us assume now that the rows of $\Phi(t_0, t)\mathrm{B}^{(\mu)}$ are linearly dependent on $[t_0, t_1]$ for any $(t_0, t_1 > t_0) \in In\mathfrak{T}_0 \times In\mathfrak{T}_0$. This ensures the existence of a nonzero vector $\mathbf{b} \in \mathfrak{R}^{1 \times n}$ such that the following holds:

$$\mathbf{b}\Phi(t_0, t)\,\mathrm{B}^{(\mu)} = \mathbf{0}_r^T, \ \forall t \in [t_0, t_1], \ (t_0, t_1 > t_0) \in In\mathfrak{T}_0 \times In\mathfrak{T}_0. \qquad (\text{D.84})$$

Let $\mathbf{S}_0$ be chosen so that

$$\mathbf{S}(t_0) = \mathbf{S}_0 = \mathbf{b}^T \neq \mathbf{0}_n.$$

The solution of the state equation (10.3) is given by Equations (10.5). We apply now the second Equation (10.5):

$$\mathbf{S}(t; t_0; \mathbf{S}_0; \mathbf{U}^\mu) = \Phi(t, t_0) \left[\mathbf{S}_0 + \int_{t_0}^t \Phi(t_0, \tau)\,\mathrm{B}^{(\mu)}\mathbf{U}^\mu(\tau)\,d\tau \right]. \qquad (\text{D.85})$$

For the chosen $\mathbf{S}_0 = \mathbf{b}^T$ the system motion reads:

$$\mathbf{S}(t; t_0; \mathbf{b}^T; \mathbf{U}^\mu) = \Phi(t, t_0) \left[\mathbf{b}^T + \int_{t_0}^t \Phi(t_0, \tau)\,\mathrm{B}^{(\mu)}\mathbf{U}^\mu(\tau)\,d\tau \right],$$

so that at the moment $t = t_1$ it becomes

$$\Phi(t_0, t_1)\,\mathbf{S}(t_1; t_0; \mathbf{b}^T; \mathbf{U}^\mu) = \mathbf{b}^T + \int_{t_0}^{t_1} \Phi(t_0, \tau)\,\mathrm{B}^{(\mu)}\mathbf{U}^\mu(\tau)\,d\tau,$$

Definition 161 of the state controllability permits arbitrary choice of the final state vector $\mathbf{S}(t_1; t_0; \mathbf{b}^T; \mathbf{U}^\mu) = \mathbf{S}(t_1)$ so that it can be at the origin: $\mathbf{S}(t_1) = \mathbf{0}_n$, which we again accept. For this choice of $\mathbf{S}(t_1)$ follows:

$$\Phi(t_0, t_1)\,\mathbf{S}(t_1; t_0; \mathbf{b}^T; \mathbf{U}^\mu) = \mathbf{0}_n = \mathbf{b}^T + \int_{t_0}^{t_1} \Phi(t_0, \tau)\,\mathrm{B}^{(\mu)}\mathbf{U}^\mu(\tau)\,d\tau,$$

or, by multiplying this equation on the left by $\mathbf{b}$:

$$0 = \mathbf{b}\mathbf{b}^T + \int_{t_0}^{t_1} \mathbf{b}\Phi\left(t_0, \tau\right) \mathrm{B}^{(\mu)} \mathbf{U}^{\mu}\left(\tau\right) d\tau.$$

Equation (D.84) simplifies the preceding equation to:

$$0 = \mathbf{b}\mathbf{b}^T,$$

which implies $\mathbf{b} = \mathbf{0}_n^T$. This is in the contradiction with the choice of $\mathbf{b} \neq \mathbf{0}_n^T$. The contradiction is the consequence of the assumption that the rows of $\Phi\left(t_0, t\right)\mathrm{B}^{(\mu)}$ are linearly dependent on $[t_0, t_1]$ for any $(t_0, t_1 > t_0) \in In\mathfrak{T}_0 \times In\mathfrak{T}_0$. This proves that the rows of the $n \times (\mu + 1)r$ matrix $\Phi\left(t_0, t\right)\mathrm{B}^{(\mu)}$ are linearly independent on $[t_0, t_1]$ for any $(t_0, t_1 > t_0) \in In\mathfrak{T}_0 \times In\mathfrak{T}_0$ and completes the proof of the necessity of 1.

2. Since the system is time-invariant it is controllable if and only if it is controllable at any $t_0 \in \mathfrak{T}$. This permits $t_0 = 0$ so that $\Phi\left(t, 0\right)\mathrm{B}^{(\mu)} = e^{At}\mathrm{B}^{(\mu)} = \Phi\left(t\right)\mathrm{B}^{(\mu)}$. Its Laplace transform is $(sI_n - A)^{-1}\mathrm{B}^{(\mu)}$. The Laplace transform is linear one-to-one operator so that for the rows of $e^{At}\mathrm{B}^{(\mu)}$ to be linearly independent it is necessary and sufficient that the rows of $(sI_n - A)^{-1}\mathrm{B}^{(\mu)}$ are linearly independent. This proves the equivalence between the criteria under 1. and 2 and the necessity of the condition 2.

3. Theorem 114 (Section 7.4) applied to the Gram matrix $G_{\Phi B}\left(t_0, t_1\right)$ (10.7) of $\Phi\left(t_0, t\right)\mathrm{B}^{(\mu)}$ proves that for $G_{\Phi B}\left(t_0, t_1\right)$ to be non-singular for any $(t_0, t_1 > t_0) \in In\mathfrak{T}_0 \times In\mathfrak{T}_0$; i.e., for $rankG_{\Phi B}\left(t_0, t_1\right) = n$ for any instants $(t_0, t_1 > t_0) \in In\mathfrak{T}_0 \times In\mathfrak{T}_0$, it is necessary and sufficient that the rows of $\Phi\left(t_0, t\right)\mathrm{B}^{(\mu)}$ are linearly independent on $[t_0, t_1]$ for any $(t_0, t_1 > t_0) \in In\mathfrak{T}_0 \times In\mathfrak{T}_0$. This proves the equivalence between the conditions 1. and 3. hence between 2. and 3., and proves the necessity of the condition 3.

4. The entries of $e^{At}\mathrm{B}^{(\mu)}$ are analytic functions. Theorem 233 implies that for the rows of $e^{At}\mathrm{B}^{(\mu)}$ to be linearly independent on $\mathfrak{T}_0$ it is necessary and sufficient that (D.80) holds, i.e.

$$rank \left[\mathrm{B}^{(\mu)} \vdots A\mathrm{B}^{(\mu)} \vdots A^2\mathrm{B}^{(\mu)} \vdots \ldots \vdots A^{n-1}\mathrm{B}^{(\mu)} \vdots \ldots\right] = n, \qquad \text{(D.86)}$$

due to the fact that the k-th derivative of $e^{At_0}\mathrm{B}^{(\mu)}$ at $t = t_0 = 0$ is $A^k\mathrm{B}^{(\mu)}$. The Cayley-Hamilton theorem ensures that A^m for $m \geq n$ can be expressed as a linear combination of $I_N, A, \ldots, A^{n-1}$ (for more details see [2, Theorem 2.1, p. 124], [76, 4.5.25 Theorem, p. 167]). This implies that the columns of

$A^m B^{(\mu)}$ for $m \geq n$ are linearly dependent of the columns of $B^{(\mu)}$, $AB^{(\mu)}$, $A^2 B^{(\mu)}$, ..., $A^{n-1} B^{(\mu)}$, which yields:

$$rank \left[B^{(\mu)} \vdots AB^{(\mu)} \vdots A^2 B^{(\mu)} \vdots ... \vdots A^{n-1} B^{(\mu)} \vdots ... \right] =$$

$$= rank \left[B^{(\mu)} \vdots AB^{(\mu)} \vdots A^2 B^{(\mu)} \vdots ... \vdots A^{n-1} B^{(\mu)} \right].$$

It follows that for all rows of $e^{At} B^{(\mu)}$ to be linearly independent on $\mathfrak{T}_0$ it is necessary and sufficient that, as shown above (Theorem 233), the condition under 4. holds. This, together with the equivalence among the conditions 1.-3., proves the equivalence among the conditions 1.-4. and the necessity of the condition 4.

  5. Let at first the system (10.3), (10.4) be assumed controllable.

  What follows is the generalization of the original necessity proof [92] by Hautus [48] accommodated to the system (10.3), (10.4) and to the notation of this book:

$$rank \left[sI_n - A \vdots B^{(\mu)} \right] < n \Longrightarrow \exists \mathbf{h} \neq \mathbf{0}_n^T \Longrightarrow \left(\mathbf{h} A = s\mathbf{h}, \ \mathbf{h} B^{(\mu)} = \mathbf{0}_{(\mu+1)r}^T \right) \Longrightarrow$$

$$\Longrightarrow \exists \mathbf{h} \neq \mathbf{0}_n^T \Longrightarrow \mathbf{h} \left[B^{(\mu)} \vdots AB^{(\mu)} \vdots A^2 B^{(\mu)} \vdots ... \vdots A^{n-1} B^{(\mu)} \right] = \mathbf{0}_{n(\mu+1)r}^T \Longrightarrow$$

$$\Longrightarrow rank \left[B^{(\mu)} \vdots AB^{(\mu)} \vdots A^2 B^{(\mu)} \vdots ... \vdots A^{n-1} B^{(\mu)} \right] < n \Longrightarrow$$

$$\Longrightarrow \left(A, B^{(\mu)} \right) \ not \ controllable.$$

The result contradicts the assume system controllability. The contradiction is the consequence of the assumption that the condition 5., i.e., (10.11) fails. The the condition 5., i.e., (10.11), is necessary for the system controllability.

  What follows is the slight generalization of the necessity proof by Chen [18, p. 206] to hold for the system (10.3), (10.4). Despite the system (10.3), (10.4) state controllability let be assumed that

$$rank \left[sI_n - A \vdots B^{(\mu)} \right] < n, \ for \ some \ s_i = s_i (A) \in \mathbb{C}. \qquad (D.87)$$

This implies the existence of a nonzero row vector $\mathbf{d} \in \mathbb{C}^{1 \times n}$ such that (statement under 3) of Lemma 100, Section 7.4):

$$\mathbf{d} \left[s_i I_n - A \vdots B^{(\mu)} \right] = \mathbf{0}_{n+(\mu+1)r}^T,$$

or

$$\mathbf{d}s_i = s_i\mathbf{d} = \mathbf{d}A \ and \ \mathbf{d}B^{(\mu)} = \mathbf{0}^T_{(\mu+1)r},$$

which imply

$$\mathbf{d}A^2 = s_i\mathbf{d}A = s_i^2\mathbf{d}, \ ...\mathbf{d}A^k = s_i^k\mathbf{d}, \ k = 1, 2, \ ...$$

Hence,

$$\mathbf{d}\left[B^{(\mu)} \vdots AB^{(\mu)} \vdots A^2B^{(\mu)} \vdots ... \vdots A^{n-1}B^{(\mu)}\right] =$$

$$= \left[\mathbf{d}B^{(\mu)} \vdots s_i\mathbf{d}B^{(\mu)} \vdots s_i^2\mathbf{d}B^{(\mu)} \vdots ... \vdots s_i^{n-1}\mathbf{d}B^{(\mu)}\right] =$$

$$= \left[\mathbf{0}^T_{(\mu+1)r} \vdots s_i\mathbf{0}^T_{(\mu+1)r} \vdots s_i^2\mathbf{0}^T_{(\mu+1)r} \vdots ... \vdots s_i^{n-1}\mathbf{0}^T_{(\mu+1)r}\right] = .\mathbf{0}^T_{n(\mu+1)r}.$$

This means that

$$rank\left[B^{(\mu)} \vdots AB^{(\mu)} \vdots A^2B^{(\mu)} \vdots ... \vdots A^{n-1}B^{(\mu)}\right] < n,$$

which implies the state noncontrollability of the system (10.3), (10.4) due to the condition under 4. This contradicts the assumed system state controllability, which is the consequence of the assumption that (D.87) holds. Therefore,

$$rank\left[sI - A \vdots B^{(\mu)}\right] = n, \ \forall s_i = s_i\left(A\right) \in \mathbb{C}, \ i = 1, 2, ..., n,$$

$$i.e., \ \forall s \in \mathbb{C}.$$

This proves that the condition 4. implies the condition 5. and proves the necessity of the condition under 5. In order to complete the proof of the equivalence of the condition 5. with the conditions 1.-4. for the sufficiency we accept the validity of the condition 5., i.e., $rank\left[sI - A \vdots B^{(\mu)}\right] < n$ for every $s \in \mathbb{C}$. Let us refer at first to Hautus [48]. What follows is the generalization of the original sufficiency proof [92] by Hautus [48] accommodated to the system (10.3), (10.4) and to the notation of this book:

$$\left(A, \mathrm{B}^{(\mu)}\right) \; not \; controllable \implies \exists \mathbf{h} \neq \mathbf{0}_n^T \implies$$

$$\implies \mathbf{h} \left[\mathrm{B}^{(\mu)} \vdots A\mathrm{B}^{(\mu)} \vdots A^2\mathrm{B}^{(\mu)} \vdots ...\vdots A^{n-1}\mathrm{B}^{(\mu)}\right] = \mathbf{0}_{n(\mu+1)r}^T.$$

Let $\psi(z)$ be minimal degree polynomial of A such that $\mathbf{h}\psi(A) = \mathbf{0}_n^T$.
Then $deg(\psi) \geq 1$. Factorize $\psi(z) = \varphi(z)(z - s)$.

Then $\mathbf{0}_n^T = \mathbf{h}\psi(A) = \mathbf{h}\varphi(A)(A - sI)$. Define $\zeta = \mathbf{h}\varphi(A)$, $\zeta \neq \mathbf{0}_n^T \implies$

$$\zeta A = s\zeta \; and \; \zeta\mathrm{B}^{(\mu)} = \mathbf{h}\varphi(A)\mathrm{B}^{(\mu)} = \mathbf{0}_{n(\mu+1)r}^T \implies$$

$$\implies \zeta\left[sI - A \vdots \mathrm{B}^{(\mu)}\right] = \mathbf{0}_{n+(\mu+1)r}^T \implies rank\left[sI - A \vdots \mathrm{B}^{(\mu)}\right] < n.$$

The obtained result $rank\left[sI - A \vdots \mathrm{B}^{(\mu)}\right] < n$ contradicts the condition

5., i.e., $rank\left[sI - A \vdots \mathrm{B}^{(\mu)}\right] < n$ for every $s \in \mathbb{C}$. The assumed noncontrollability of the system caused the contradiction. The condition 5., i.e.,
$rank\left[sI - A \vdots \mathrm{B}^{(\mu)}\right] < n$ for every $s \in \mathbb{C}$ is sufficient for the system controllability.

Let us generalize also the proof by Chen [18, p. 206]. It is accommodated
to the system (10.3), (10.4) and to the notation of this book. Let us assume
that the condition 5. holds but that the system is state uncontrollable, which
means that every condition 1.-4. fails. The supposed system state uncontrol-
lability guarantees the existence of the equivalence transformation T such
that

$$TAT^{-1} = \begin{bmatrix} A_C & A_{12} \\ O_{n-k,k} & A_{NC} \end{bmatrix} = \overline{A} \in \mathfrak{R}^{n \times n}, \; A_C \in \mathfrak{R}^{k \times k}, \; 0 < k < n,$$

$$TB^{(\mu)} = \begin{bmatrix} \mathrm{B}_C^{(\mu)} \\ O_{n-k,(\mu+1)r} \end{bmatrix} = \overline{\mathrm{B}^{(\mu)}} \in \mathfrak{R}^{n \times (\mu+1)r}, \; \mathrm{B}_C^{(\mu)} \in \mathfrak{R}^{k \times (\mu+1)r},$$

where A_C corresponds to the state controllable part of the system, and
$A_{NC} \in \mathfrak{R}^{(n-k) \times (n-k)}$ is determined by the state uncontrollable part of the
system. The rows of the submatrix A_{NC} are linearly dependent. For any
eigenvalue $s_i = s_i(A)$ of the matrix A, $i \in \{1, 2, ..., n\}$, let a nonzero row
vector $\mathbf{b} \in \mathfrak{R}^{1 \times (n-k)}$, $\mathbf{b} \neq \mathbf{0}_{n-k}^T$, obey $\mathbf{b}A_{NC} = s_i\mathbf{b}$. Let another subsidiary

$1 \times n$ nonzero vector $\mathbf{a} = \begin{bmatrix} \mathbf{0}_k^T & \vdots & \mathbf{b} \end{bmatrix}$ multiply $\begin{bmatrix} s_i I - \overline{A} & \vdots & \overline{B^{(\mu)}} \end{bmatrix}$:

$$\mathbf{a} \begin{bmatrix} s_i I - \overline{A} & \vdots & \overline{B^{(\mu)}} \end{bmatrix} = \begin{bmatrix} \mathbf{0}_k^T & \vdots & \mathbf{b} \end{bmatrix} \begin{bmatrix} s_i I - \overline{A} & \vdots & \overline{B^{(\mu)}} \end{bmatrix} =$$

$$= \begin{bmatrix} \mathbf{0}_k^T & \vdots & \mathbf{b} \end{bmatrix} \begin{bmatrix} s_i I_k - A_C & A_{12} & \vdots & B_C^{(\mu)} \\ O_{n-k,k} & s_i I_{n-k} - A_{NC} & \vdots & O_{n-k,(\mu+1)r} \end{bmatrix} = \mathbf{0}_{n+(\mu+1)r}^T.$$

This yields

$$\mathbf{a} \begin{bmatrix} T \left(s_i I - A \right) T^{-1} & \vdots & TV \end{bmatrix} = \mathbf{a} T \begin{bmatrix} \left(s_i I - A \right) T^{-1} & \vdots & B^{(\mu)} \end{bmatrix} = \mathbf{0}_{n+(\mu+1)r}^T.$$

The nonsingularity of T and $\mathbf{a} = \begin{bmatrix} \mathbf{0}_k^T & \vdots & \mathbf{b} \end{bmatrix} \neq \mathbf{0}_n^T$ ensure $\mathbf{c} = \mathbf{a} T \neq \mathbf{0}_n^T$ so that $\mathbf{c} \left(s_i I - A \right) T^{-1} = \mathbf{0}_n^T$ implies $\mathbf{c} \left(s_i I - A \right) = \mathbf{0}_n^T$ in view of $det T \neq 0$. Hence,

$$\mathbf{a} T \begin{bmatrix} \left(s_i I - A \right) T^{-1} & \vdots & B^{(\mu)} \end{bmatrix} = \mathbf{0}_{n+(\mu+1)r}^T \implies \mathbf{c} \begin{bmatrix} \left(s_i I - A \right) & \vdots & B^{(\mu)} \end{bmatrix} = \mathbf{0}_{n+(\mu+1)r}^T.$$

The last equation means that there is an eigenvalue s_i of the matrix A such that

$$rank \begin{bmatrix} \left(s_i I - A \right) & \vdots & B^{(\mu)} \end{bmatrix} < n.$$

This contradicts the condition 5. that $rank \begin{bmatrix} \left(s_i I - A \right) & \vdots & B^{(\mu)} \end{bmatrix} = n$ for every $s \in \mathbb{C}$, i.e., for every eigenvalue $s_i = s_i(A)$ of the matrix A. The contradiction is the consequence of the assumption that the system is output uncontrollable despite $rank \begin{bmatrix} \left(s I - A \right) & \vdots & B^{(\mu)} \end{bmatrix} = n$ for every $s \in \mathbb{C}$, i.e., for every eigenvalue $s_i = s_i(A)$ of the matrix A. It follows that the condition 5. guarantees the system state controllability and the complete equivalence of the condition 5. to the conditions 1.-4.

    6.   a) If $\mu = 0$ then the first Equation (10.5) simplifies to:

$$\mathbf{S}(t; t_0; \mathbf{S}_0; \mathbf{U}) = \Phi(t, t_0) \mathbf{S}_0 + \int_{t_0}^{t} \Phi(t, \tau) B_0 \mathbf{U}(\tau) d\tau. \tag{D.88}$$

The system state controllability means that there is control $\mathbf{U}(t)$ that an arbitrary initial state $\mathbf{S}_0$ steers to an arbitrary final state $\mathbf{S}_1$ at a moment

$t_1 \in In\mathfrak{T}_0$ so that Equation (D.88) gives:

$$\int_{t_0}^{t_1} \Phi\left(t_1, \tau\right) \mathrm{B}_0 \mathbf{U}\left(\tau\right) d\tau = \mathbf{S}_1 - \Phi\left(t_1, t_0\right) \mathbf{S}_0. \qquad \text{(D.89)}$$

The linear independence of all rows of $\Phi\left(t_1, \tau\right) \mathrm{B}_0$ in $\tau \in [t_0, t_1]$ for $t_1 \in In\mathfrak{T}_0$, (the condition 1.) guarantees the nonsingularity of the grammian $G_{\Phi B}\left(t_1, t_0\right)$ of $\Phi\left(t_1, \tau\right)$ in $\tau \in [t_0, t_1]$ for $t_1 \in In\mathfrak{T}_0$,

$$G_{\Phi B}\left(t_1, t_0\right) = \int_{t_0}^{t} \Phi\left(t_1, \tau\right) \mathrm{B}_0 \mathrm{B}_0^T \Phi^T\left(t_1, \tau\right) d\tau.$$

This and the condition 1. imply the following solution to Equation (D.89):

$$\mathbf{U}\left(t\right) = \left(\Phi\left(t_1, t\right) \mathrm{B}_0\right)^T G_{\Phi B}^{-1}\left(t_1, t_0\right) \left[\mathbf{S}_1 - \Phi\left(t_1, t_0\right) \mathbf{S}_0\right],$$
$$\forall t \in [t_0, t_1], \ t_1 \in In\mathfrak{T}_0. \qquad \text{(D.90)}$$

Equation (D.90) is Equation (10.12). This proves the necessity of the condition 6.a) and its equivalence with the conditions 1.-5. for the necessity.

b) The condition 1. guarantees the linear independence of the rows of $\Phi\left(t_1, t\right) \mathrm{B}^{(\mu)}$ that further implies the linear independence of the rows of $\Phi\left(t_1, t\right) \mathrm{B}^{(\mu)} T$ on $[t_0, t_1]$, for any nonsingular matrix $T \in \mathfrak{R}^{(\mu+1)r \times (\mu+1)r}$ and for any $(t_0, t_1 > t_0) \in In\mathfrak{T}_0 \times In\mathfrak{T}_0$ (due to the statement 3) of Lemma 107 in Section 7.4). It follows that the Gram matrix $G_{\Phi BT}\left(t_1, t_0\right)$, Equation (10.14), of $\Phi\left(t_1, t\right) \mathrm{B}^{(\mu)} T$ is nonsingular (Theorem 114). The system motion, Equation (10.5), can be presented in the following form:

$$\mathbf{S}\left(t; t_0; \mathbf{S}_0; \mathbf{U}^\mu\right) = \Phi\left(t, t_0\right) \mathbf{S}_0 + \int_{t_0}^{t} \Phi\left(t, \tau\right) \mathrm{B}^{(\mu)} T T^{-1} \mathbf{U}^\mu\left(\tau\right) d\tau,$$

which for $t = t_1$ can be shown as

$$\int_{t_0}^{t_1} \Phi\left(t_1, \tau\right) \mathrm{B}^{(\mu)} T T^{-1} \mathbf{U}^\mu\left(\tau\right) d\tau = \mathbf{S}\left(t_1; t_0; \mathbf{S}_0; \mathbf{U}^\mu\right) - \Phi\left(t_1, t_0\right) \mathbf{S}_0,$$

or, for an arbitrary initial state $\mathbf{S}_0$ and final state $\mathbf{S}_1$ at $t = t_1$, $\mathbf{S}_1 = \mathbf{S}\left(t_1; t_0; \mathbf{S}_0; \mathbf{U}^\mu\right)$, becomes

$$\int_{t_0}^{t_1} \Phi\left(t_1, \tau\right) \mathrm{B}^{(\mu)} T T^{-1} \mathbf{U}^\mu\left(\tau\right) d\tau = \mathbf{S}_1 - \Phi\left(t_1, t_0\right) \mathbf{S}_0. \qquad \text{(D.91)}$$

The system state controllability, the linear independence of the rows of $\Phi(t_1, \tau) \mathrm{B}^{(\mu)} T$ on $[t_0, t_1]$, for any $(t_0, t_1 > t_0) \in In\mathfrak{T}_0 \times In\mathfrak{T}_0$, and the nonsingularity of the Gram matrix $G_{\Phi BT}(t_1, t_0)$, Equation (10.14), of $\Phi(t_1, t) \mathrm{B}^{(\mu)} T$, imply the existence of the solution $\mathbf{U}^\mu(t)$ to Equation (D.91) in the following form:

$$T^{-1} \mathbf{U}^\mu(t) = T^T \mathrm{B}^{(\mu)T} \Phi^T(t_1, t) G_{\Phi BT}(t_1, t_0) [\mathbf{S}_1 - \Phi(t_1, t_0) \mathbf{S}_0].$$

This proves the necessity of Equation (10.13) and its equivalence with the conditions 1.-5. for the necessity.

c) The condition 1. ensures the linear independence of the rows of $\Phi(t_1, \tau) \mathrm{B}^{(\mu)}$ on $[t_0, t_1]$, for any $(t_0, t_1 > t_0) \in In\mathfrak{T}_0 \times In\mathfrak{T}_0$, which implies the linear independence of the rows of $\Phi(t_1, t)$ on $[t_0, t_1]$, for any $(t_0, t_1 > t_0) \in In\mathfrak{T}_0 \times In\mathfrak{T}_0$ (Lemma 108, Section 7.4). It follows that the Gram matrix $G_\Phi(t_1, t_0)$ (10.16) of $\Phi(t_1, t)$ is nonsingular (Theorem 114). The system motion, Equation (10.5), can be presented in the following form:

$$\mathbf{S}(t; t_0; \mathbf{S}_0; \mathbf{U}^\mu) = \Phi(t, t_0) \mathbf{S}_0 + \int_{t_0}^{t} \Phi(t, \tau) \left(\mathrm{B}^{(\mu)} \mathbf{U}^\mu(\tau) \right) d\tau,$$

which for $t = t_1$ can be expressed as

$$\int_{t_0}^{t_1} \Phi(t_1, \tau) \left(\mathrm{B}^{(\mu)} \mathbf{U}^\mu(\tau) \right) d\tau = \mathbf{S}(t_1; t_0; \mathbf{S}_0; \mathbf{U}^\mu) - \Phi(t_1, t_0) \mathbf{S}_0,$$

and for an arbitrary initial state $\mathbf{S}_0$ and final state $\mathbf{S}_1$ at $t = t_1$, $\mathbf{S}_1 = \mathbf{S}(t_1; t_0; \mathbf{S}_0; \mathbf{U}^\mu)$, becomes

$$\int_{t_0}^{t_1} \Phi(t_1, \tau) \left(\mathrm{B}^{(\mu)} \mathbf{U}^\mu(\tau) \right) d\tau = \mathbf{S}_1 - \Phi(t_1, t_0) \mathbf{S}_0.$$

This, the system state controllability and the linear independence of the rows of $\Phi(t_1, t)$ on $[t_0, t_1]$, for any $(t_0, t_1 > t_0) \in In\mathfrak{T}_0 \times In\mathfrak{T}_0$, and the nonsingularity of the Gram matrix $G_\Phi(t_1, t_0)$ (10.16) of $\Phi(t_1, t)$ imply the existence of the solution $\mathbf{U}^\mu(t)$ to the preceding equation in the following form:

$$\mathrm{B}^{(\mu)} \mathbf{U}^\mu(t) = \Phi^T(t_1, t) G_\Phi^{-1}(t_1, t_0) [\mathbf{S}_1 - \Phi(t_1, t_0) \mathbf{S}_0].$$

This proves the necessity of Equation (10.15) and completes the proof of the necessity of the condition 6c and its equivalence with the conditions 1.-5. for the necessity. It completes the proof of the necessity part.

Proof of the sufficiency of condition 6.

Let $\mathbf{S}_0 \in \mathfrak{R}^n$ and $\mathbf{S}_1 \in \mathfrak{R}^n$ be arbitrarily selected. Let any of the equivalent conditions 1. through 6a if $\mu = 0$ or 1. through 6b or 1. through 6c if $\mu > 0$ be valid. Then, each of them holds. We treat separately the case 6.a) $\mu = 0$ and the case 6.b) or 6.c) in the proof of the sufficiency.

Case $\mu = 0$.

6.a) The condition 6.a) guarantees the existence of the control vector function $\mathbf{U}(.)$ defined by Equation (10.12):

$$\mathbf{U}(t) = (\Phi(t_1, t) \, \mathbf{B}_0)^T \, G_{\Phi B}^{-1}(t_1, t_0) \, [\mathbf{S}_1 + \Phi(t_1, t_0) \, \mathbf{S}_0],$$
$$\forall t \in [t_0, t_1], \ t_1 \in In\mathfrak{T}_0.$$

We use this equation to eliminate the control vector from the first Equation (10.5) for $t = t_1$:

$$\mathbf{S}(t_1; t_0; \mathbf{S}_0; \mathbf{U}) = \Phi(t_1, t_0) \, \mathbf{S}_0 + \int_{t_0}^{t_1} \Phi(t_1, \tau) \, \mathbf{B}_0 \mathbf{U}(\tau) \, d\tau =$$

$$= \Phi(t_1, t_0) \, \mathbf{S}_0 + \left\{ \begin{array}{c} \displaystyle\int_{t_0}^{t_1} \Phi(t_1, \tau) \, \mathbf{B}_0 \, (\Phi(t_1, \tau) \, \mathbf{B}_0)^T \, d\tau \bullet \\ \\ \bullet G_{\Phi B}^{-1}(t_1, t_0) \, [\mathbf{S}_1 - \Phi(t_1, t_0) \, \mathbf{S}_0] \end{array} \right\} =$$

$$= \Phi(t_1, t_0) \, \mathbf{S}_0 + G_{\Phi B}(t_1, t_0) \, G_{\Phi B}^{-1}(t_1, t_0) \, [\mathbf{S}_1 - \Phi(t_1, t_0) \, \mathbf{S}_0] =$$
$$= \Phi(t_1, t_0) \, \mathbf{S}_0 + \mathbf{S}_1 - \Phi(t_1, t_0) \, \mathbf{S}_0 = \mathbf{S}_1.$$

The beginning and the end of these equations read:

$$\mathbf{S}(t_1; t_0; \mathbf{S}_0; \mathbf{U}) = \mathbf{S}_1.$$

The control $\mathbf{U}(t)$ (10.12) steers the system state from an arbitrarily chosen $\mathbf{S}_0 \in \mathfrak{R}^n$ to any chosen $\mathbf{S}_1 \in \mathfrak{R}^n$ over the finite time interval $[t_0, t_1]$. The system (10.3), (10.4) is state controllable in view of Definition 161 if $\mu = 0$.

This completes the necessity and equivalency of the conditions 1.-6.a).

Case $\mu > 0$

6.b) Let any of the equivalent conditions 1. through 5., 6.b) be valid. Then, each of them holds. We apply Equation (10.13) to the system motion, Equation (10.5), for $t = t_1$:

$$\mathbf{S}(t_1; t_0; \mathbf{S}_0; \mathbf{U}^\mu) = \Phi(t_1, t_0) \, \mathbf{S}_0 + \int_{t_0}^{t_1} \Phi(t_1, \tau) \, \mathbf{B}^{(\mu)} \mathbf{U}^\mu(\tau) \, d\tau =$$

$$= \Phi\left(t_1, t_0\right) \mathbf{S}_0 + \int_{t_0}^{t_1} \Phi\left(t_1, \tau\right) B^{(\mu)} TT^{-1} \mathbf{U}^\mu\left(\tau\right) d\tau =$$

$$= \Phi\left(t_1, t_0\right) \mathbf{S}_0 + \left\{ \begin{array}{c} \int_{t_0}^{t_1} \Phi\left(t_1, \tau\right) B^{(\mu)} TT^T B^{(\mu)T} \Phi^T\left(t_1, \tau\right) d\tau \bullet \\ \bullet G_{\Phi BT}^{-1}\left(t_1, t_0\right) \left[\mathbf{S}_1 - \Phi\left(t_1, t_0\right) \mathbf{S}_0\right] \end{array} \right\} =$$

$$= \Phi\left(t_1, t_0\right) \mathbf{S}_0 + \left\{ \begin{array}{c} G_{\Phi BT}\left(t_1, t_0\right) \bullet \\ \bullet G_{\Phi BT}^{-1}\left(t_1, t_0\right) \left[\mathbf{S}_1 - \Phi\left(t_1, t_0\right) \mathbf{S}_0\right] \end{array} \right\} =$$

$$= \Phi\left(t_1, t_0\right) \mathbf{S}_0 + \mathbf{S}_1 - \Phi\left(t_1, t_0\right) \mathbf{S}_0 = \mathbf{S}_1.$$

The result

$$\mathbf{S}\left(t_1; t_0; \mathbf{S}_0; \mathbf{U}^\mu\right) = \mathbf{S}_1$$

shows that the control $\mathbf{U}^\mu(t)$ (10.13) steers the system state from an arbitrarily chosen $\mathbf{S}_0 \in \mathfrak{R}^n$ to any chosen $\mathbf{S}_1 \in \mathfrak{R}^n$ over the finite time interval $[t_0, t_1]$. The system (10.3), (10.4) is state controllable in view of Definition 161 if $\mu > 0$. This completes the proof of the sufficiency and equivalency of every condition 1. -5. and of the condition 6.b)

6.c) Let Equation (10.15) hold. We replace $B^{(\mu)} \mathbf{U}^\mu(t)$ by the right-hand side of Equation (10.15) into Equation (10.5), for $t = t_1$:

$$\mathbf{S}\left(t_1; t_0; \mathbf{S}_0; \mathbf{U}^\mu\right) = \Phi\left(t_1, t_0\right) \mathbf{S}_0 + \int_{t_0}^{t_1} \Phi\left(t_1, \tau\right) \left(B^{(\mu)} \mathbf{U}^\mu\left(\tau\right)\right) d\tau =$$

$$= \Phi\left(t_1, t_0\right) \mathbf{S}_0 + \left\{ \begin{array}{c} \int_{t_0}^{t_1} \Phi\left(t_1, \tau\right) \Phi^T\left(t_1, \tau\right) d\tau \bullet \\ \bullet G_\Phi^{-1}\left(t_1, t_0\right) \left[\mathbf{S}_1 - \Phi\left(t_1, t_0\right) \mathbf{S}_0\right] \end{array} \right\} =$$

$$= \Phi\left(t_1, t_0\right) \mathbf{S}_0 + \left\{ \begin{array}{c} G_\Phi\left(t_1, t_0\right) \bullet \\ \bullet G_\Phi^{-1}\left(t_1, t_0\right) \left[\mathbf{S}_1 - \Phi\left(t_1, t_0\right) \mathbf{S}_0\right] \end{array} \right\} =$$

$$= \Phi\left(t_1, t_0\right) \mathbf{S}_0 + \mathbf{S}_1 - \Phi\left(t_1, t_0\right) \mathbf{S}_0 = \mathbf{S}_1.$$

The result

$$\mathbf{S}\left(t_1; t_0; \mathbf{S}_0; \mathbf{U}^\mu\right) = \mathbf{S}_1$$

proves that the control $\mathbf{U}^\mu(t)$ (10.15) steers the system state from an arbitrarily chosen $\mathbf{S}_0 \in \mathfrak{R}^n$ to any chosen $\mathbf{S}_1 \in \mathfrak{R}^n$ over the finite time interval $[t_0, t_1]$. The system (10.3), (10.4) is state controllable in view of Definition 161 if $\mu > 0$. This completes the proof of the sufficiency and equivalency of every condition 1.-5. and 6.c) and the whole proof. ■

D.11   Proof of Theorem 163

Proof. General necessary condition

Equation (10.4) reads at the initial moment t_0 :

$$\mathbf{Y}_0 = C\mathbf{X}_0 + Q\mathbf{U}_0 = \begin{bmatrix} C \vdots Q \end{bmatrix} \begin{bmatrix} \mathbf{X}_0 \\ \mathbf{U}_0 \end{bmatrix}.$$

In order to satisfy the condition of Definition 165 that the initial output vector $\mathbf{Y}_0$ can be any vector in the output space $\mathfrak{R}^N$ the vector $\begin{bmatrix} \mathbf{X}_0^T \vdots \mathbf{U}_0^T \end{bmatrix}^T$ should span the whole output space $\mathfrak{R}^N$. For the vector $\begin{bmatrix} \mathbf{X}_0^T \vdots \mathbf{U}_0^T \end{bmatrix}^T$ to span the whole output space $\mathfrak{R}^N$ it is necessary and sufficient that the matrix $\begin{bmatrix} C \vdots Q \end{bmatrix}$ has the full rank N, i.e., that Equation (10.29) holds. If this condition is not satisfied then there is not a control that can satisfy Definition 165. If the rank condition (10.29) is fulfilled then there can be control that satisfies Definition 165, but the rank condition (10.29) does not guarantee its existence. The rank condition (10.29) is not sufficient, but it is necessary condition for the system output controllability. From now on it is accepted that the system obeys the rank condition (10.29).

Necessity

Let the system (10.3), (10.4) be output controllable. Definition 165 is fulfilled, i.e., for every initial output vector $\mathbf{Y}_0 \in \mathfrak{R}^N$ at any $t_0 \in \mathfrak{T}$ and for any final output vector $\mathbf{Y}_1 \in \mathfrak{R}^N$ there exist a moment $t_1 \in In\mathfrak{T}_0$ and an extended control $\mathbf{U}_{[t_0,t_1]}^\mu$ on the *time* interval $[t_0,t_1]$ such that

$$\mathbf{Y}(t_1;t_0;\mathbf{Y}_0;\mathbf{U}_{[t_0,t_1]}^\mu) = \mathbf{Y}_1.$$

1. Let $\mu > 0$. Let us assume that the rows of the system matrix $H_S(t,t_0)$ (10.21) are linearly dependent on $[t_0,t_1]$ for any $(t_0,t_1 > t_0) \in In\mathfrak{T}_0 \times In\mathfrak{T}_0$. There is a nonzero $1xN$ vector $\mathbf{w}$ such that (Lemma 100, Section 7.4):

$$\mathbf{w} \in \mathfrak{R}^{1\times N}, \ \mathbf{w} \neq \mathbf{0}_N^T \ \ and \ \mathbf{w}H_S(t,t_0) = \mathbf{0}_{(\mu+1)r}^T, \ \forall t \in [t_0,t_1], \ t_1 \in In\mathfrak{T}_0.$$

Let the initial state vector $\mathbf{S}_0$ and the final output vector $\mathbf{Y}_1$ be chosen as follows:

a) *if* $\mathbf{w}C \neq \mathbf{0}_n^T \Longrightarrow \mathbf{S}_0 = \Phi^T(t_1,t_0)C^T\mathbf{w}^T \in \mathfrak{R}^n$ *and* $\mathbf{Y}_1 = Q^{(\mu-1)}\mathbf{U}_0^{\mu-1}$,

b) $\mathbf{w}C = \mathbf{0}_n^T \Longrightarrow \mathbf{S}_0$ *is any vector in* $\mathfrak{R}^n$ *and* $\mathbf{Y}_1 = \mathbf{w}^T + Q^{(\mu-1)}\mathbf{U}_0^{\mu-1}$.

Let $\mathbf{w}$ multiplies on the left the system response (10.28):

$$\mathbf{w}\mathbf{Y}\left(t; t_0; \mathbf{S}_0; \mathbf{U}^\mu\right) =$$

$$= \mathbf{w}C\Phi\left(t, t_0\right)\mathbf{S}_0 + \mathbf{w}\int_{t_0}^{t} H_S(t, \tau)\mathbf{U}^\mu\left(\tau\right)d\tau + \mathbf{w}Q^{(\mu-1)}\mathbf{U}_0^{\mu-1},$$

which at $t = t_1$ becomes

$$\mathbf{w}\mathbf{Y}\left(t_1; t_0; \mathbf{S}_0; \mathbf{U}^\mu\right) = \mathbf{w}\mathbf{Y}_1 =$$

$$= \mathbf{w}C\Phi\left(t_1, t_0\right)\mathbf{S}_0 + \mathbf{w}\int_{t_0}^{t_1} H_S(t_1, \tau)\mathbf{U}^\mu\left(\tau\right)d\tau + \mathbf{w}Q^{(\mu-1)}\mathbf{U}_0^{\mu-1} \implies$$

In the case a) : $\mathbf{w}\mathbf{Y}_1 = \mathbf{w}Q^{(\mu-1)}\mathbf{U}_0^{\mu-1} =$

$$= \mathbf{w}C\Phi\left(t_1, t_0\right)\Phi^T\left(t_1, t_0\right)C^T\mathbf{w}^T + \mathbf{w}\int_{t_0}^{t_1} H_S(t_1, \tau)\mathbf{U}^\mu\left(\tau\right)d\tau + \mathbf{w}Q^{(\mu-1)}\mathbf{U}_0^{\mu-1}$$

$$\implies$$

$$0 = \mathbf{w}C\Phi\left(t_1, t_0\right)\Phi^T\left(t_1, t_0\right)C^T\mathbf{w}^T \neq 0$$

because $\Phi\left(t_1, t_0\right)\Phi^T\left(t_1, t_0\right)$ is positive definite matrix at every $t \in \mathfrak{T}$ due to its symmetricity and nonsingularity at every $t \in \mathfrak{T}$, and because $C^T\mathbf{w}^T \neq \mathbf{0}_N$. The result $0 \neq 0$ shows that the assumed vector $\mathbf{w} \neq \mathbf{0}_N^T$ does not exist.

In the case b) : $\mathbf{w}\mathbf{Y}_1 = \mathbf{w}\mathbf{w}^T + \mathbf{w}Q^{(\mu-1)}\mathbf{U}_0^{\mu-1} =$

$$= \mathbf{w}C\Phi\left(t_1, t_0\right)\mathbf{S}_0 + \mathbf{w}\int_{t_0}^{t_1} H_S(t_1, \tau)\mathbf{U}^\mu\left(\tau\right)d\tau + \mathbf{w}Q^{(\mu-1)}\mathbf{U}_0^{\mu-1} \implies$$

$$\mathbf{w}\mathbf{w}^T = 0 \iff \mathbf{w} = \mathbf{0}_N^T;$$

Again, the assumed vector $\mathbf{w} \neq \mathbf{0}_N^T$ does not exist. The conclusions for the cases a) and b) prove that the rows of the system matrix $H_S(t_1, \tau)$ (10.28) are linearly independent on $[t_0, t_1]$, for any $(t_0, t_1 > t_0) \in In\mathfrak{T}_0 \times In\mathfrak{T}_0$. This proves the necessity of the condition 1. if $\mu > 0$.

Let now $\mu = 0$. We slightly modify the preceding proof valid for $\mu > 0$. Let us assume that the rows of the system matrix function $H_S(t_1, \tau)$

(10.21) are linearly dependent on $[t_0, t_1]$ for any $(t_0, t_1 > t_0) \in In\mathfrak{T}_0 \times In\mathfrak{T}_0$. There is a nonzero $1xN$ vector $\mathbf{w}$ such that

$$\mathbf{w} \in \mathfrak{R}^{1 \times N}, \ \mathbf{w} \neq \mathbf{0}_N^T \ and \ \mathbf{w}H_S(t_1, t) = \mathbf{0}_N^T, \ \forall t \in [t_0, t_1], \ t_1 \in In\mathfrak{T}_0.$$

Let the initial state vector $\mathbf{S}_0$ and the final output vector $\mathbf{Y}_1$ be chosen as follows:

$$a) \ if \ \mathbf{w}C \neq \mathbf{0}_n^T \Longrightarrow \mathbf{S}_0 = \Phi^T (t_1, t_0) C^T \mathbf{w}^T \in \mathfrak{R}^n \ and \ \mathbf{Y}_1 = \mathbf{0}_N,$$

$$b) \ \mathbf{w}C = \mathbf{0}_n^T \Longrightarrow \mathbf{S}_0 \ is \ any \ vector \ in \ \mathfrak{R}^n \ and \ \mathbf{Y}_1 = \mathbf{w}^T$$

Let $\mathbf{w}$ multiplies on the left the system response (10.28):

$$\mathbf{w}\mathbf{Y}(t; t_0; \mathbf{S}_0; \mathbf{U}) =$$

$$= \mathbf{w}C\Phi(t, t_0) \mathbf{S}_0 + \mathbf{w} \int_{t_0}^{t} H_S(t, \tau)\mathbf{U}(\tau) d\tau,$$

which at $t = t_1$ becomes

$$\mathbf{w}\mathbf{Y}(t_1; t_0; \mathbf{S}_0; \mathbf{U}) = \mathbf{w}\mathbf{Y}_1 =$$

$$= \mathbf{w}C\Phi(t_1, t_0) \mathbf{S}_0 + \mathbf{w} \int_{t_0}^{t_1} H_S(t_1, \tau)\mathbf{U}(\tau) d\tau \Longrightarrow$$

$$In \ the \ case \ a): \ \mathbf{w}\mathbf{Y}_1 = \mathbf{w}\mathbf{0}_N = 0 =$$

$$= \mathbf{w}C\Phi(t_1, t_0) \Phi^T (t_1, t_0) C^T \mathbf{w}^T + \mathbf{w} \int_{t_0}^{t_1} H_S(t_1, \tau)\mathbf{U}(\tau) d\tau$$

$$\Longrightarrow$$

$$0 = \mathbf{w}C\Phi(t_1, t_0) \Phi^T (t_1, t_0) C^T \mathbf{w}^T \neq 0$$

because $\Phi(t, t_0) \Phi^T (t, t_0)$ is positive definite matrix at every $t \in \mathfrak{T}$ due to its symmetricity and nonsingularity at every $t \in \mathfrak{T}$, and because $C^T \mathbf{w}^T \neq \mathbf{0}_N$. The result $0 \neq 0$ shows that the assumed vector $\mathbf{w} \neq \mathbf{0}_N^T$ does not exist.

$$In \ the \ case \ b): \ \mathbf{w}\mathbf{Y}_1 = \mathbf{w}\mathbf{w}^T =$$

$$= \mathbf{w}C\Phi(t_1, t_0) \mathbf{S}_0 + \mathbf{w} \int_{t_0}^{t_1} H_S(t_1, \tau)\mathbf{U}(\tau) d\tau \Longrightarrow$$

$$\mathbf{w}\mathbf{w}^T = 0 \Longleftrightarrow \mathbf{w} = \mathbf{0}_N^T;$$

Again, the assumed vector $\mathbf{w} \neq \mathbf{0}_N^T$ does not exist. The conclusions for the cases a) and b) prove that the rows of the system matrix $H_S(t_1, \tau)$ (10.21) are linearly independent on $[t_0, t_1]$, for any $(t_0, t_1 > t_0) \in In\mathfrak{T}_0 \times In\mathfrak{T}_0$, also if $\mu = 0$. This completes the proof of the necessity of the condition 1.

2. $H_S(s)$ (10.22) is the Laplace transform of $H_S(t, t_0)$ (10.21). The Laplace transform is linear one-to-one operator so that for the rows of $H_S(s)$ to be linearly independent it is necessary and sufficient that the rows of $H_S(t, t_0)$ are linearly independent. This and the criterion under 1. prove the necessity of the criterion under 2.

3. The matrix $H_S(t_1, t)$ (10.21) is integrable on $\mathfrak{T}$ for every $\tau \in \mathfrak{T}$ which implies the existence of its Gram matrix $G_{H_S}(t_1, t_0)$ (10.30). The output controllability of the system (10.3), (10.4) implies the linear independence of the rows of the system matrix $H_S(t, t_0)$ (10.21) (due to the condition 1.), which proves the nonsingularity of its Gram matrix $G_{H_S}(t, t_0)$ (10.30) due to Theorem 114 (Section 7.4). The condition 3. is necessary condition.

4. The output controllability of the system guarantees the linear independence of the rows of $H_S(t, t_0)$ (10.21) on $\mathfrak{T}_0$, condition 1. It is known that the exponential matrix function $\Phi(.) = e^{A(.)}$, where $A \in \mathfrak{R}^{n \times n}$, is defined in terms of an infinite series,

$$\Phi(t, \tau) = e^{A(t-\tau)} = \sum_{i=0}^{i=\infty} \alpha_i(t, \tau) A^i,$$

$$\alpha_i(t, \tau) = \frac{t^i}{i!}, \quad i = 0, 1, ..., n-1, ... \implies$$

$$\alpha_0(t, \tau) \equiv 1, \quad \alpha_i(0) = 0, \quad i = 1, 2, ..., n-1, ... \tag{D.92}$$

[2, Equation (2.101), p. 125, Definition 4.1, p. 149], [76, 4.11.49 Definition, p. 251], and that it can be expressed, by applying Cayley-Hamilton theorem [2, Theorem 2.1, p. 124], [76, 4.5.25 Theorem, p. 167], as a finite series [2, Equation (2.101), p. 125], [66, Equation (11-8), p. 491], [81, p. 386],

$$A \in \mathfrak{R}^{n \times n}, \quad \Phi(t, \tau) = e^{A(t-\tau)} = \sum_{i=0}^{i=n-1} \beta_i(t, \tau) A^i,$$

$$\beta_i(t, \tau) \in \mathfrak{C}^{\infty}, \quad i = 0, 1, ..., n-1 \implies$$

$$\beta_0(t, \tau) \equiv 1, \quad \beta_i(0, 0) = 0, \quad i = 1, 2, ..., n-1. \tag{D.93}$$

The functions $\beta_i(t, \tau)$ are linearly independent on $\mathfrak{T}$ for every $\tau \in \mathfrak{T}$. We may set $H_S(t,\tau)$ (10.21) into the following form:

$$H_S(t,\tau) = \left\{ \begin{array}{l} \left[C\Phi(t,\tau)\mathrm{B}^{(\mu)} \vdots \widetilde{Q} \right] \left[\begin{array}{c} I_{(\mu+1)r} \\ I_{(\mu+1)r} \end{array} \right], \ \mu > 0, \\[12pt] \left[C\Phi(t,\tau)\mathrm{B}_0 \vdots Q \right] \left[\begin{array}{c} I_{(\mu+1)r} \\ \delta(t,\tau) I_{(\mu+1)r} \end{array} \right], \ \mu = 0, \end{array} \right\} \qquad \text{(D.94)}$$

so that the linear independence of the rows of $H_S(t_1,\tau)$ implies, due to 1) of Lemma 107 (Section 7.4), the linear independence of the rows of

$$\left[C\Phi(t_1,\tau)\mathrm{B}^{(\mu)} \vdots \widetilde{Q} \right], \ \mu > 0,$$

$$\left[C\Phi(t_1,\tau)\mathrm{B}_0 \vdots Q \right], \ \mu = 0, . \qquad \text{(D.95)}$$

Equations (D.93) enable us to set $\left[C\Phi(t_1,\tau)\mathrm{B}^{(\mu)} \vdots \widetilde{Q} \right]$ and $\left[C\Phi(t_1,\tau)\mathrm{B}_0 \vdots Q \right]$ in the following forms:

$$\mu > 0 \Longrightarrow \left[C\Phi(t_1,\tau)\mathrm{B}^{(\mu)} \vdots \widetilde{Q} \right] =$$

$$= \left[\beta_0(t_1,\tau) C\mathrm{B}^{(\mu)} \vdots \beta_1(t_1,\tau) CA\mathrm{B}^{(\mu)} \vdots ... \vdots \beta_{n-1}(t_1,\tau) CA^{n-1}\mathrm{B}^{(\mu)} \vdots \widetilde{Q} \right]. \quad \text{(D.96)}$$

$$\mu = 0 \Longrightarrow \left[C\Phi(t_1,\tau)\mathrm{B}_0 \vdots Q \right] =$$

$$= \left[\beta_0(t_1,\tau) C\mathrm{B}_0 \vdots \beta_1(t_1,\tau) CA\mathrm{B}_0 \vdots ... \vdots \beta_{n-1}(t_1,\tau) CA^{n-1}\mathrm{B}_0 \vdots Q \right]. \quad \text{(D.97)}$$

The rows of the matrix functions on the right-hand sides of these equations are linearly independent in $\tau \in [t_0,t_1]$, $t_1 \in In\mathfrak{T}_0$, due to the linear independence of the rows of the matrix functions $\left[C\Phi(t_1,.)\mathrm{B}^{(\mu)} \vdots \widetilde{Q} \right]$ and $\left[C\Phi(t_1,.)\mathrm{B}_0 \vdots Q \right]$ on the left-hand sides of the same equations, respectively. In spite of that, let us assume the matrix $\mathcal{C}_{Sout}$ has rank defective, i.e., that $rank\mathcal{C}_{Sout} < N$. This implies (Theorem 86, Section 7.2) that the rows of the constant matrices

$$\left[CB^{(\mu)} \vdots CAB^{(\mu)} \vdots CA^2B^{(\mu)} \vdots ... \vdots CA^{n-1}B^{(\mu)} \vdots \widetilde{Q} \right] \in \mathfrak{R}^{N \times (n+1)(\mu+1)r},$$

(D.98)

$$\left[CB_0 \vdots CAB_0 \vdots CA^2B_0 \vdots ... \vdots CA^{n-1}B_0 \vdots Q \right] \in \mathfrak{R}^{N \times (n+1)r}, \qquad \text{(D.99)}$$

are linearly dependent so that there is a constant non zero, $1 \times N$ vector $\mathbf{a}$, $\mathbf{a} \in \mathfrak{R}^{1 \times N}$, $\mathbf{a} \neq \mathbf{0}_N^T$, such that (due to Lemma 100, Section 7.4):

$$\mathbf{a} \left[CB^{(\mu)} \vdots CAB^{(\mu)} \vdots CA^2B^{(\mu)} \vdots ... \vdots CA^{n-1}B^{(\mu)} \vdots \widetilde{Q} \right] =$$

$$= \left[\mathbf{a}CB^{(\mu)} \vdots \mathbf{a}CAB^{(\mu)} \vdots \mathbf{a}CA^2B^{(\mu)} \vdots ... \vdots \mathbf{a}CA^{n-1}B^{(\mu)} \vdots \mathbf{a}\widetilde{Q} \right] = \mathbf{0}_{(n+1)(\mu+1)r}^T$$

(D.100)

$$\Longleftrightarrow$$

$$\mathbf{a}CB^{(\mu)} = \mathbf{a}CAB^{(\mu)} = \mathbf{a}CA^2B^{(\mu)} = .. = \mathbf{a}CA^{n-1}B^{(\mu)} = \mathbf{a}\widetilde{Q} = \mathbf{0}_{(\mu+1)r}^T.$$

These equations and Equation (D.97) multiplied on the left by $\mathbf{a}$ yield

$$\mathbf{a} \left[\beta_0(t_1, \tau) CB^{(\mu)} \vdots \beta_1(t_1, \tau) CAB^{(\mu)} \vdots ... \vdots \beta_{n-1}(t_1, \tau) CA^{n-1}B^{(\mu)} \vdots \widetilde{Q} \right] =$$

$$= \left[\beta_0(t_1, \tau) \mathbf{a}CB^{(\mu)} \vdots \beta_1(t_1, \tau) \mathbf{a}CAB^{(\mu)} \vdots ... \vdots \beta_{n-1}(t_1, \tau) \mathbf{a}CA^{n-1}B^{(\mu)} \vdots \mathbf{a}\widetilde{Q} \right] =$$

$$= \left[\beta_0(t_1, \tau) \mathbf{0}_{(\mu+1)r}^T \vdots ... \vdots \beta_{n-1}(t_1, \tau) \mathbf{0}_{(\mu+1)r}^T \vdots \mathbf{0}_r^T \right] = \mathbf{0}_{(n+1)(\mu+1)r}^T.$$

This means that the rows of the matrix function

$$\left[\beta_0(t_1, \tau) CB^{(\mu)} \vdots \beta_1(t_1, \tau) CAB^{(\mu)} \vdots ... \vdots \beta_{n-1}(t_1, \tau) CA^{n-1}B^{(\mu)} \vdots \widetilde{Q} \right]$$

are linearly dependent that contradicts their independence. The contradiction is the consequence of the assumption that the rows of the constant matrix

$$\left[CB^{(\mu)} \vdots CAB^{(\mu)} \vdots CA^2B^{(\mu)} \vdots ... \vdots CA^{n-1}B^{(\mu)} \vdots \widetilde{Q} \right]$$

are linearly dependent. The assumption failure implies the linear independence of the rows of the constant matrix

$$\left[CB^{(\mu)} \vdots CAB^{(\mu)} \vdots CA^2B^{(\mu)} \vdots ... \vdots CA^{n-1}B^{(\mu)} \vdots \widetilde{Q} \right] \in \mathfrak{R}^{N \times (n+1)(\mu+1)r}.$$

This means that the matrix has the full row rank N:

$$rank \left[CB^{(\mu)} \vdots CAB^{(\mu)} \vdots CA^2B^{(\mu)} \vdots ... \vdots CA^{n-1}B^{(\mu)} \vdots \widetilde{Q} \right] = N. \quad (D.101)$$

By repeating the preceding presentation starting with Equation (D.100) in which Q is replaced by $\widetilde{Q}$ we prove the following:

$$rank \left[CB_0 \vdots CAB_0 \vdots CA^2B_0 \vdots ... \vdots CA^{n-1}B_0 \vdots Q \right] = N. \quad (D.102)$$

Equation (D.101) and Equation (D.102) prove that the condition for the matrix $\mathcal{C}_{Sout0}$ (10.32) to have the full rank N is necessary for the system output controllability, i.e., that Equation (10.33) holds. This completes the proof of the necessity of the condition 4.

5. Let be assumed that

$$rank \left[C(sI_n - A) \vdots CB^{(\mu)} \vdots \widetilde{Q} \right] < N, \; for \; some \; s_i = s_i(A) \in \mathbb{C}, \quad (D.103)$$

if $\mu > 0$. This implies the existence of a nonzero row vector $\mathbf{a} \in \mathbb{C}^{1 \times N}$ such that

$$\mathbf{a} \left[C(sI_n - A) \vdots CB^{(\mu)} \vdots \widetilde{Q} \right] = \mathbf{0}_{n+2(\mu+1)r}^T,$$

or, for $s \neq 0$:

$$s\mathbf{a}C = \mathbf{a}CA \; and \; \mathbf{a}CB^{(\mu)} = \mathbf{a}\widetilde{Q} = \mathbf{0}_r^T,$$

which imply

$$\mathbf{a}CA^2 = s\mathbf{a}CA = s^2\mathbf{a}C, ...\mathbf{a}CA^k = s^k\mathbf{a}C, \; k = 1, 2, ...$$

Hence,

$$\mathbf{a} \left[CAB^{(\mu)} \vdots CA^2B^{(\mu)} \vdots ... \vdots CA^{n-1}B^{(\mu)} \vdots CB^{(\mu)} \vdots \widetilde{Q} \right] =$$

$$= \left[s\mathbf{a}CB^{(\mu)} \vdots s^2\mathbf{a}CB^{(\mu)} \vdots ... \vdots s^{n-1}\mathbf{a}CB^{(\mu)} \vdots \mathbf{a}CB^{(\mu)} \vdots \mathbf{a}\widetilde{Q} \right] =$$

$$= \left[s\mathbf{0}_{(\mu+1)r}^T \vdots s^2\mathbf{0}_{(\mu+1)r}^T \vdots ... \vdots s^{n-1}\mathbf{0}_{(\mu+1)r}^T \vdots \mathbf{0}_{(\mu+1)r}^T \vdots \mathbf{0}_{(\mu+1)r}^T \right] = .$$

$$= \mathbf{0}_{(n+2)(\mu+1)r}^T.$$

This means that

$$rank \left[CAB^{(\mu)} \vdots CA^2B^{(\mu)} \vdots ... \vdots CA^{n-1}B^{(\mu)} \vdots CB^{(\mu)} \vdots \widetilde{q} \right] < N,$$

which implies the output noncontrollability of the system (10.3), (10.4) due to the condition under 4. This contradicts the assumed system output controllability, which is the consequence of the assumption that (D.87) holds. Therefore,

$$rank \left[C\left(sI_n - A\right) \vdots CB^{(\mu)} \vdots \widetilde{Q} \right] = N, \ \forall s_i = s_i\left(A\right) \in \mathbb{C},$$

$$i = 1, 2, ..., n, \ i.e., \ \forall s \in \mathbb{C}.$$

If $\mu = 0$ then we repeat the preceding proof of the necessity of the conditions under 4. with $\widetilde{Q}$ replaced by Q. The result is the following:

$$rank \left[C\left(sI_n - A\right) \vdots CB_0 \vdots Q \right] = N, \ \forall s_i = s_i\left(A\right) \in \mathbb{C},$$

$$i = 1, 2, ..., n, \ i.e., \ \forall s \in \mathbb{C}.$$

This completes the proof of the necessity of the condition under 5.

6.   a) In the case $\mu = 0$ Equation (10.28) for $t = t_1$, i.e., the following equation:

$$\mathbf{Y}\left(t_1; t_0; \mathbf{S}_0; \mathbf{U}\right) = C\Phi\left(t_1, t_0\right)\mathbf{S}_0 + \int_{t_0}^{t_1} H_S\left(t_1, \tau\right)\mathbf{U}\left(\tau\right)d\tau, \qquad \text{(D.104)}$$

together with $\mathbf{Y}\left(t_1; t_0; \mathbf{S}_0; \mathbf{U}\right) = \mathbf{Y}_1$ becomes

$$\int_{t_0}^{t_1} H_S\left(t_1, \tau\right)\mathbf{U}\left(\tau\right)d\tau = \mathbf{Y}_1 - C\Phi\left(t_1, t_0\right)\mathbf{S}_0. \qquad \text{(D.105)}$$

All rows of $H_S(t_1, \tau)$ are linearly independent on $[t_0, t_1]$, for any moment $t_1 \in In\mathfrak{T}_0$ (the condition 1.). This guarantees the nonsingularity of the Gram matrix $G_{H_S}\left(t_1, t_0\right)$ of $H_S(t_1, \tau)$ is nonsingular. These facts imply the following solution $\mathbf{U}\left(t\right)$ to Equation (D.105):

$$\mathbf{U}\left(t\right) = \left(H_S(t_1, \tau)\right)^T G_{H_S}^{-1}\left(t_1, t_0\right)\left[\mathbf{Y}_1 - C\Phi\left(t_1, t_0\right)\mathbf{S}_0\right], \qquad \text{(D.106)}$$

Equation (D.106) is Equation (10.36). In order to verify whether this solves Equation (D.105) we replace $\mathbf{U}\left(t\right)$ by the right-hand side of Equation (D.106) into Equation (D.105):

$$\int_{t_0}^{t_1} H_S\left(t_1, \tau\right)\mathbf{U}\left(\tau\right)d\tau = \left\{ \begin{array}{c} \left\{ \displaystyle\int_{t_0}^{t_1} H_S\left(t_1, \tau\right)\left(H_S\left(t_1, \tau\right)\right)^T d\tau \right\} \bullet \\ \bullet G_{H_S}^{-1}\left(t_1, t_0\right) \bullet \\ \bullet \left[\mathbf{Y}_1 - C\Phi\left(t_1, t_0\right)\mathbf{S}_0\right] \end{array} \right\} =$$

$$= G_{H_S}\left(t_1, t_0\right) G_{H_S}^{-1}\left(t_1, t_0\right)\left[\mathbf{Y}_1 - C\Phi\left(t_1, t_0\right)\mathbf{S}_0\right] =$$
$$= \mathbf{Y}\left(t_1; t_0; \mathbf{S}_0; \mathbf{U}\right) - C\Phi\left(t_1, t_0\right)\mathbf{S}_0,$$

i.e.,

$$\int_{t_0}^{t_1} H_S\left(t_1, \tau\right)\mathbf{U}\left(\tau\right) d\tau = \mathbf{Y}_1 - C\Phi\left(t_1, t_0\right)\mathbf{S}_0,$$

which is Equation (D.105). This completes successfully the verification.

b) Let $\mu > 0$ and $R \in \mathfrak{R}^{(\mu+1)r \times (\mu+1)r}$ be nonsingular constant matrix, $det R \neq 0$. This enables us to present the system response (10.28) in the following form:

$$\mathbf{Y}\left(t; t_0; \mathbf{S}_0; \mathbf{U}^\mu\right) = C\Phi\left(t, t_0\right)\mathbf{S}_0 +$$

$$+ \int_{t_0}^{t} H_S\left(t, \tau\right) R R^{-1}\mathbf{U}^\mu\left(\tau\right) d\tau + Q^{(\mu-1)}\mathbf{U}_0^{\mu-1}, \ \mu > 0. \qquad (\text{D.107})$$

The condition 1. implies the linear independence of the rows of $H_S(t, \tau)R$ in τ on $[t_0, t_1]$, for $t_1 \in In\mathfrak{T}_0$ (the statement 3) of Lemma 107 in Section 7.4). The linear independence of the rows of $H_S(t, \tau)R$ in τ on $[t_0, t_1]$, for $t_1 \in In\mathfrak{T}_0$, implies the nonsingularity of its Gram matrix $G_{H_SR}\left(t_1, t_0\right)$, Equation (10.38), and the following solution $\mathbf{U}\left(t\right)$ to Equation (D.107) for $t = t_1$:

$$R^{-1}\mathbf{U}^\mu\left(t\right) = \left\{ \begin{array}{c} \left(H_S\left(t_1, \tau\right) R\right)^T \bullet G_{H_SR}^{-1}\left(t_1, t_0\right) \bullet \\ \bullet \left[\mathbf{Y}_1 - C\Phi\left(t_1, t_0\right)\mathbf{S}_0 - Q^{(\mu-1)}\mathbf{U}_0^{\mu-1}\right] \end{array} \right\}, \ \mu > 0. \quad (\text{D.108})$$

This proves the necessity of Equation (10.37). It also completes the proof of the necessity of the condition 6b) and of the necessity part.

Sufficiency

Let $\mathbf{S}_0 \in \mathfrak{R}^n$, $\mathbf{Y}_0$, $\mathbf{Y}_1 \in \mathfrak{R}^N$ be any vectors. Let any of the equivalent conditions 1. - 4. holds, which means that each of them is valid.

a) Let $\mu = 0$. Let any of the equivalent conditions 1. - 4. and 6.a) holds, which means that each of them is valid if $\mu = 0$. We exploit the condition 6.a). Equation (10.36) transforms the system response at $t = t_1$, i.e., Equation (10.28) for $t = t_1$:

$$\mathbf{Y}\left(t_1; t_0; \mathbf{S}_0; \mathbf{U}\right) = C\Phi\left(t_1, t_0\right)\mathbf{S}_0 +$$

$$+ \int_{t_0}^{t_1} H_S\left(t_1, \tau\right)\left(H_S\left(t_1, \tau\right)\right)^T d\tau G_{H_S}^{-1}\left(t_1, t_0\right)\left[\mathbf{Y}_1 - C\Phi\left(t_1, t_0\right)\mathbf{S}_0\right] =$$

$$= C\Phi\left(t_1, t_0\right) \mathbf{S}_0 + G_{H_S}\left(t_1, t_0\right) G_{H_S}^{-1}\left(t_1, t_0\right)\left[\mathbf{Y}_1 - C\Phi\left(t_1, t_0\right) \mathbf{S}_0\right] =$$

$$= C\Phi\left(t_1, t_0\right) \mathbf{S}_0 + \mathbf{Y}_1 - C\Phi\left(t_1, t_0\right) \mathbf{S}_0 = \mathbf{Y}_1.$$

This result, $\mathbf{Y}\left(t_1; t_0; \mathbf{S}_0; \mathbf{U}\right) = \mathbf{Y}_1$, proves that the control $\mathbf{U}\left(t\right)$ defined by the right-hand side of Equation (10.36) steers the system output from an arbitrary initial output $\mathbf{Y}_0 = C\mathbf{S}_0$ to an arbitrary final output $\mathbf{Y}_1$ at a moment $t = t_1$. Definition 165 is fulfilled. The system (10.3), (10.4) is output controllable in the case a).

b) Let $\mu > 0$. Let any of the equivalent conditions 1. - 4., 6.b) holds, which means that each of them is valid if $\mu > 0$. Equation (10.37) and the system response Equation (10.28) for $t = t_1$ result in the following:

$$\mathbf{Y}\left(t_1; t_0; \mathbf{S}_0; \mathbf{U}^\mu\right) =$$

$$= C\Phi\left(t_1, t_0\right) \mathbf{S}_0 + \int_{t_0}^{t_1} H_S\left(t_1, \tau\right) RR^{-1}\mathbf{U}^\mu\left(\tau\right) d\tau + Q^{(\mu-1)}\mathbf{U}_0^{\mu-1} =$$

$$= C\Phi\left(t_1, t_0\right) \mathbf{S}_0 + Q^{(\mu-1)}\mathbf{U}_0^{\mu-1}+$$

$$+ \int_{t_0}^{t_1} H_S\left(t_1, \tau\right) R\left\{ \begin{array}{c} \left(H_S\left(t_1, \tau\right) R\right)^T \bullet G_{H_S R}^{-1}\left(t_1, t_0\right) \bullet \\ \bullet\left[\mathbf{Y}_1 - C\Phi\left(t_1, t_0\right) \mathbf{S}_0 - Q^{(\mu-1)}\mathbf{U}_0^{\mu-1}\right] \end{array} \right\} d\tau =$$

$$= C\Phi\left(t_1, t_0\right) \mathbf{S}_0 + Q^{(\mu-1)}\mathbf{U}_0^{\mu-1}+$$

$$+ G_{H_S R}\left(t_1, t_0\right) G_{H_S R}^{-1}\left(t_1, t_0\right)\left[\mathbf{Y}_1 - C\Phi\left(t_1, t_0\right) \mathbf{S}_0 - Q^{(\mu-1)}\mathbf{U}_0^{\mu-1}\right] =$$

$$= C\Phi\left(t_1, t_0\right) \mathbf{S}_0 + Q^{(\mu-1)}\mathbf{U}_0^{\mu-1} + \mathbf{Y}_1 - C\Phi\left(t_1, t_0\right) \mathbf{S}_0 - Q^{(\mu-1)}\mathbf{U}_0^{\mu-1} = \mathbf{Y}_1.$$

This proves

$$\mathbf{Y}\left(t_1; t_0; \mathbf{S}_0; \mathbf{U}^\mu\right) = \mathbf{Y}_1.$$

The control vector function solution $\mathbf{U}\left(.\right)$ of the *time*-invariant linear vector differential Equation (10.37) steers an arbitrary system initial output $\mathbf{Y}_0$ at the initial moment $t = t_0$ to an arbitrary final output $\mathbf{Y}_1 = \mathbf{Y}\left(t_1\right)$ at a moment $t_1 \in In\mathfrak{T}_0$. Definition 165 is satisfied. The system (10.3), (10.4) is output controllable. This completes the proof for the case b) and the proof in the whole. ∎

D.12 Proof of Theorem 173

Proof. Let the *IO* system (2.15) with $A_0 = O_N$ be robust state controllable relative to arbitrary matrices A_i, $i = 1, ..., \nu - 1$. Anyone of the conditions of Theorem 180 is satisfied for $\mu \geq 0$. We should now prove that the rank condition (11.23) is also necessary and sufficient.

Necessity.

In the case $A_0 = O_N$ for the *IO* system (2.15) to be state controllable the condition (11.23) becomes also necessary due to the fact that the system state controllability implies:

$$rank \left[(sI_n - A) \vdots B^{(\mu)} \right] = \left[sI_n - A \vdots \begin{array}{c} O_{(\nu-1)N,N} \\ A_\nu^{-1} B^{(\mu)} \end{array} \right] = n, \ \forall s \in \mathbb{C},$$

i.e.,

$$\nu > 1 \Longrightarrow \left[sI_n - A \vdots B^{(\mu)} \right] =$$

$$\begin{bmatrix}
sI_N & -I_N & .. & O_N & O_N & O_N & O_N \\
O_N & sI_N & .. & O_N & O_N & O_N & O_N \\
O_N & O_N & .. & O_N & O_N & O_N & O_N \\
O_N & O_N & .. & O_N & O_N & O_N & O_N \\
O_N & O_N & .. & O_N & O_N & O_N & O_N \\
\cdots & \cdots & .. & \cdots & \cdots & \cdots & \cdots \\
O_N & O_N & .. & sI_N & -I_N & O_N & O_N \\
O_N & O_N & .. & O_N & sI_N & -I_N & O_N \\
A_\nu^{-1}A_0 & A_\nu^{-1}A_1 & .. & A_\nu^{-1}A_{\nu-2} & A_\nu^{-1}A_{\nu-2} & sI_N + A_\nu^{-1}A_{\nu-1} & A_\nu^{-1}B^{(\mu)}
\end{bmatrix}$$

$$\nu = 1 \Longrightarrow A = -A_1^{-1}A_0 \in \mathfrak{R}^N, \left[sI_n - A \vdots B^{(1)} \right] = \left[sI_N + A_1^{-1}A_0 \vdots A_1^{-1}B^{(1)} \right].$$

which for $A_0 = O_N$ and $s = 0$ becomes

$$\left[(sI_n - A)_{s=0} \vdots B^{(\mu)} \right] =$$

$$= \left\{ \begin{array}{c} \begin{bmatrix}
O_N & -I_N & \cdots & O_N & O_N & O_N \\
O_N & O_N & \cdots & O_N & O_N & O_N \\
O_N & O_N & \cdots & O_N & O_N & O_N \\
\cdots & \cdots & \cdots & \cdots & \cdots & \cdots \\
O_N & O_N & \cdots & O_N & -I_N & O_N \\
O_N & A_\nu^{-1}A_1 & \cdots & A_\nu^{-1}A_{\nu-2} & A_\nu^{-1}A_{\nu-1} & A_\nu^{-1}B^{(\mu)}
\end{bmatrix}, \nu > 1, \\[4mm]
\left[O_N \vdots A_1^{-1}B^{(1)} \right], \nu = 1. \end{array} \right.$$

This yields

$$\forall s \in \mathbb{C} \implies n = rank \left[(sI_n - A)_{s=0} \vdots \mathrm{B}^{(\mu)} \right] =$$

$$= \left\{ rank \begin{bmatrix} O_N & -I_N & \dots & O_N & O_N & O_N \\ O_N & O_N & \dots & O_N & O_N & O_N \\ O_N & O_N & \dots & O_N & O_N & O_N \\ \dots & \dots & \dots & \dots & \dots & \dots \\ O_N & O_N & \dots & O_N & -I_N & O_N \\ O_N & A_\nu^{-1}A_1 & \dots & A_\nu^{-1}A_{\nu-2} & A_\nu^{-1}A_{\nu-1} & A_\nu^{-1}\mathrm{B}^{(\mu)} \end{bmatrix}, \\ if \ \nu > 1, \\ rank \left[O_N \vdots A_1^{-1}B^{(1)} \right], \ \nu = 1 \right\} =$$

$$= \left\{ rank \begin{bmatrix} -I_N & O_N & \dots & O_N & O_N & O_N \\ O_N & -I_N & \dots & O_N & O_N & O_N \\ O_N & O_N & \dots & O_N & O_N & O_N \\ \dots & \dots & \dots & \dots & \dots & \dots \\ O_N & O_N & \dots & O_N & -I_N & O_N \\ A_\nu^{-1}A_1 & A_\nu^{-1}A_2 & \dots & A_\nu^{-1}A_{\nu-2} & O_N & A_\nu^{-1}\mathrm{B}^{(\mu)} \end{bmatrix}, \\ if \ \nu > 1, \\ rank A_1^{-1}B^{(1)} = rank B^{(1)} due \ to \ det A_\nu^{-1} \neq 0, \ \nu = 1 \right\} =$$

$$= \left\{ \begin{array}{c} (\nu - 1) N + rank A_\nu^{-1}B^{(\mu)} = (\nu - 1) N + rank B^{(\mu)} = \nu N = n \\ \iff rank B^{(\mu)} = N, \ \nu > 1, \\ N \iff rank B^{(1)} = N, \ \nu = 1. \end{array} \right\}$$

Since $rank \left[(sI_n - A) \vdots \mathrm{B}^{(\mu)} \right] = n = \nu N$ for every $s \in \mathbb{C}$ is necessary and sufficient condition for the state controllability of the system, then it is necessary that $rank \left[(sI_n - A)_{s=0} \vdots \mathrm{B}^{(\mu)} \right] = n$ that implies $rank B^{(\mu)} = N$ independently of the arbitrary matrices A_i, $i = 1, ..., \nu - 1$, which proves the necessity of the condition (11.23) for any matrices A_i, $i = 1, ..., \nu - 1$.

Sufficiency.

Let the rank condition (11.23) hold in order to prove its sufficiency. It is valid independently of the arbitrary matrices A_i, $i = 1, ..., \nu - 1$. We repeat that $A_0 = O_N$ and

$$\nu > 1 \Longrightarrow \left[sI_n - A \vdots \mathrm{B}^{(\mu)} \right] =$$

$$\begin{bmatrix}
sI_N & -I_N & \cdots & O_N & O_N & O_N & O_N \\
O_N & sI_N & \cdots & O_N & O_N & O_N & O_N \\
O_N & O_N & \cdots & O_N & O_N & O_N & O_N \\
O_N & O_N & \cdots & O_N & O_N & O_N & O_N \\
O_N & O_N & \cdots & O_N & O_N & O_N & O_N \\
\cdots & \cdots & \cdots & \cdots & \cdots & \cdots & \cdots \\
O_N & O_N & \cdots & sI_N & -I_N & O_N & O_N \\
O_N & O_N & \cdots & O_N & sI_N & -I_N & O_N \\
O_N & A_\nu^{-1}A_1 & \cdots & A_\nu^{-1}A_{\nu-2} & A_\nu^{-1}A_{\nu-2} & sI_N + A_\nu^{-1}A_{\nu-1} & A_\nu^{-1}\mathrm{B}^{(\mu)}
\end{bmatrix}$$

$$\nu = 1 \Longrightarrow A = -A_1^{-1}A_0 \in \mathfrak{R}^N, \quad \left[sI_n - A \vdots \mathrm{B}^{(1)} \right] = \left[sI_N \vdots A_1^{-1}\mathrm{B}^{(1)} \right].$$

Its submatrix

$$\nu > 1 \Longrightarrow$$

$$\begin{bmatrix}
-I_N & \cdots & O_N & O_N & O_N & O_N \\
sI_N & \cdots & O_N & O_N & O_N & O_N \\
O_N & \cdots & O_N & O_N & O_N & O_N \\
O_N & \cdots & O_N & O_N & O_N & O_N \\
O_N & \cdots & O_N & O_N & O_N & O_N \\
\cdots & \cdots & \cdots & \cdots & \cdots & \cdots \\
O_N & \cdots & sI_N & -I_N & O_N & O_N \\
O_N & \cdots & O_N & sI_N & -I_N & O_N \\
A_\nu^{-1}A_1 & \cdots & A_\nu^{-1}A_{\nu-2} & A_\nu^{-1}A_{\nu-2} & sI_N + A_\nu^{-1}A_{\nu-1} & A_\nu^{-1}\mathrm{B}^{(\mu)}
\end{bmatrix}$$

$$\nu = 1 \Longrightarrow A_1^{-1}\mathrm{B}^{(1)} \text{ and } n = N.$$

has the full rank n due to the full rank N of $B^{(\mu)}$. This proves

$$rank \left[(sI_n - A) \vdots \mathrm{B}^{(\mu)} \right] = n \ on \ \mathbb{C}$$

and the sufficiency of the condition (11.23) that permits the following determination of the control vector function $\mathbf{U}(.)$ as an alternative to Equation (11.21). Let the control vector $\mathbf{U}(t)$ be the solution of the following *time-invariant* linear vector differential equation:

$$\mathrm{B}^{(\mu)}\mathbf{U}^\mu(t) = A_\nu \left(\Phi(t_0, t)\, \mathrm{B}_{inv} \right)^T G_{ctBinv}^{-1}(t_1, t_0) \left[\Phi(t_0, t_1)\, \mathbf{X}_1 - \mathbf{X}_0 \right],$$
$$\tag{D.109}$$

where $\mathbf{X}_0$ and $\mathbf{X}_1$ are arbitrary vectors in $\mathfrak{R}^n$. We use $\mathrm{B}^{(\mu)} = \mathrm{B}_{inv}A_{\nu}^{-1}B^{(\mu)}$ and replace $\mathrm{B}^{(\mu)}\mathbf{U}^{\mu}(t)$ by the right hand side of Equation (D.109) into the second Equation (11.10) that for the IO system (2.15) at $t = t_1$ then reads:

$$\mathbf{X}(t_1; t_0; \mathbf{X}_0; \mathbf{U}^{\mu}) =$$

$$= \Phi(t_1, t_0) \left[\mathbf{X}_0 + \int_{t_0}^{t_1} \left\{ \begin{array}{c} \Phi(t_0, \tau) \, \mathrm{B}_{inv} A_{\nu}^{-1} A_{\nu} \left(\Phi(t_0, t) \, \mathrm{B}_{inv} \right)^T \bullet \\ \bullet G_{ctBinv}^{-1}(t_1, t_0) \left[\Phi(t_0, t_1) \mathbf{X}_1 - \mathbf{X}_0 \right] \end{array} \right\} d\tau \right] =$$

$$= \Phi(t_1, t_0) \left\{ \mathbf{X}_0 + G_{ctBinv}(t_1, t_0) \bullet G_{ctBinv}^{-1}(t_1, t_0) \left[\Phi(t_0, t_1) \mathbf{X}_1 - \mathbf{X}_0 \right] \right\} =$$

$$= \Phi(t_1, t_0) \left[\mathbf{X}_0 + \Phi^{-1}(t_1, t_0) \mathbf{X}_1 - \mathbf{X}_0 \right] = \mathbf{X}_1.$$

The beginning and the end of these equations prove

$$\mathbf{X}(t_1; t_0; \mathbf{X}_0; \mathbf{U}^{\mu}) = \mathbf{X}_1.$$

The control vector $\mathbf{U}(t)$ (D.109) steers the system state from an arbitrary initial state $\mathbf{X}_0$ into an arbitrary final state $\mathbf{X}_1$ at $t_1 \in In\mathfrak{T}_0$. The IO system (2.15) with $A_0 = O_N$ is robust state controllable relative to arbitrary matrices A_i, $i = 1, ..., \nu - 1$. ■

D.13 Proof of Theorem 181

Lemma 234 Full rank of $\mathcal{C}_{IOout}$ and of $B^{(\mu)}$

If $\nu \geq 1$ and $\mu \geq 0$ then the full rank N of the matrix $B^{(\mu)}$ implies the full rank N of the matrix $\mathcal{C}_{IOout}$ (11.37),

$$\mathcal{C}_{IOout} = \left[C\mathrm{B}^{(\mu)} \vdots CA\mathrm{B}^{(\mu)} \vdots CA^2\mathrm{B}^{(\mu)} \vdots ... \vdots CA^{n-1}\mathrm{B}^{(\mu)} \right], \qquad \text{(D.110)}$$

$$full \; rank B^{(\mu)} = N. \Longrightarrow full \; rank \mathcal{C}_{IOout} =$$

$$= fullrank \left[C\mathrm{B}^{(\mu)} \vdots CA\mathrm{B}^{(\mu)} \vdots CA^2\mathrm{B}^{(\mu)} \vdots ... \vdots CA^{n-1}\mathrm{B}^{(\mu)} \right] = N. \quad \text{(D.111)}$$

Proof. Simple algebraic calculations show that Equations (11.5)-(11.7) (Section 11.1) yield the following:

$$CA = \left[\underbrace{O_N \;\; I_N \;\; ... \;\; O_N \;\; O_N}_{\nu \; columns} \right] \in R^{N \times \nu N}, \; CA\mathrm{B}^{(\mu)} = O_{N,r},$$

$$CA^2 = \begin{bmatrix} O_N & O_N & I_N & \ldots & O_N & O_N \end{bmatrix} \in R^{N \times \nu N}, \; CA^2 B^{(\mu)} = O_{N,r},$$

$$\ldots\ldots$$

$$CA^{\nu-1} = \begin{bmatrix} O_N & O_N & O_N & \ldots & O_N & I_N \end{bmatrix} \in R^{N \times \nu N}$$
$$CA^{\nu-1} B^{(\mu)} = A_\nu^{-1} B^{(\mu)},$$

which imply

$$rank C A^{\nu-1} B^{(\mu)} = rank A_\nu^{-1} B^{(\mu)} = rank B^{(\mu)}$$

because the nonsingular matrix A_ν^{-1} does not influence the rank of the product $A_\nu^{-1} B^{(\mu)}$, which equals the rank of $B^{(\mu)}$:

$$rank A_\nu^{-1} B^{(\mu)} = rank B^{(\mu)}.$$

Therefore,

$$rank B^{(\mu)} = N = rank C A^{\nu-1} B^{(\mu)} \Longrightarrow$$

$$full \; rank B^{(\mu)} = N \Longrightarrow rank \mathcal{C}_{IOout} =$$

$$rank \begin{bmatrix} O_N \vdots O_N \vdots \ldots \vdots O_N \vdots CA^{\nu-1}B^{(\mu)} \vdots CA^\nu B^{(\mu)} \vdots \ldots \vdots CA^{\nu N-1}B^{(\mu)} \end{bmatrix} =$$

$$= rank \begin{bmatrix} O_N \vdots O_N \vdots \ldots \vdots O_N \vdots A_\nu^{-1}B^{(\mu)} \vdots CA^\nu B^{(\mu)} \vdots \ldots \vdots CA^{\nu N-1}B \end{bmatrix} =$$

$$= full rank A_\nu^{-1} B^{(\mu)} = full rank B^{(\mu)} = N \Longrightarrow$$

$$full \; rank B^{(\mu)} = N \Longrightarrow rank \mathcal{C}_{IOout} = full \; rank \mathcal{C}_{IOout} = N.. \qquad \text{(D.112)}$$

This proves Lemma 234. ■

Proof. This is the proof of Theorem 191.

Let the condition (11.45) be valid. Lemma 234, the equivalence of the conditions $rank \mathcal{C}_{IOout} = N$ and the linear independence of the rows of $H_S(t, t_0)$ (10.21) due to Theorem 173, together with

$$H_S(t, t_0) = C\Phi(t, \tau)B^{(\mu)} = C\Phi(t, t_0)B_{inv}A_\nu^{-1}B^{(\mu)},$$

in view of Equation (11.32), imply the nonsingularity of $G_{C\Phi Binv}^{-1}(t_1, t_0)$ on $[t_1, t_0]$ for every $t_1 > t_0$, $(t_1, t_0) \in \mathfrak{T} \times \mathfrak{T}$. The right-hand side of Equation (11.46) is well defined and its solution $\mathbf{U}(t)$. Equations (11.31), (11.6) (Section 11.1) and (11.32) result in:

$$\mathbf{Y}(t; t_0; \mathbf{X}_0; \mathbf{U}^\mu) =$$

$$= C\Phi(t, t_0)\mathbf{X}_0 + \int_{t_0}^{t} C\Phi(t, \tau)B_{inv}A_\nu^{-1}B^{(\mu)}\mathbf{U}^\mu(\tau)\, d\tau. \qquad \text{(D.113)}$$

In order to verify that $\mathbf{U}(t)$ (11.46) steers an arbitrary initial output $\mathbf{Y}_0 = C\mathbf{X}_0$ to an arbitrary final output $\mathbf{Y}_1$ at $t = t_1$ we eliminate $B^{(\mu)}\mathbf{U}^\mu(\tau)$ from Equation (D.113) by using the right-hand side of Equation (11.46):

$$\mathbf{Y}(t_1; t_0; \mathbf{X}_0; \mathbf{U}^\mu) = C\Phi(t_1, t_0)\mathbf{X}_0+$$

$$+ \int_{t_0}^{t_1} C\Phi(t_1, \tau)B_{inv}A_\nu^{-1}A_\nu \left\{ \begin{array}{c} (C\Phi(t_1, t)B_{inv})^T G^{-1}_{C\Phi Binv}(t_1, t_0) \bullet \\ \bullet [\mathbf{Y}_1 - C\Phi(t_1, t_0)\mathbf{S}_0] \end{array} \right\} d\tau =$$

$$= C\Phi(t_1, t_0)\mathbf{X}_0 + \left\{ \begin{array}{c} G_{C\Phi Binv}(t_1, t_0) G^{-1}_{C\Phi Binv}(t_1, t_0) \bullet \\ \bullet [\mathbf{Y}_1 - C\Phi(t_1, t_0)\mathbf{X}_0] \end{array} \right\} = \mathbf{Y}_1.$$

The result $\mathbf{Y}(t_1; t_0; \mathbf{X}_0; \mathbf{U}^\mu) = \mathbf{Y}_1$ proves that the control $\mathbf{U}(t)$ (11.46) steers an arbitrary initial output $\mathbf{Y}_0 = C\mathbf{X}_0$ to an arbitrary final output $\mathbf{Y}_1$ at $t = t_1$. The *IO* system (2.15) is output controllable (Definition 165). ∎

D.14 Proof of Theorem 185

Proof. *Necessity.* Let the *IO* system be output controllable.

1) Let $\mathbf{Y}_{0-}^{\nu-1}$ and $\mathbf{U}_{0-}^{\mu-1}$ be arbitrary. The proof is by contradiction. Let be supposed that the rows of $\Gamma_{IOU}(.)$ are not linearly independent on $[t_0, t_1]$, i.e., that they are linearly dependent on $[0, t_1]$, $t_1 \in In\mathfrak{T}_0$. Then there is a nonzero, constant, $1 \times N$ vector $\mathbf{a}$,

$$\mathbf{a} \in \Re^{1 \times N}, \ \mathbf{a} \neq \mathbf{0}_N^T, \tag{D.114}$$

such that

$$\mathbf{a}\Gamma_{IOU}(t_1, t) = \mathbf{0}_r^T, \ \forall t \in [0, t_1], \ t_1 \in In\mathfrak{T}_0. \tag{D.115}$$

Premultiplying both sides of the first equation (2.40) by $\mathbf{a}$ for $t = t_1$ the result is

$$\mathbf{a}\mathbf{Y}(t_1; \mathbf{Y}_{0-}^{\nu-1}; \mathbf{U}) =$$

$$= \int_{0-}^{t_1} [\mathbf{a}\Gamma_{IOU}(t_1, \tau)\mathbf{U}(\tau)d\tau] + \mathbf{a}\Gamma_{IOu_0}(t_1)\mathbf{U}_{0-}^{\mu-1} + \mathbf{a}\Gamma_{IOy_0}(t_1)\mathbf{Y}_{0-}^{\nu-1},$$

i.e.,

$$\mathbf{a}\mathbf{Y}(t_1; \mathbf{Y}_{0-}^{\nu-1}; \mathbf{U}) = \mathbf{a}\Gamma_{IOu_0}(t_1)\mathbf{U}_{0-}^{\mu-1} + \mathbf{a}\Gamma_{IOy_0}(t_1)\mathbf{Y}_{0-}^{\nu-1}, \ t_1 \in In\mathfrak{T}_0,$$

due to Equation (D.115). Let, for any chosen $t_1 \in In\mathfrak{T}_0$, $\mathbf{Y}(t_1) = \mathbf{0}_N$ so that

$$\mathbf{aY}(t_1; \mathbf{Y}_{0-}^{\nu-1}; \mathbf{U}) = 0 = \mathbf{a}\Gamma_{IOu_0}(t_1)\mathbf{U}_{0-}^{\mu-1} + \mathbf{a}\Gamma_{IOy_0}(t_1)\mathbf{Y}_{0-}^{\nu-1},$$
$$\forall(\mathbf{U}_{0-}^{\mu-1}, .\mathbf{Y}_{0-}^{\nu-1}) \in \mathfrak{R}^{\mu r} \times \mathfrak{R}^{\nu N}.$$

Since this holds for any $t_1 \in In\mathfrak{T}_0$, $\mathbf{Y}_{0-}^{\nu-1}$ and $\mathbf{U}_{0-}^{\mu-1}$ then it implies $\mathbf{a} = \mathbf{0}_N^T$ that contradicts (D.114). Hence, the rows of $\Gamma_{IOU}(t_1, t)$ are linearly independent on $[t_0, t_1]$ for $t_1 \in In\mathfrak{T}_0$.

2) The linear independence of the rows of $\Gamma_{IOU}(t_1, \tau)$ on $[t_0, t_1]$ for $t_1 \in In\mathfrak{T}_0$ implies the linear independence of the rows of $\Gamma_{IOU}(t)$ on $[t_0, t_1]$ for $t_1 \in In\mathfrak{T}_0$ due to the following fact implied by Equation (D.115):

$$\int_{0-}^{t_1} [\mathbf{a}\Gamma_{IOU}(\tau)\mathbf{U}(t_1, \tau)d\tau] = \int_{0-}^{t_1} [\mathbf{a}\Gamma_{IOU}(t_1, \tau)\mathbf{U}(\tau)d\tau] = 0.$$

The Laplace transform $\mathcal{L}\{\Gamma_{IOU}(t)\}$ is the linear operator and represents the system transfer function $G_{IOU}(s)$ due to Equations (2.46). Altogether, The linear independence of the rows of $\Gamma_{IOU}(t_1, \tau)$ on $[t_0, t_1]$ for $t_1 \in In\mathfrak{T}_0$ implies the linear independence of the rows of the transfer function matrix $G_{IOU}(s)$ of the *IO* system (2.15) (see Equation (2.46)) on $\mathbb{C}$. This proves the necessity of the condition 2.

3) The equivalency of the conditions 1) and 3) results from Theorem 114, Section 7.4. It follows also from 2) of Lemma 194.

4) Another form of the system response $\mathbf{Y}(t; \mathbf{Y}_{0-}^{\nu-1}; \mathbf{U}^\mu)$ (11.58) is the following for $t = t_1$:

$$\mathbf{Y}(t_1; t_0; \mathbf{Y}_{0-}^{\nu-1}; \mathbf{U}^\mu) =$$
$$= \int_{t_0^-}^{t_1} \Theta_{IO}(t_1, \tau)A_\nu^{-1}B^{(\mu)}\mathbf{U}^\mu(\tau)d\tau + \Gamma_{IOy_0}(t_1)\mathbf{Y}_{0-}^{\nu-1},$$
$$\Gamma_{IOy_0}(t_1) = \int_{t_0^-}^{t_1} \Theta_{IO}(t, \tau)A_\nu^{-1}A^{(\nu)}\mathcal{L}^{-1}\left\{Z_N^{(\nu-1)}(s)\right\}d\tau, \qquad \text{(D.116)}$$

The proof is by contradiction. Let be supposed that the rows of the matrix product $\Theta_{IO}(t_1, t_0)A_\nu^{-1}B^{(\mu)}$ are linearly dependent on $[t_0, t_1]$, $t_1 \in In\mathfrak{T}_0$. Then there is a nonzero, constant, $1 \times N$ vector $\mathbf{g}$,

$$\mathbf{g} \in \mathfrak{R}^{1 \times N}, \ \mathbf{g} \neq \mathbf{0}_N^T \qquad \text{(D.117)}$$

such that

$$\mathbf{g}\Theta_{IO}(t_1, t)A_\nu^{-1}B^{(\mu)} = \mathbf{0}_{(\mu+1)r}^T, \ \forall t \in [t_0, t_1]. \qquad \text{(D.118)}$$

Premultiplying both sides of the first equation (D.116) by $\mathbf{g}$ the result is the following:

$$\mathbf{g}\mathbf{Y}(t_1; t_0; \mathbf{Y}_{0-}^{\nu-1}; \mathbf{U}) = \mathbf{g}\Gamma_{IOy_0}(t_1)\mathbf{Y}_{0-}^{\nu-1} +$$

$$+ \int_{t_0^-}^{t_1} \left[\mathbf{g}\Theta_{IO}(t_1, \tau)A_\nu^{-1}B^{(\mu)}\mathbf{U}^\mu(\tau)d\tau \right],$$

i.e.,

$$\mathbf{g}\mathbf{Y}(t_1; \mathbf{Y}_{0-}^{\nu-1}; \mathbf{U}) = \mathbf{g}\Gamma_{IOy_0}(t_1)\mathbf{Y}_{0-}^{\nu-1}, \ t_1 \in In\mathfrak{T}_0,$$

due to (D.118). Let any $t_1 \in In\mathfrak{T}_0$ be chosen and let $\mathbf{Y}(t_1) = \mathbf{0}_N$.so that

$$\mathbf{g}\mathbf{Y}(t_1; \mathbf{Y}_{0-}^{\nu-1}; \mathbf{U}) = 0 = \mathbf{g}\Gamma_{IOy_0}(t_1)\mathbf{Y}_{0-}^{\nu-1},$$

$$\forall(\mathbf{U}_{0-}^{\mu-1}, .\mathbf{Y}_{0-}^{\nu-1}) \in \mathfrak{R}^{\mu r} \times \mathfrak{R}^{\nu N}.$$

This holds for any $t_1 \in In\mathfrak{T}_0$ and any $\mathbf{Y}_{0-}^{\nu-1}$, which implies $\mathbf{g} = \mathbf{0}_N^T$ that contradicts (D.117). Hence, the rows of $\Theta_{IO}(t, t_0)A_\nu^{-1}B^{(\mu)}$ are linearly independent on $[t_0, t_1]$ for $t_1 \in In\mathfrak{T}_0$. This and the statement 1) of Lemma 107, (Section 7.4), prove that the rows of $\Theta_{IO}(t, t_0)$ are linearly independent on $[t_0, t_1]$ for $t_1 \in In\mathfrak{T}_0$.

5. The necessity of the condition 5) follows from the condition 4) of Lemma 194, (Section 11.2), and Theorem 114.

6)   a) We apply the system response $\mathbf{Y}(t_1; t_0; \mathbf{Y}_{0-}^{\nu-1}; \mathbf{U})$ in the form (2.40) for $t = t_1$:

$$\mathbf{Y}(t_1; t_0; \mathbf{Y}_{0-}^{\nu-1}; \mathbf{U}) = \left\{ \begin{array}{c} \int_{t_0^-}^{t_1} \Gamma_{IOU}(t_1, \tau)\mathbf{U}(\tau)d\tau + \\ +\Gamma_{IOu_0}(t_1)\mathbf{U}_{0-}^{\mu-1} + \Gamma_{IOy_0}(t_1)\mathbf{Y}_{0-}^{\nu-1} \end{array} \right\}. \quad (D.119)$$

It and the conditions 1) and 2) imply Equation

$$\mathbf{U}(t) = \Gamma_{IOU}(t_1, t)G_{\Gamma IO}^{-1}(t_1, t_0) \left[\mathbf{Y}_1 - \Gamma_{IOu_0}(t_1)\mathbf{U}_{0-}^{\mu-1} - \Gamma_{IOy_0}(t_1)\mathbf{Y}_{0-}^{\nu-1} \right], \quad (D.120)$$

which is Equation (11.75).

b) We start with the introduction of any nonsingular square matrix $R \in \mathfrak{R}^{(\mu+1)r \times (\mu+1)r}$, $detR \neq 0$. The system response $\mathbf{Y}(t; \mathbf{Y}_{0-}^{\nu-1}; \mathbf{U})$ (2.40) can be set in the following equivalent form that incorporates the matrix R and its inverse R^{-1}, i.e., their product $RR^{-1} = I_{(\mu+1)r}$:

$$\mathbf{Y}(t; t_0; \mathbf{Y}_{0-}^{\nu-1}; \mathbf{U}^\mu) =$$

$$= \int_0^t \left\{ \bullet \left[\begin{array}{c} \Theta_{IO}(t, \tau)\bullet \\ A_\nu^{-1}B^{(\mu)}RR^{-1}\mathbf{U}^\mu(\tau) + \\ +A_\nu^{-1}A^{(\nu)}\mathcal{L}^{-1}\left\{Z_N^{(\nu-1)}(s)\right\}\mathbf{Y}_{0\mp}^{\nu-1}. \end{array} \right] d\tau \right\}. \quad (D.121)$$

This and the conditions 4) and 5), i.e., Equation (11.74) (Section 11.2), imply the control   vector function $\mathbf{U}(.)$ defined by the following time-invariant linear differential equation:

$$R^{-1}\mathbf{U}^{\mu}(t) = \left(\Theta_{IO}(t_1, t)A_{\nu}^{-1}B^{(\mu)}R\right)^T G_{\Theta IOABR}^{-1}(t_1, t_0) \bullet$$

$$\bullet \left[\mathbf{Y}_1 - A_{\nu}^{-1}A^{(\nu)}\mathcal{L}^{-1}\left\{Z_N^{(\nu-1)}(s)\right\}\mathbf{Y}_{0\mp}^{\nu-1}\right]. \tag{D.122}$$

This is Equation (11.76).

c) The system response $\mathbf{Y}(t; t_0; \mathbf{Y}_{0-}^{\nu-1}; \mathbf{U})$ (11.58) can be set also in the following equivalent form:$^{\mu}$

$$\mathbf{Y}(t; t_0; \mathbf{Y}_{0-}^{\nu-1}; \mathbf{U}^{\mu}) =$$

$$= \int_{t_0}^{t} \left\{ \begin{array}{c} \Theta_{IO}(t, \tau)A_{\nu}^{-1}B^{(\mu)}\mathbf{U}^{\mu}(\tau) + \\ +\Theta_{IO}(t, \tau)A_{\nu}^{-1}A^{(\nu)}\mathcal{L}^{-1}\left\{Z_N^{(\nu-1)}(s)\right\}\mathbf{Y}_{0\mp}^{\nu-1} \end{array} \right\} d\tau.$$

It yields for $t = t_1$ and $\mathbf{Y}(t_1; t_0; \mathbf{Y}_{0-}^{\nu-1}; \mathbf{U}) = \mathbf{Y}_1$:

$$\int_{t_0}^{t_1} \Theta_{IO}(t_1, \tau)A_{\nu}^{-1}B^{(\mu)}\mathbf{U}^{\mu}(\tau)d\tau = \mathbf{Y}_1 -$$

$$- \int_{t_0}^{t_1} \Theta_{IO}(t_1, \tau)A_{\nu}^{-1}A^{(\nu)}\mathcal{L}^{-1}\left\{Z_N^{(\nu-1)}(s)\right\}\mathbf{Y}_{0\mp}^{\nu-1}d\tau.$$

This and the condition 5), i.e., Equation (11.71), imply

$$B^{(\mu)}\mathbf{U}^{\mu}(t) = A_{\nu}\Theta_{IO}^T(t_1, t)G_{\Theta IO}^{-1}(t_1, t_0) \bullet$$

$$\bullet \left[\mathbf{Y}_1 - \int_{t_0}^{t_1} \Theta_{IO}(t_1, \tau)A_{\nu}^{-1}A^{(\nu)}\mathcal{L}^{-1}\left\{Z_N^{(\nu-1)}(s)\right\}\mathbf{Y}_{0\mp}^{\nu-1}d\tau,\right]$$

which is Equation (11.77).

Sufficiency. Let the initial conditions $\mathbf{U}_{0-}^{\mu-1}$ and $\mathbf{Y}_{0-}^{\nu-1}$, and the final $\mathbf{Y}_1$ be arbitrary and fixed. Let any $t_1 \in In\mathfrak{T}_0$ be chosen. Let any of the conditions 1) - 5) holds.

a) Let the control vector be defined by Equation (11.75):

$$\mathbf{U}(t) = \Gamma_{IOU}(t_1, t)G_{\Gamma IO}^{-1}(t_1, t_0)\left[\mathbf{Y}_1 - \Gamma_{IOu_0}(t_1)\mathbf{U}_{0-}^{\mu-1} - \Gamma_{IOy_0}(t_1)\mathbf{Y}_{0-}^{\nu-1}\right].$$

We use this equation to eliminate $\mathbf{U}(\tau)$ from Equation (D.119):

$$\mathbf{Y}(t_1; t_0; \mathbf{Y}_{0-}^{\nu-1}; \mathbf{U}) =$$

$$= \int_{0-}^{t_1} \left\{ \begin{array}{c} \Gamma_{IOU}(t_1, \tau)\Gamma_{IOU}(t_1, t)G_{\Gamma IO}^{-1}(t_1, t_0) \bullet \\ \bullet \left[\mathbf{Y}_1 - \Gamma_{IOu_0}(t_1)\mathbf{U}_{0-}^{\mu-1} - \Gamma_{IOy_0}(t_1)\mathbf{Y}_{0-}^{\nu-1}\right] \end{array} \right\} d\tau +$$

$$+ \Gamma_{IOu_0}(t_1)\mathbf{U}_{0-}^{\mu-1} + \Gamma_{IOy_0}(t_1)\mathbf{Y}_{0-}^{\nu-1} =$$

$$\mathbf{Y}(t_1; t_0; \mathbf{Y}_{0-}^{\nu-1}; \mathbf{U}) = \left\{ \begin{array}{c} G_{\Gamma IO}(t_1, t_0)\, G_{\Gamma IO}^{-1}(t_1, t_0)\, \bullet \\ \bullet\left[\mathbf{Y}_1 - \Gamma_{IOu_0}(t_1)\mathbf{U}_{0-}^{\mu-1} - \Gamma_{IOy_0}(t_1)\mathbf{Y}_{0-}^{\nu-1}\right] \end{array} \right\} + $$
$$+\Gamma_{IOu_0}(t_1)\mathbf{U}_{0-}^{\mu-1} + \Gamma_{IOy_0}(t_1)\mathbf{Y}_{0-}^{\nu-1} = $$

$$\mathbf{Y}(t_1; t_0; \mathbf{Y}_{0-}^{\nu-1}; \mathbf{U}) = \mathbf{Y}_1 - \Gamma_{IOu_0}(t_1)\mathbf{U}_{0-}^{\mu-1} - \Gamma_{IOy_0}(t_1)\mathbf{Y}_{0-}^{\nu-1} + $$
$$+\Gamma_{IOu_0}(t_1)\mathbf{U}_{0-}^{\mu-1} + \Gamma_{IOy_0}(t_1)\mathbf{Y}_{0-}^{\nu-1} = \mathbf{Y}_1.$$

The chosen control vector function $\mathbf{U}(.)$ defined by (11.75) steers the system output from any chosen initial output $\mathbf{Y}_{0-}$ to any accepted final output $\mathbf{Y}_1$ at the chosen moment $t_1 \in In\mathfrak{T}_0$. Definition 165 is satisfied. The IO system (2.15) is output controllable.

     b) Let the control vector function be defined by Equation (11.76) that reads

$$R^{-1}\mathbf{U}^\mu(t) = \left(\Theta_{IO}(t_1, t) A_\nu^{-1} B^{(\mu)} R\right)^T G_{\Theta IOABR}^{-1}(t_1, t_0)\, \bullet$$
$$\bullet\left[\mathbf{Y}_1 - A_\nu^{-1} A^{(\nu)} \mathcal{L}^{-1}\left\{Z_N^{(\nu-1)}(s)\right\}\mathbf{Y}_{0-}^{\nu-1}\right] \qquad (D.123)$$

Equation (D.123) transforms Equation (D.116) into the following for $t = t_1$:

$$\mathbf{Y}(t_1; t_0; \mathbf{Y}_{0-}^{\nu-1}; \mathbf{U}^\mu) = $$

$$= \int_{t_0^-}^{t_1} \left\{ \begin{array}{c} \Theta_{IO}(t_1, \tau) A_\nu^{-1} B^{(\mu)} R \left(\Theta_{IO}(t_1, \tau) A_\nu^{-1} B^{(\mu)} R\right)^T \bullet \\ \bullet G_{\Theta IOABR}^{-1}(t_1, t_0)\left[\mathbf{Y}_1 - A_\nu^{-1} A^{(\nu)} \mathcal{L}^{-1}\left\{Z_N^{(\nu-1)}(s)\right\}\mathbf{Y}_{0-}^{\nu-1}\right] \end{array} \right\} d\tau + $$
$$+\Gamma_{IOy_0}(t_1)\mathbf{Y}_{0-}^{\nu-1} = $$

$$= \left\{ \begin{array}{c} G_{\Theta IOABR}(t_1, t_0)\, G_{\Theta IOABR}^{-1}(t_1, t_0)\, \bullet \\ \bullet\left[\mathbf{Y}_1 - A_\nu^{-1} A^{(\nu)} \mathcal{L}^{-1}\left\{Z_N^{(\nu-1)}(s)\right\}\mathbf{Y}_{0-}^{\nu-1}\right] \end{array} \right\} + $$
$$+A_\nu^{-1} A^{(\nu)} \mathcal{L}^{-1}\left\{Z_N^{(\nu-1)}(s)\right\}\mathbf{Y}_{0-}^{\nu-1} = $$

$$= \mathbf{Y}_1 - A_\nu^{-1} A^{(\nu)} \mathcal{L}^{-1}\left\{Z_N^{(\nu-1)}(s)\right\}\mathbf{Y}_{0-}^{\nu-1} + $$
$$+A_\nu^{-1} A^{(\nu)} \mathcal{L}^{-1}\left\{Z_N^{(\nu-1)}(s)\right\}\mathbf{Y}_{0-}^{\nu-1} = \mathbf{Y}_1.$$

The chosen control vector function $\mathbf{U}(.)$ defined by (11.76) steers the system output from any chosen initial output $\mathbf{Y}_{0-}$ to any accepted final output $\mathbf{Y}_1$ at the chosen moment $t_1 \in In\mathfrak{T}_0$. Definition 165 is satisfied. The IO system (2.15) is output controllable.

c) We replace $B^{(\mu)}\mathbf{U}^\mu(t)$ by the right hand side of Equation (11.77) in Equation (11.58) for $t = t_1$:

$$\mathbf{Y}(t_1; t_0; \mathbf{Y}_{0-}^{\nu-1}; \mathbf{U}^\mu) =$$

$$= \int_{t_0}^{t_1} \left\{ \begin{array}{c} \Theta_{IO}(t_1, \tau)\bullet \\ \bullet \left[\begin{array}{c} A_\nu^{-1} B^{(\mu)}\mathbf{U}^\mu(\tau) - A_\nu^{-1} B^{(\mu)} Z_r^{(\mu-1)}(\tau)\mathbf{U}_{0\mp}^{\mu-1} + \\ + A_\nu^{-1} A^{(\nu)} \mathcal{L}^{-1}\left\{ Z_N^{(\nu-1)}(s)\right\}\mathbf{Y}_{0\mp}^{\nu-1}. \end{array} \right] \end{array} d\tau \right\} \Longrightarrow .$$

$$\mathbf{Y}(t_1; t_0; \mathbf{Y}_{0-}^{\nu-1}; \mathbf{U}^\mu) =$$

$$= \left\{ \left\{ \begin{array}{c} \int_{t_0}^{t_1} \Theta_{IO}(t_1, \tau) A_\nu^{-1} A_\nu \Theta_{IO}^T(t_1, t) G_{\Theta IO}^{-1}(t_1, t_0)\, d\tau \bullet \\ \bullet \left[\begin{array}{c} \mathbf{Y}_1 - \int_{t_0}^{t_1} \Theta_{IO}(t_1, \tau) A_\nu^{-1} A^{(\nu)}\bullet \\ \bullet \mathcal{L}^{-1}\left\{ Z_N^{(\nu-1)}(s)\right\}\mathbf{Y}_{0\mp}^{\nu-1} d\tau, \end{array} \right] \\ + \int_{t_0}^{t_1} \Theta_{IO}(t_1, \tau) A_\nu^{-1} A^{(\nu)} \mathcal{L}^{-1}\left\{ Z_N^{(\nu-1)}(s)\right\}\mathbf{Y}_{0\mp}^{\nu-1} d\tau \end{array} \right\} + \right\},$$

or,

$$\mathbf{Y}(t_1; t_0; \mathbf{Y}_{0-}^{\nu-1}; \mathbf{U}^\mu) =$$

$$= \left\{ \left\{ \bullet \left[\begin{array}{c} G_{\Theta IO}(t_1, t_0)\, G_{\Theta IO}^{-1}(t_1, t_0)\bullet \\ \mathbf{Y}_1 - \\ -\int_{t_0}^{t_1} \Theta_{IO}(t_1, \tau) A_\nu^{-1} A^{(\nu)}\bullet \\ \bullet \mathcal{L}^{-1}\left\{ Z_N^{(\nu-1)}(s)\right\}\mathbf{Y}_{0\mp}^{\nu-1} d\tau, \end{array} \right] \right\} + \right\} =$$

$$\qquad + \int_{t_0}^{t_1} \Theta_{IO}(t_1, \tau) A_\nu^{-1} A^{(\nu)} Z_N^{(\nu-1)}(\tau)\mathbf{Y}_{0\mp}^{\nu-1} d\tau$$

$$= \mathbf{Y}_1.$$

The result $\mathbf{Y}(t_1; t_0; \mathbf{Y}_{0-}^{\nu-1}; \mathbf{U}^\mu) = \mathbf{Y}_1$ proves that the control vector function $\mathbf{U}(.)$ (11.77) steers any initial output $\mathbf{Y}_{0-}$ to any final output $\mathbf{Y}_1$ at a moment $t_1 \in In\mathfrak{T}_0$. Definition 165 is satisfied. The IO system (2.15) is output controllable. ∎

D.15   Proof of Theorem 215

Proof. Let $(t_0, t_1 > t_0) \in In\mathfrak{T}_0 \times In\mathfrak{T}_0$ be any chosen.

Necessity. Let the IIO system (11.179), (11.180) be output controllable.

1) Let $\mathbf{Y}_{0-}^{\nu-1}$ and $\mathbf{U}_{0-}^{\mu-1}$ be arbitrary. The proof is by contradiction. Let be supposed that the rows of $\Gamma_{IIO}(t, t_0)$ are not linearly independent on

$[t_0, t_1]$, i.e., that they are linearly dependent on $[t_0, t_1]$, $t_1 \in In\mathfrak{T}_0$. Then there is a nonzero, constant ,$1 \times N$ vector $\mathbf{a} \in \mathfrak{R}^{1 \times N}$,

$$\mathbf{a} \neq \mathbf{0}_N^T \tag{D.124}$$

such that

$$\mathbf{a}\Gamma_{IIO}(t, t_0) = \mathbf{0}_r^T, \ \forall t \in [t_0, t_1]. \tag{D.125}$$

Premultiplying Equation (11.241) by $\mathbf{a}$ the result is for $t_0 = 0$:

$$\mathbf{a}\mathbf{Y}(t; \mathbf{R}_{0-}^{\alpha-1}; \mathbf{Y}_{0-}^{\nu-1}; \mathbf{U}) =$$

$$= \int_{0-}^{t} [\mathbf{a}\Gamma_{IIO}(t, \tau)\mathbf{U}(\tau)d\tau] + \mathbf{a}\Gamma_{IIOU_0}(t)\mathbf{U}_{0-}^{\mu-1} + \mathbf{a}\Gamma_{IIOR_0}(t)\mathbf{R}_{0-}^{\alpha-1} +$$

$$+ \left\{ \begin{array}{c} \mathbf{a}\Gamma_{IIOY_0}(t)\mathbf{Y}_{0-}^{\nu-1}, \ \nu \geq 1, \\ \mathbf{0}_N \ , \ \nu = 0 \end{array} \right\}, \ \forall t \in \mathfrak{T}_0,$$

i.e.,

$$\mathbf{a}\mathbf{Y}(t; \mathbf{R}_{0-}^{\alpha-1}; \mathbf{Y}_{0-}^{\nu-1}; \mathbf{U}) = \mathbf{a}\Gamma_{IIOU_0}(t)\mathbf{U}_{0-}^{\mu-1} + \mathbf{a}\Gamma_{IIOR_0}(t)\mathbf{R}_{0-}^{\alpha-1} +$$

$$+ \left\{ \begin{array}{c} \mathbf{a}\Gamma_{IIOY_0}(t)\mathbf{Y}_{0-}^{\nu-1}, \ \nu \geq 1, \\ \mathbf{0}_N \ , \ \nu = 0 \end{array} \right\}, \ \forall t \in \mathfrak{T}_0,$$

due to (D.125). Let $\mathbf{Y}(t_1) = \mathbf{0}_N$.so that

$$\mathbf{a}\mathbf{Y}(t_1; \mathbf{R}_{0-}^{\alpha-1}; \mathbf{Y}_{0-}^{\nu-1}; \mathbf{U}) = \mathbf{0}_N = \mathbf{a}\Gamma_{IIOU_0}(t_1)\mathbf{U}_{0-}^{\mu-1} + \mathbf{a}\Gamma_{IIOR_0}(t_1)\mathbf{R}_{0-}^{\alpha-1} +$$

$$+ \left\{ \begin{array}{c} \mathbf{a}\Gamma_{IIOY_0}(t_1)\mathbf{Y}_{0-}^{\nu-1}, \ \nu \geq 1, \\ \mathbf{0}_N \ , \ \nu = 0 \end{array} \right\},$$

$$\forall (t_1, \mathbf{U}_{0-}^{\mu-1}, \mathbf{R}_{0-}^{\alpha-1}, \mathbf{Y}_{0-}^{\nu-1}) \in In\mathfrak{T}_0 \times \mathfrak{R}^{\mu r} \times \mathfrak{R}^{\alpha \rho} \times \mathfrak{R}^{\nu N}.$$

Since this holds for any t_1, $\mathbf{Y}_{0-}^{\nu-1}$, $\mathbf{R}_{0-}^{\alpha-1}$ and $\mathbf{U}_{0-}^{\mu-1}$ then it implies $\mathbf{a} = \mathbf{0}_N^T$ that contradicts (D.124). Hence, the rows of $\Gamma_{IIO}(t, t_0)$ are linearly independent on $[t_0, t_1]$.

2) Let $t_0 = 0$. Since the Laplace transform $\mathcal{L}\{\Gamma_{IIO}(t)\}$ of $\Gamma_{IIO}(t)$ is the transfer function matrix $G_{IIO}(s)$ of the *IIO* system (11.179), (11.180) (see Equation (6.32)),

$$\mathcal{L}\{\Gamma_{IIO}(t)\} = G_{IIO}(s). \tag{D.126}$$

then the necessity of the condition 2) is implied by the necessity of the condition 1).

3) The equivalence of the conditions 1) and 2) for the necessity results from Theorem 114, Section 7.4.

4) Equation (11.241) for $t = t_1$ reads:

$$\int_{0-}^{t_1} \Gamma_{IIO}(t_1, \tau)\mathbf{U}(\tau)d\tau = \mathbf{Y}(t_1; \mathbf{R}_{0-}^{\alpha-1}; \mathbf{Y}_0^{\nu-1}; \mathbf{U}^\mu) -$$

$$-\Gamma_{IIOI_0}(t_1)\mathbf{U}_{0-}^{\mu-1} - \Gamma_{IIOR_0}(t_1)\mathbf{R}_{0-}^{\alpha-1} - \Gamma_{IIOY_0}(t_1)\mathbf{Y}_{0-}^{\nu-1},$$

$$t_1 \in In\mathfrak{T}_0.$$

The conditions 1) and 2) imply the following solution $\mathbf{U}(t)$ to the preceding equation for $\mathbf{Y}(t_1; \mathbf{R}_{0-}^{\alpha-1}; \mathbf{Y}_0^{\nu-1}; \mathbf{U}^\mu) = \mathbf{Y}_1$ and $t_0 = 0$:

$$\mathbf{U}(t) = \Gamma_{IIO}^T(t_1, t)G_{\Gamma IIO}^{-!}(t_1) \begin{bmatrix} \mathbf{Y}_1 - \Gamma_{IIOI_0}(t_1)\mathbf{U}_{0-}^{\mu-1} - \\ -\Gamma_{IIOR_0}(t_1)\mathbf{R}_{0-}^{\alpha-1} - \Gamma_{IIOY_0}(t_1)\mathbf{Y}_{0-}^{\nu-1} \end{bmatrix}.$$

This is Equation 11.244, which proves its necessity.

Sufficiency. Let the initial conditions $\mathbf{U}_{0-}^{\mu-1}$ and $\mathbf{Y}_{0-}^{\nu-1}$, and the final $\mathbf{Y}_1$ be arbitrary and fixed. Let any $t_1 \in In\mathfrak{T}_0$ be chosen. Let any of the necessary conditions 1) - 3) holds. They guarantee the existence of the control vector function $\mathbf{U}(.)$ defined by Equation (11.244). We replace $\mathbf{U}(t)$ by the right-hand side of Equation (11.244) in the right-hand side of Equation (11.241) for $t = t_1$:

$$\mathbf{Y}(t_1; \mathbf{R}_{0-}^{\alpha-1}; \mathbf{Y}_0^{\nu-1}; \mathbf{U}^\mu) = \int_{0-}^{t_1} \Gamma_{IIO}(t_1, \tau)\mathbf{U}(\tau)d\tau +$$

$$+\Gamma_{IIOI_0}(t_1)\mathbf{U}_{0-}^{\mu-1} + \Gamma_{IIOR_0}(t_1)\mathbf{R}_{0-}^{\alpha-1} + \Gamma_{IIOY_0}(t_1)\mathbf{Y}_{0-}^{\nu-1} =$$

$$= \int_{0-}^{t_1} \left\{ \bullet \begin{bmatrix} \Gamma_{IIO}(t_1, \tau)\Gamma_{IIO}^T(t_1, t)d\tau G_{\Gamma IIO}^{-1}(t_1) \bullet \\ \mathbf{Y}_1 - \Gamma_{IIOI_0}(t_1)\mathbf{U}_{0-}^{\mu-1} - \\ -\Gamma_{IIOR_0}(t_1)\mathbf{R}_{0-}^{\alpha-1} - \Gamma_{IIOY_0}(t_1)\mathbf{Y}_{0-}^{\nu-1} \end{bmatrix} \right\} +$$

$$+\Gamma_{IIOI_0}(t_1)\mathbf{U}_{0-}^{\mu-1} + \Gamma_{IIOR_0}(t_1)\mathbf{R}_{0-}^{\alpha-1} + \Gamma_{IIOY_0}(t_1)\mathbf{Y}_{0-}^{\nu-1} =$$

$$= \left\{ \bullet \begin{bmatrix} G_{\Gamma IIO}(t_1)\, G_{\Gamma IIO}^{-1}(t_1) \bullet \\ \mathbf{Y}_1 - \Gamma_{IIOI_0}(t_1)\mathbf{U}_{0-}^{\mu-1} - \\ -\Gamma_{IIOR_0}(t_1)\mathbf{R}_{0-}^{\alpha-1} - \Gamma_{IIOY_0}(t_1)\mathbf{Y}_{0-}^{\nu-1} \end{bmatrix} \right\} +$$

$$+\Gamma_{IIOI_0}(t_1)\mathbf{U}_{0-}^{\mu-1} + \Gamma_{IIOR_0}(t_1)\mathbf{R}_{0-}^{\alpha-1} + \Gamma_{IIOY_0}(t_1)\mathbf{Y}_{0-}^{\nu-1} =$$

$$= \begin{bmatrix} \mathbf{Y}_1 - \Gamma_{IIOI_0}(t_1)\mathbf{U}_{0-}^{\mu-1} - \\ -\Gamma_{IIOR_0}(t_1)\mathbf{R}_{0-}^{\alpha-1} - \Gamma_{IIOY_0}(t_1)\mathbf{Y}_{0-}^{\nu-1} \end{bmatrix} +$$

$$+\Gamma_{IIOI_0}(t_1)\mathbf{U}_{0-}^{\mu-1} + \Gamma_{IIOR_0}(t_1)\mathbf{R}_{0-}^{\alpha-1} + \Gamma_{IIOY_0}(t_1)\mathbf{Y}_{0-}^{\nu-1} = \mathbf{Y}_1.$$

The chosen control vector function $\mathbf{U}(.)$ defined by (11.244) steers the system output from any chosen initial output $\mathbf{Y}_0$- to any accepted final output $\mathbf{Y}_1$ at the chosen moment $t_1 \in In\mathfrak{T}_0$. Definition 165 is satisfied. The *IIO* system (11.179), (11.180) is output controllable. ∎

Bibliography

[1] B. D. O. Anderson and J. B. Moore, *Linear Optimal Control*, Englewood Cliffs: Prentice Hall, 1971.

[2] P. J. Antsaklis and A. N. Michel, *Linear Systems*, New York: The McGraw Hill Companies, Inc., 1997, Boston: Birkhaüser, 2006.

[3] P. J. Antsaklis and A. N. Michel, *A Linear Systems Primer*, Boston: Boston: Birkhaüser, 2007.

[4] S. Barnett, *Introduction to Mathematical Control Theory*, Oxford: Clarendon Press, 1975.

[5] A. Benzaoiua, F. Mesquine and M. Benhayoun, *Saturated Control of Linear Systems*, Berlin: Springer, 2018.

[6] L. D. Berkovitz, *Optimal Control Theory*, New York: Springer Verlag, 2010.

[7] L. D. Berkovitz and N. G. Medhin, *Nonlinear Optimal Control*, Boca Raton, FL: Taylor & Francis-CRC, 2013.

[8] J. E. Bertram and P. E. Sarachik, "On optimal computer control", *Proc. of the First International Congress of the Federation of Automatic Control*, London: Butterworths, pp. 419-422, 1961.

[9] S. P. Bhattacharyya, A. Datta and L. H. Keel, *Linear Control Theory: Structure, Robustness, and Optimization*, Boca Raton, FL: CRC Press, Taylor & Francis Group, 2009.

[10] D. Biswa, *Numerical Methods for Linear Control Systems*, London: Elsevier Inc., 2004.

[11] P. Borne, G. Dauphin-Tanguy, J.-P. Richard, F. Rotella and I. Zambet-takis, *Commande et Optimisation des Processus*, Paris: Éditions TECH-NIP, 1990.

[12] R. W. Brockett and M. D. Mesarović, "The reproducibility of multi-variable systems", *J. Mathematics Analysis and Applications*, Vol. 1, pp. 548-563, 1965.

[13] W. L. Brogan, *Modern Control Theory*, New York: Quantum Publish-ers, Inc., 1974.

[14] Z. M. Buchevats and Ly. T. Gruyitch, *Linear Discrete-time Systems*, Boca Raton, FL: CRC Press, 2018.

[15] F. M. Callier and C. A. Desoer, *Linear System Theory*, New York: Springer-Verlag, 1991.

[16] F. M. Callier and C. A. Desoer, *Multivariable Feedback Systems*, New York: Springer-Verlag, 1982.

[17] G. E. Carlson, *Signal and Linear Systems Analysis and Matlab*, second edition, New York: Wiley, 1998.

[18] C.-T. Chen, *Linear System Theory and Design*, New York: Holt, Rine-hart and Winston, Inc., 1984; Oxford: Oxford University Press, 2013.

[19] H. Chestnut and R. W. Mayer, *Servomechanisms and Regulating Sys-tem Design*, New York: Wiley, 1955.

[20] M. J. Corless and A. E. Frazho, *Linear Systems and Control*, Boca Raton, FL: CRC Press, 2003.

[21] J. J. D'Azzo and C. H. Houpis, *Linear Control System Analysis & Design*, New York: McGraw-Hill Book Company, 1988.

[22] J. J. D'Azzo, C. H. Houpis and S. N. Sheldon, *Linear Control System Analysis & Design with Matlab*, Boca Raton, FL: CRC Press, 2003.

[23] L. Debnath, *Integral Transformations and Their Applications*, Boca Ra-ton, FL: CRC Press, 1995.

[24] C. A. Desoer, *Notes for A Second Course on Linear Systems*, New York: Van Nostrand Reinhold Company, 1970.

[25] C. A. Desoer and M. Vidyasagar, *Feedback Systems: Input-Output Properties*, New York: Academic Press, 1975.

[26] J.L. Domínguez-García, M.I. García-Planas, "Output Controllability and Steady-Output Controllability Analysis of Fixed Speed Wind Turbine", PHYSCON 2011, León, Spain, September, 5 – September, 8, pp. 1-5, 2011.

[27] F. W. Fairman, *Linear Control Theory: The State Space Approach*, Chichester, England: John Wiley & Sons, 1998.

[28] P. Falb, *Methods of Algebraic Geometry in Control Theory: Multivariable Linear Systems and Projective Algebraic Geometry Part II*, Boston: Birkhauser, 1999.

[29] T. E. Fortmann and K. L. Hitz, *An Introduction to Linear Control Systems*, New York: Marcel Dekker, Inc., 1977.

[30] F. R. Gantmacher, *The Theory of Matrices*, Vol. 1, New York: Chelsea Publishing Co., 1960, 1974.

[31] F. R. Gantmacher, *The Theory of Matrices*, Vol. 2, New York: Chelsea Publishing Co., 1960, 1974.

[32] E. G. Gilbert, "Controllability and Observability in Multivariable Control Systems", *SIAM Journal of Control*, Ser.A, Vol. 1, 1963, pp. 128 - 151.

[33] I. Gohberg, P. Lancaster, L. Rodman, *Matrix Polynomials*, New York: Academic Press, 1982.

[34] G. C. Goodwin, S. F. Graebe and M. E. Salgado, *Control System Design*, New Jersey USA, London UK: Prentice Hall-Pearson, 2001.

[35] Lj. T. Grujić (Ly. T. Gruyitch), *Continuous Time Control Systems*, Lecture notes for the course "DNEL4CN2: Control Systems", Durban: Department of Electrical Engineering, University of Natal, South Africa, 1993.

[36] Ly. T. Gruyitch, *Advances in the Linear Dynamic Systems Theory*, Tamarac, FL: Llumina Press, 2013.

[37] Ly. T. Gruyitch, *Control of Linear Systems. II: Tracking and Trackability of General Linear Systems*, Boca Raton, FL: CRC Press, 2018.

[38] Ly. T. Gruyitch, *Einstein's Relativity Theory. Correct, Paradoxical, and Wrong*, ISBN 1-4122-0211-6, Trafford, Victoria, Canada, http://www.trafford.com/06-2239, 2006.

[39] Ly. T. Gruyitch, *Galilean-Newtonean Rebuttal to Einsteins Relativity Theory*, Cambridge, UK: Cambridge International Science Publishing, 2015.

[40] Ly. T. Gruyitch, *Linear Continuous-time Systems*, Boca Raton, FL: CRC Press/Taylor & Francis Group, 2017.

[41] Ly. T. Gruyitch, *Nonlinear Systems Tracking*, Boca Raton, FL: CRC Press/Taylor & Francis Group, 2016.

[42] Ly. T. Gruyitch, *Time and Consistent Relativity. Physical and Mathematical Fundamentals*, Waretown NJ and Oakville, ON: Apple Academic Press, Inc., 2015.

[43] Ly. T. Gruyitch, *Time. Fields, Relativity, and Systems*, ISBN 1-59526-671-2, LCCN 2006909437, Coral Springs, FL: Llumina, 2006.

[44] Ly. T. Gruyitch, *Time and Time Fields. Modeling, Relativity, and Systems Control*, ISBN 1-4251-0726-5, Victoria, Canada: Trafford, 2006.

[45] Ly. T. Gruyitch, *Tracking Control of Linear Systems*, Boca Raton, FL: CRC Press/Taylor & Francis Group, 2013.

[46] M. Haidekker, *Linear Feedback Controls*, London: Elsevier, 2013

[47] S. C. Hamilton and E. M. Broucke, *Geometric Control of Patterned Linear Systems*, Berlin: Springer, 2012.

[48] M. L. J. Hautus, "Controllability and observability conditions of linear autonomous systems", *Nederlandse Akademie Vanwettenschappen*, Series A, V72, 1969, pp. 443-448.

[49] E. Hendricks, O. SØrensen and H. Paul, *Linear Systems Control*, Berlin: Springer, 2008.

[50] J. P. Hespanh, *Linear Systems Theory*, Princeton, NJ: Princeton University Press, 2009.

[51] R. M. Hirschorn, "Output tracking in multivariable nonlinear systems", *IEEE Transactions on Automatic Control*, Vol, AC-26, No. 6, pp. 593-595, 1981.

[52] C. H. Houpis and S. N. Sheldon, *Linear Control System Analysis and Design with MATLAB®*, Boca Raton, FL: CRC Press, Taylor and Francis Group, 2014.

[53] D. G. Hull, *Optimal Control Theory for Applications*, New York: Springer Verlag, 2003.

[54] R. Johnson, R. Obaya, S. Novo, C. Núñez and R. Fabbri, *Nonautonomous Linear Hamiltonian Systems: Oscillation, Spectral Theory and Control*, Berlin: Springer, 2016.

[55] M. K.-J. Johansson, *Picewise Linear Control Systems*, Berlin: Springer, 2003.

[56] T. Kaczorek, *Polynomial and Rational Matrices: Applications in Dynamical Systems Theory*, Berlin: Springer, 2007.

[57] T. Kailath, *Linear Systems*, Englewood Cliffs, NJ: Prentice-Hall, Inc., 1980.

[58] R. E. Kalman, "Algebraic structure of linear dynamical systems, I. The module of Σ," *Proc. National Academy of Science: Mathematics*, USA NAS, Vol. 54, pp. 1503-1508, 1965.

[59] R. E. Kalman, "Canonical structure of linear dynamical systems", *Proceedings of the National Academy of Science: Mathematics*, USA NAS, Vol. 48, pp. 596-600, 1962.

[60] R. E. Kalman, "Mathematical description of linear dynamical systems", *J.S.I.A.M. Control*, Ser. A, Vol. 1, No. 2, pp. 152-192, 1963.

[61] R. E. Kalman, "On the General Theory of Control Systems", *Proceedings of the First International Congress on Automatic Control*, London: Butterworth, 1960, pp. 481-491.

[62] R. E. Kalman, P. L. Falb and M. A. Arbib, *Topics on Mathenatical System Theory*, New York: McGraw Hill, 1969.

[63] R. E. Kalman, Y. C. Ho and K. S. Narendra, "Controllability of Linear Dynamical Systems", *Contributions to Differential Equations*, Vol.*1*, No. 2, pp. 189-213, 1963.

[64] D. E. Kirk, *Optimal Control Theory: An Introduction.* Englewood Cliffs, NJ: Prentice-Hall, 1970, Dover republication, 2004.

[65] B. Kisačanin.and G. C. Agarwal, *Linear Control Systems: With Solved Problems and MATLAB Examples*, New York: Kluwer Academic Press/Plenum Publishers, 2001.

[66] B. C. Kuo, *Automatic Control Systems*, Englewood Cliffs, NJ: Prentice-Hall, Inc., 1967.

[67] B. C. Kuo, *Automatic Control Systems*, Englewood Cliffs, NJ, CA: Prentice-Hall, Inc., 1987.

[68] H. Kwakernaak and R. Sivan, *Linear Optimal Control Systems*, New York: Wiley-Interscience, 1972.

[69] P. Lancaster and M. Tismenetsky, *The Theory of Matrices*, San Diego, CA: Academic Press, 1985.

[70] J. R. Leigh, *Functional Analysis and Linear Control Theory*, Mineola, NY: Dover Publications, Inc., 1980, 2007.

[71] J. M. Maciejowski, *Multivariable Feedback Systems*, Wokingham: Addison-Wesley Publishing Company, 1989.

[72] G. Marsaglla, *Bounds for the rank of the Sum of two Matrices*, Seattle: Mathematics Research Laboratory, Boeing Scientific Research Laboratories: Mathematical Note No, 3bA, D1-82-0343, pp. 1-13, April 1964. http://www.dtic.mil/dtic/tr/fulltext/u2/600471.pdf

[73] G. Marsaglla, *Bounds for the Rank of the Sum of Matrices*, Seattle: Mathematics Research Laboratory, Boeing Scientific Research Laboratories, pp. 455-462, April 1964. http://www.ic.unicamp.br/~meidanis/PUB/Doutorado/2012-Biller/Marsaglia1964.pdf

[74] E. J. McShane, *Integration*, Princeton, NJ: Princeton University Press, 1944.

[75] J. L. Melsa and D. G, Schultz, *Linear Control Systems*, New York: McGraw-Hill Book Company, 1969.

[76] A. N. Michel and C. J. Herget, *Algebra and Analysis for Engineers and Scientists*, Boston, MA: Birkhäuser, 2007.

[77] R. K. Miller and A. N. Michel, *Ordinary Differential Equations*, New York, NY: Academic Press, 1982.

[78] B. R. Milojković and Lj. T. Grujić, *Automatic Control*, the textbook in Serbo-Croatian, Belgrade: Faculty of Mechanical Engineering, University of Belgrade, 1977.

[79] B. R. Milojković and L. T. Grujić, *Automatic Control*, in Serbo-Croatian, Belgrade: Faculty of Mechanical Engineering, 1981.

[80] Sir Isaac Newton, *Mathematical Principles of Natural Philosophy - BOOK I. The Motion of Bodies*, Chicago, IL: William Benton, Publisher, Ecyclopaedia Britannica, Inc., *(The first publication: 1687)* 1952.

[81] K. Ogata, *State Space Analysis of Control Systems*, Englewood Cliffs, NJ: Prentice Hall, 1967.

[82] K. Ogata, *Modern Control Engineering*, Englewood Cliffs, NJ: Prentice Hall, 1970.

[83] D. H. Owens, *Feedback and Multivariable Systems*, Stevenage, Herts: Peter Peregrinus Ltd., 1978.

[84] *Polynomial and Polynomial Matrix Glossary*, http://www.polyx.cz/glossary.htm, 2017.

[85] B. Porter, "Reachability and Controllability of Discrete-Time Linear Dynamical Systems", *Int. J. General Systems*, B. Porter, Vol. 2, 1975, p. 115.

[86] H. M. Power and R. J. Simpson, *Introduction to Dynamics and Control*, London: McGraw-Hill Book Company (UK) Limited, 1978.

[87] H. H. Rosenbrock, "Some properties of relatively prime polynomial matrices", *Electronic Letters*, Vol. 4, 1968, pp. 374 - 375.

[88] H. H. Rosenbrock, *State-space and Multivariable Theory*, London: Thomas Nelson and Sons Ltd., 1970.

[89] A. Sinha, *Linear Systems: Optimal and Robust Control*, Boca Raton, FL: CRC Press, Taylor & Francis Group, 2007.

[90] R. E. Skelton, *Dynamic Systems Control: Linear Systems Analysis and Synthesis*, New York: John Wiley & Sons, 1988.

[91] E. D. Sontag, *Mathematical Control Theory: Deterministic Finite Dimensional Systems*, New York: Springer, 1990.

[92] H. L. Trentelman, *Sense and Simplicity*, Talk at the Occasion of the Afscheidscollege of M.L.J. Hautus on April 8, 2005 at the Eindhoven University of Technology, available at: http://www.math.rug.nl/~trentelman/seminars/Hautus.pdf

[93] H. L. Trentelman, A. A. Stoorvogel, M. Hautus, *Control Theory for Linear Systems*, London, Berlin: Springer, 2001, 2012.

[94] J. Tsiniaas and J. Karafyllis, "ISS property for time-varying systems and application to partial-static feedback stabilization and asymptotic tracking," *IEEE Trans. Automatic Control*, Vol. 44, No. 11, pp. 2179-2184, November 1999.

[95] D. M. Wiberg, *State Space and Linear Systems*, New York: McGraw-Hill Book Company, 1971.

[96] R. L. Williams II and D. A. Lawrence, *Linear State-Space Control Systems*, Hoboken, NJ: John Wiley & Sons, Inc., 2007.

[97] W. A. Wolovich, *Linear Multivariable Systems*, New York: Springer-Verlag, 1974.

[98] W. M. Wonham, *Linear Multivariable Control. A Geometric Approach*, Berlin: Springer-Verlag, 1974.

[99] B.-T. Yazdan, *Introduction to Linear Control Systems*, New York: Academic Press, 2017.

Part V

INDEX

Author Index

Subject Index